FOURIER SERIES, TRANSFORMS, AND BOUNDARY VALUE PROBLEMS

Second Edition

J. RAY HANNA

JOHN H. ROWLAND

DOVER PUBLICATIONS, INC.
MINEOLA, NEW YORK

Bibliographical Note

This Dover edition, first published in 2008, is an unabridged republication of the first work originally published in 1990 by John Wiley & Sons, Inc., New York.

Library of Congress Cataloging-in-Publication Data

Hanna, J. Ray
    Fourier series, transforms, and boundary value problems. — 2nd ed. / J. Ray Hanna and John H. Rowland.
        p. cm.
    "This Dover edition, first published in 2008, is an unabridged republication of the work originally published in 1990 by John Wiley & Sons, Inc., New York."
    Includes index.
    ISBN-13: 978-0-486-46673-6
    ISBN-10: 0-486-46673-6
    1. Boundary value problems. 2. Fourier series. I. Rowland, John H. II. Title.

QA379.H36 2008
515'.35—dc22

                                            2007051617

Printed in Canada
46673603   2025
www.doverpublications.com

PREFACE

The basic philosophy remains the same as in the first edition. The primary changes consist of the addition of new material on integral transforms, discrete and fast Fourier transforms, series solutions, harmonic analysis, spherical harmonics, and a glance at some of the numerical techniques for the solution of boundary value problems. The order of presentation of some of the material from the first edition has been rearranged to provide more flexibility in arranging courses based on this text.

The book contains more than enough material for a one semester course. For this reason we have attempted to keep the later chapters relatively self-contained. The first three chapters contain basic material which would ordinarily be covered in a course of this nature. These could be followed by any combination of Chapters 4, 5, 6, and 8, except that Sections 8.11–8.13 depend on Chapter 5 and Sections 8.14 and 8.15 depend on Chapter 6. Chapter 7 depends somewhat on the theory presented in Chapter 4 and the Hankel and Legendre transforms depend on Chapters 5 and 6, respectively. Chapter 9, taken in its entirety, is dependent on all of the preceding chapters. However, instructors who prefer to interweave applications with the development of tools will find that it is possible to select pertinent topics from Chapters 8 and 9 as the necessary mathematics is developed.

A one-semester course given at the University of Wyoming covers substantial portions of Chapters 1, 2, 3, 4, 7, and 8. This course has upper class and graduate students from fields such as geophysics, physics, engineering, computer science, and mathematics.

We are indebted to Maria Taylor and Bob Hilbert for valuable editorial assistance and to many students for catching errors and suggesting improvements. Finally, we express our special appreciation to Janet Netzel and Mitzi Stephens for their skillful typing.

J. Ray Hanna

Aurora, Colorado

John H. Rowland

Laramie, Wyoming
May, 1990

PREFACE TO THE FIRST EDITION

This book is a result of the development of a set of notes for a course in boundary value problems, using Fourier series and integrals. Its primary objective is to acquaint students with the solutions of boundary value problems associated with natural phenomena. It is therefore necessary for the reader to understand basic concepts and manipulations of elementary calculus. Although some mathematical ideas from advanced calculus are beneficial, many of the concepts are contained in this book. A minimal background in physics will aid one in understanding the modeling of a few problems concerning heat, wave, and potential theory.

This book refers to the main process for solving boundary value problems as the Fourier method. To understand the details of this procedure, topics of orthogonality, Fourier series, and integrals precede the discussion of the Fourier method. Of necessity such topics as convergence, existence, and uniqueness are included. Emphasis is placed clearly on the use of basic concepts and techniques rather than the details of developing the theory. There are many completely solved examples. These are followed by exercises that allow the reader ample opportunity to test his/her understanding of the material. Most exercises are accompanied with answers. Some answers are implied in the problems, while others are given in an answer section. The abbreviations used are listed in the index.

Content similar to that of this book has been used in a course of three semester hours with several classes. If a prerequisite of ordinary differential equations is prescribed, much of Chapter 1 may be omitted. To shorten the course chapters on either Bessel functions or Legendre polynomials may be deleted. Work with operators may be reduced, or other sections may be omitted without seriously affecting the continuity of the course. The preference of the instructor, the background and interests of the students, and the intensity of the course should govern the choice of subject matter.

Numerous colleagues and students have influenced the form of this book. It is my pleasure to thank everyone who has offered suggestions for its improvement. I am particularly indebted to my department head, Joseph

Martin, for his faithful support of the project, and to Daniel Katz, a former student, for his helpful ideas and his solutions for many of the problems. To Beatrice Shube for valuable editorial assistance, and to all of the Wiley publication staff, I am deeply grateful. Finally, I express a special appreciation for the skillful typing of the manuscript by Laureda Dolan, Paula Melcher, and Pat Twitchell.

Humbly I acknowledge the volumes of literature, extending from a time before J. Fourier to the present, which have influenced the composition of this book. It would be an endless task to mention each one. A list of references, by no means exhaustive, is included to aid the reader.

In spite of careful proofreading, some errors are elusive and not discovered. I encourage readers to inform me of mistakes and to offer suggestions for the improvement of the book.

J. RAY HANNA

Laramie, Wyoming
January 1982

CONTENTS

FOURIER SERIES, TRANSFORMS, AND BOUNDARY VALUE PROBLEMS

1

LINEAR DIFFERENTIAL EQUATIONS

The primary objective of this book is to develop procedures for obtaining solutions to boundary value problems (abbreviated BVPs) containing partial differential equations (abbreviated PDEs). One of the principal solution procedures for PDEs requires some knowledge of ordinary differential equations (abbreviated ODEs) and their solutions. This chapter begins with a review of basic concepts for solving ODEs. The topics considered will include the superposition principle, characteristic equations, power series solutions, the Frobenius method, and numerical solutions. The remainder of the chapter will involve definitions, classifications, and solutions of PDEs.

1.1. LINEAR OPERATORS

An *operator* is a mathematical transformation applied to a function to produce another function. If Q is an operator, the notation Qy means that Q acts upon the function y to produce a new function Qy. For $Qy = y^2$, Q is a squaring operator. The new function is y^2. When $Qy = Dy$, Q is a derivative operator D and the transformed function is the derivative of y.

A *linear operator* L changes each function so that for two functions y_1 and y_2 of a class

$$L(c_1 y_1 + c_2 y_2) = c_1 L y_1 + c_2 L y_2 \tag{1.1}$$

if c_1 and c_2 are constants. One finds by using (1.1) that the differential operator D is linear but the squaring operator is not linear.

The sum of two linear operators L and M is defined by

$$(L + M)y = Ly + My$$

A product of two linear operators LM is the linear operator M acting upon y and sequentially L operating on My. This is expressed by

$$LMy = L(My)$$

In (1.1), $c_1y_1 + c_2y_2$ is called a *linear combination* of y_1 and y_2. The linear operator acting upon the linear combination of y_1 and y_2 is the linear combination of Ly_1 and Ly_2. *A linear combination of a set of n functions* is defined by

$$\sum_{k=1}^{n} c_k y_k$$

1.2. ORDINARY DIFFERENTIAL EQUATIONS

For a linear differential operator

$$L = D^n + \alpha_1(x)D^{n-1} + \cdots + \alpha_{n-1}(x)D + \alpha_n(x)$$

and a function $f(x)$

$$Ly = f \tag{1.2}$$

is a *linear ODE of order n*. If $f \equiv 0$, the equation

$$Ly = 0 \tag{1.3}$$

is a *linear homogeneous ODE of order n*.

An *initial value problem* (abbreviated IVP) is composed of a differential equation (1.2) or (1.3) and set of n restrictions at a single point. These restrictions, called *initial conditions*, have the form

$$y(x_0) = y_0, \quad y'(x_0) = y_0', \ldots, y^{(n-1)}(x_0) = y_0^{(n-1)} \tag{1.4}$$

where $y_0, y_0', \ldots, y_0^{(n-1)}$ are constants.

A *boundary value problem* (abbreviated BVP) contains a differential equation and a set of n constants called *boundary conditions*. These conditions are given at two or more points. At this time definitions for both the IVP and the BVP are relative to ODEs.

The equation $y'' + 4y = 0$ with restrictions $y(0) = y'(0) = 1$ is an IVP. The ODE $y'' + 4y = 0$ accompanied by $y(0) = 1$, $y(\pi/4) = 0$ is a BVP.

Although our main emphasis is the determination of solutions, questions of existence and uniqueness of solutions for differential equations with constraints are important. If one succeeds in finding a solution for an IVP or a BVP then a solution exists. To ascertain whether other solutions exist for the same problem may be as necessary as knowing a solution. We state without proof an existence-uniqueness theorem for an IVP.

Theorem 1.1. Let $x_0 \in (a, b)$ and $\alpha_1(x)$, $\alpha_2(x), \ldots, \alpha_n(x)$, and $f(x)$ be continuous on (a, b) for the IVP composed of (1.2) with initial conditions (1.4). Then there *exists* a *unique* solution $y(x)$ for the problem.

A set of functions

$$\{y_1, y_2, \ldots, y_n\} \tag{1.5}$$

defined on (a, b) is *linearly independent* if the linear combination for the set

$$c_1 y_1 + c_2 y_2 + \cdots + c_n y_n = 0 \tag{1.6}$$

implies that for all x on the interval the only c solution is

$$c_1 = c_2 = \cdots = c_n = 0$$

If some constants $c_k \neq 0$ exist in (1.6) then the set is *linearly dependent*.

Example 1.1. Are the functions e^x and e^{-x} linearly dependent?
Examine the equation

$$c_1 e^x + c_2 e^{-x} = 0 \tag{1.7}$$

If both sides of (1.7) are multiplied by e^x

$$c_1 e^{2x} = -c_2$$

The two members are identical for all x only if

$$c_1 = c_2 = 0$$

The Wronskian of the set of functions (1.5), assumed differentiable $n - 1$ times, is defined by

$$W(y_1, y_2, \ldots, y_n) = \begin{vmatrix} y_1 & y_2 & \cdots & y_n \\ y_1' & y_2' & \cdots & y_n' \\ \vdots & & & \vdots \\ y_1^{(n-1)} & y_2^{(n-1)} & \cdots & y_n^{(n-1)} \end{vmatrix}$$

It can be shown that if $W(y_1, y_2, \ldots, y_n)$ is not zero for any $x \in (a, b)$, then (1.5) is *linearly independent*. If (1.5) is a solution set for (1.3) the linear combination of (1.5)

$$y = c_1y_1 + c_2y_2 + \cdots + c_ny_n \tag{1.8}$$

is a solution of (1.3). This idea is referred to as the *superposition principle*. The n solutions of (1.3) form a *fundamental solution set* if every solution of (1.3) can be formed as a linear combination of (1.5) as shown in (1.8). $W(y_1, y_2, \ldots, y_n)$ for the solution set (1.5) can be shown either to vanish identically or else never be zero. The following theorem describes conditions for a fundamental solution set and defines a general solution.

Theorem 1.2. Let $\alpha_1(x), \alpha_2(x), \ldots, \alpha_n(x)$ be continuous on (a, b), let (1.5) be the solution set of the linear homogeneous ODE (1.3), and let $W(y_1, y_2, \ldots, y_n) \neq 0$ for one point on (a, b). Then it is possible to form any solution of (1.3) as a linear combination of (1.5). The solution set is a *fundamental set*. The linear combination (1.8) is called the *general solution*.

Exercises 1.1

1. Show that the two conditions $L(y_1 + y_2) = Ly_1 + Ly_2$ and $L(cy) = cLy$ taken together imply linearity.

2. Show that the sum of two linear operators L and M is linear.

3. If both L and M are linear operators, is LM linear?

4. Assume that L and M are linear operators. Show by contradiction that LM and ML are not always the same.

5. (a) Verify that $y_1 = 1/(x + 1)$ and $y_2 = 1/(x + 2)$ are both solutions of $y' + y^2 = 0$.

   (b) Compute the Wronskian $W(y_1, y_2)$. Is the set $\{y_1, y_2\}$ linearly independent?

   (c) Is there a c_1 so that c_1y_1 is a solution?

   (d) Is there a c_2 so that c_2y_2 is a solution?

   (e) If nonzero values of c_1 and c_2 are used from (c) and (d) then is $c_1y_1 + c_2y_2$ a solution of the differential equation? Do your results violate Theorem 1.2? Explain.

6. (a) If y_1 and y_2 are solutions of the differential equation $y'' + x^2y = 0$, is $y = c_1y_1 + c_2y_2$ a solution? Why?

   (b) If y_1 and y_2 are solutions of $y'' + y = x^2$, is $y = c_1y_1 + c_2y_2$ a solution?

7. (a) Assume that y_1 and y_2 satisfy the differential equation $y'' + \sin y = 0$. Is $y = c_1y_1 + c_2y_2$ a solution?

   (b) If y_1 and y_2 are solutions of $y'' + \sin x = 0$, is $y = c_1y_1 + c_2y_2$ a solution? Explain basic differences in the differential equations of (a) and (b).

8. (a) Is the set of functions $\{e^x, e^{2x}\}$ linearly independent?

(b) Test the set $\{e^x, e^{2x}, xe^{2x}\}$ for linear independence.

(c) For the differential equation $y''' - 5y'' + 8y' - 4y = 0$ verify that e^x and e^{2x} are solutions.

(d) Is $c_1 e^x + c_2 e^{2x}$ a solution of the ODE in (c)? Is it a general solution? Why?

(e) Is xe^{2x} a solution of the ODE in (c)? Is $c_1 e^x + c_2 e^{2x} + c_3 xe^{2x}$ a general solution? Explain.

9. (a) Verify that the equation $y'' + 4y = 0$ is satisfied by the two solutions $y_1 = 2\cos^2 x - 1$ and $y_2 = 1 - 2\sin^2 x$.

(b) Determine the Wronskian $W(2\cos^2 x - 1, 1 - 2\sin^2 x)$. Is the set $\{y_1, y_2\}$ of (a) linearly independent?

(c) Is $c_1 y_1 + c_2 y_2$ a general solution for $y'' + 4y = 0$?

10. The differential equation $y'' + 4y = 4$ has two solutions $y_1 = \cos 2x + 1$ and $y_2 = \sin 2x + 1$. Is $c_1 y_1 + c_2 y_2$ a solution of the ODE? Is Theorem 1.2 violated?

The ODE

$$y'' + \alpha^2 y = 0$$

and its general solution

$$y = c_1 \cos \alpha x + c_2 \sin \alpha x$$

play a prominent role in our study of BVPs of PDEs. Usually our linear ODEs will be homogeneous of order no greater than 2. Since the discussion of the nth order case is as easy basically as the second, the nth order equation is our choice.

1.3. HOMOGENEOUS LINEAR ODE WITH CONSTANT COEFFICIENTS

The equation described under this heading has the form

$$Ly = (D^n + \alpha_1 D^{n-1} + \cdots + \alpha_{n-1}D + \alpha_n)y = 0 \qquad (1.9)$$

where $\alpha_1, \ldots, \alpha_n$ are real constants. Let $y = e^{mx}$ be a proposed solution for (1.9). Actual substitution of e^{mx} in (1.9) implies that

$$m^n + \alpha_1 m^{n-1} + \cdots + \alpha_{n-1}m + \alpha_n = 0 \qquad (1.10)$$

This polynomial equation is called the *auxiliary* or *characteristic equation* for (1.9). Of the n roots of (1.10) (a) all may be real and distinct, (b) some may be imaginary, or (c) some may be multiple roots.

1. *Real and Distinct Roots.* If roots of (1.10) are $m_1, m_2, \ldots, m_n$, then the fundamental solution set is $e^{m_1 x}, e^{m_2 x}, \ldots, e^{m_n x}$, and by superposition

$$y = c_1 e^{m_1 x} + c_2 e^{m_2 x} + \cdots + c_n e^{m_n x}$$

is the general solution.

2. *Imaginary Roots.* If $\alpha_1, \alpha_2, \ldots, \alpha_n$ are all real and $a + ib$ (a and b real numbers) is a root of (1.10), then $a - ib$ is also a root. A solution corresponding to the conjugate pair of roots $a \pm ib$, $b \neq 0$, is

$$y_k = e^{ax}(c_1 \cos bx + c_2 \sin bx)$$

3. *Multiple Roots.* If one root of the characteristic equation m_k is repeated r times, then the solution corresponding to the multiple root is

$$y_k = (c_1 + c_2 x + \cdots + c_r x^{r-1})e^{m_k x}$$

If the ODE has the differential operator of (1.9) but the form

$$Ly = f \tag{1.11}$$

then the equation is nonhomogeneous if $f \neq 0$. To solve (1.11) one first finds a general solution y_c for the equation $Ly = 0$. Next find a function y_p that satisfies (1.11). The general solution for (1.11) is $y = y_c + y_p$. We refrain from discussing procedures for determining y_p. For readers having a need for this information see Boyce and DiPrima [6, pp. 143–162 and 270–278].

Example 1.2. Find the general solution for the ODE $y'' - 4y' + 13y = 0$. The characteristic equation is

$$m^2 - 4m + 13 = 0$$

with roots $2 \pm 3i$. The general solution is written

$$y = e^{2x}(c_1 \cos 3x + c_2 \sin 3x)$$

Example 1.3. Determine the solution for the BVP

$$y''' - 6y'' + 12y' - 8y = 0, \quad y(0) = 0, \quad y(1) = 0, \quad y'(0) = 1$$

The ODE has a characteristic equation

$$m^3 - 6m^2 + 12m - 8 = 0 \tag{1.12}$$

or

$$(m-2)^3 = 0$$

A root 2 of multiplicity 3 is the solution of (1.12). The general solution of the ODE is expressed by the linear combination

$$y = (c_1 + c_2 x + c_3 x^2)e^{2x}$$

If $y(0) = 0$, then $c_1 = 0$. If $y(1) = 0$, $c_2 + c_3 = 0$. Note that

$$y' = 2(c_2 x + c_3 x^2)e^{2x} + (c_2 + 2c_3 x)e^{2x}$$

If $y'(0) = 1$, $c_2 = 1$. Therefore, $c_3 = -1$. The BVP has the solution

$$y = (x - x^2)e^{2x}$$

1.4. EULER'S ODE

The operator of the Euler (or Cauchy) ODE is the operator of (1.9) with an added factor x^n inserted in each coefficient, where n is the order of the derivative. For the ODE

$$Ly = (x^n D^n + \alpha_1 x^{n-1} D^{n-1} + \cdots + \alpha_{n-1} x D + \alpha_n)y = f \tag{1.13}$$

a transformation $x = e^t$ is employed to change the independent variable x to t. This transformation converts (1.13) to a new ODE with constant coefficients.

Example 1.4. Solve the differential equation

$$x^2 y'' + 7xy' + 9y = 0 \tag{1.14}$$

Let $x = e^t$ and $t = \ln x$. Then

$$\frac{dy}{dx} = \frac{1}{x}\frac{dy}{dt} \quad \text{and} \quad \frac{d^2 y}{dx^2} = \frac{1}{x^2}\left[\frac{d^2 y}{dt^2} - \frac{dy}{dt}\right]$$

The new ODE with the independent variable t is

$$\frac{d^2y}{dt^2} + 6\frac{dy}{dt} + 9y = 0 \qquad (1.15)$$

The characteristic equation

$$m^2 + 6m + 9 = 0$$

has a double root -3. Equation (1.15) has a general solution

$$y(t) = (c_1 + c_2 t)e^{-3t}$$

Using the transformation again, one obtains

$$y(x) = (c_1 + c_2 \ln x)x^{-3} \qquad (1.15a)$$

Euler equations appear in solutions of BVPs involving spherical geometry.

Exercises 1.2

1. Determine a general solution for the equation $y'' + 5y' + 6y = 0$.
2. Find a general solution for the equation $y'' - 4y' + 4y = 0$.
3. Solve the differential equation $y'' + 2y' + 2y = 0$.
4. Show that the characteristic equation for $y''' - 2y' - 4y = 0$ has a root of 2. Then solve the differential equation.
5. Solve the boundary value problem $y'' - y = 0$, $y(0) = 0$, $y'(\pi) = 1$.
6. Find a general solution for $y^{(4)} - y = 0$.
7. Solve the differential equation $y''' - 5y'' + 6y' = 0$.
8. Determine a general solution for the equation $x^2 y'' - 3xy' + 3y = 0$.
9. Solve the BVP $x^2 y'' - 3xy' + 4y = 0$, $y(1) = 0$, $y(e) = e^2$.
10. Find a general solution for $x^2 y'' - xy' + 5y = 0$.
11. Find a solution for the BVP $x^2 y'' + xy' + y = 0$, $y(1) = 0$, $y(e^{\pi/2}) = \pi$.

1.5. SERIES SOLUTIONS

In Section 1.3 we have seen how to determine solutions for linear ODEs with constant coefficients. These closed form solutions are expressed by elementary functions. If the coefficients of (1.9) or (1.11) are not constants, then with a few exceptions such as the Euler equation it is not possible to find closed form solutions for second and higher order equations. For these situations we introduce the *power series and numerical methods* for solving differential equations. It is assumed that the reader has an acquaintance with the elementary theory of power series such as that described in Boyce and DiPrima [6, Section 4.1].

If a function is represented by a power series on the interior of its interval of convergence, then termwise differentiation of that series produces a power series which converges on that same interval to the derivative of the function. Frequently, the ratio test is useful for determining the radius of convergence. The foregoing fact along with the usual algebraic operations are necessary when a series is substituted into a differential equation. To illustrate the mechanics of this method, we consider a simple example.

Example 1.5. Determine a power series solution for the initial value problem

$$y' - 2y = 0, \quad y(0) = 3 \tag{1.16}$$

Assuming that the ODE has a power series solution about $x = 0$, we have

$$y = \sum_{k=0}^{\infty} C_k x^k \tag{1.17}$$

$$y' = \sum_{k=1}^{\infty} k C_k x^{k-1} \tag{1.18}$$

Insert (1.17) and (1.18) into the differential equation (1.16). This gives

$$\sum_{k=1}^{\infty} k C_k x^{k-1} - 2 \sum_{k=0}^{\infty} C_k x^k = 0 \tag{1.19}$$

The index of summation in the series of (1.19) is a parameter, much the same as a variable of integration in a definite integral. Changes of indices are possible without changing the actual sum. If k is replaced by $k + 1$ in the first sum of (1.19) and the factor 2 is moved inside the second sum, we have

$$\sum_{k=0}^{\infty} (k + 1) C_{k+1} x^k - \sum_{k=0}^{\infty} 2 C_k x^k = 0$$

This may be expressed as a single sum

$$\sum_{k=0}^{\infty} [(k + 1) C_{k+1} - 2 C_k] x^k = 0 \tag{1.20}$$

Now if the power series in (1.20) is zero for all x in an interval about zero, the coefficient of each power of x must be zero; so

$$(k + 1) C_{k+1} - 2 C_k = 0 \quad \text{or} \quad C_{k+1} = \frac{2}{k+1} C_k, \quad k = 0, 1, 2, 3, \ldots \tag{1.21}$$

Successively substituting $k = 0, 1, 2, 3, \ldots$, in the *recursion formula* (1.21), we obtain

$$C_1 = 2C_0, \quad C_2 = \frac{2}{2} C_1 = \frac{2^2}{2!} C_0, \quad C_3 = \frac{2}{3} C_2 = \frac{2^3}{3!} C_0, \ldots$$

The solution y is given by

$$y = C_0 + 2C_0 x + \frac{2^2}{2!} C_0 x^2 + \frac{2^3}{3!} C_0 x^3 + \cdots$$

or

$$y = C_0 \left[1 + 2x + \frac{2^2}{2!} x^2 + \frac{2^3}{3!} x^3 + \cdots \right] = C_0 \sum_{k=0}^{\infty} \frac{(2x)^k}{k!}$$

But $y(0) = 3$; so

$$3 = C_0 \left(1 + \sum_{k=1}^{\infty} \frac{0^k}{k!} \right) \quad \text{and} \quad C_0 = 3 \tag{1.22}$$

It follows from (1.22) that the solution is

$$y = 3 \sum_{k=0}^{\infty} \frac{(2x)^k}{k!} = 3e^{2x} \tag{1.23}$$

This solution is the same as we obtain using the procedure of Section 1.3. The series in (1.23) converges for all values of x.

A function $f(x)$ is *analytic* at x_0 if it can be represented by a power series of the form $\sum_{k=0}^{\infty} C_k(x - x_0)^k$ with positive radius of convergence. Consider the second order homogeneous differential equation

$$y'' + p(x)y' + q(x)y = 0 \tag{1.24}$$

When both $p(x)$ and $q(x)$ are analytic at $x = x_0$, then x_0 is an *ordinary point* of the differential equation. If either or both $p(x)$ and $q(x)$ fail to be analytic at $x = x_0$, then x_0 is a *singular point*.

If $x = x_0$ is an ordinary point of (1.24), then each solution can be expressed as a power series

$$y = \sum_{k=0}^{\infty} C_k(x - x_0)^k$$

which converges on the interval $(x_0 - R, x_0 + R)$. Here R is the smaller of the radii of convergence of the power series (in powers of $x - x_0$) representing $p(x)$ and $q(x)$.

Example 1.6. Determine a power series solution for the ODE

$$y'' + xy = 0 \tag{1.25}$$

The differential equation (1.25) has an ordinary point at $x = 0$, and we may assume a power series solution of the type (1.17). Substituting the series for y and y'' into (1.25), we obtain

$$\sum_{k=2}^{\infty} k(k-1)C_k x^{k-2} + x \sum_{k=0}^{\infty} C_k x^k = 0$$

Replacing k by $k+2$ in the first series, k by $k-1$ in the second, and multiplying inside the second series termwise by x, we obtain

$$\sum_{k=0}^{\infty} (k+2)(k+1)C_{k+2} x^k + \sum_{k=1}^{\infty} C_{k-1} x^k = 0$$

At this point it is desirable to have the indices of the sums begin with the same number; this can be accomplished by displaying the first term of the first sum separately. Then the sume can be combined to yield

$$2C_2 + \sum_{k=1}^{\infty} [(k+2)(k+1)C_{k+2} + C_{k-1}]x^k = 0$$

Therefore,

$$C_2 = 0 \quad \text{and} \quad (k+2)(k+1)C_{k+2} + C_{k-1} = 0$$

or

$$C_{k+2} = -\frac{C_{k-1}}{(k+2)(k+1)}, \quad k = 1, 2, 3, \ldots$$

This implies that

$$C_3 = -\frac{C_0}{3 \cdot 2} = -\frac{C_0}{3!}, \quad C_4 = -\frac{C_1}{4 \cdot 3} = -\frac{2}{4!}C_1,$$

Similarly, we obtain

$$C_5 = 0, \quad C_6 = \frac{4C_0}{6!}, \quad C_7 = \frac{5 \cdot 2}{7!}C_1$$

$$C_8 = 0, \quad C_9 = -\frac{7 \cdot 4}{9!}C_0, \quad C_{10} = -\frac{8 \cdot 5 \cdot 2}{10!}C_1$$

where C_0 and C_1 are arbitrary. Therefore,

$$y = C_0\left[1 - \frac{x^3}{3!} + \frac{4}{6!}x^6 - \frac{7\cdot 4}{9!}x^9 + \cdots\right]$$
$$+ C_1\left[x - \frac{2}{4!}x^4 + \frac{5\cdot 2}{7!}x^7 - \frac{8\cdot 5\cdot 2}{10!}x^{10} + \cdots\right] \tag{1.26}$$

The solution (1.26) is a linear combination of two series which can each be shown to be convergent for all x.

The equation

$$(1 - x^2)y'' - 2xy' + n(n+1)y = 0 \tag{1.27}$$

is known as the *Legendre differential equation of degree n*. Since $x = 0$ is an ordinary point of the equation, we can use a power series to solve the equation. Let

$$y = \sum_{k=0}^{\infty} C_k x^k$$

be inserted in (1.27). After several summation simplifications, the result may be written

$$\sum_{k=0}^{\infty} \{(k+2)(k+1)C_{k+2} + [(n-k)(n+k+1)]C_k\}x^k = 0 \tag{1.28}$$

Since (1.28) is an identity with zero,

$$C_{k+2} = -\frac{(n-k)(n+k+1)}{(k+2)(k+1)}C_k \tag{1.29}$$

Two arbitrary constants, C_0 and C_1, appear in the series solution. If $k = n$ in (1.29) the coefficient $C_{k+2} = 0$ and all successive coefficients C_{k+2m} will be zeros also. Therefore, if n is a positive integer the series truncates and becomes a polynomial. We include the results of (1.29) for C_0 and C_1 for a few values of the index k.

$$C_2 = -\frac{n(n+1)}{2!}C_0, \quad C_3 = -\frac{(n-1)(n+2)}{3!}C_1$$

$$C_4 = \frac{(n-2)(n+3)n(n+1)}{4!}C_0, \quad C_5 = \frac{(n-3)(n+4)(n-1)(n+2)}{5!}C_1$$

$$C_6 = -\frac{(n-4)(n+5)(n-2)(n+3)(n)(n+1)}{6!}C_0,$$

$$C_7 = -\frac{(n-5)(n+6)(n-3)(n+4)(n-1)(n+2)}{7!}C_1$$

If $y_0(x)$ represents the part of the solution associated with C_0 and $y_1(x)$ associated with C_1, then

$$y(x) = C_0 y_0(x) + C_1 y_1(x) \tag{1.30}$$

where

$$y_0(x) = 1 - \frac{n(n+1)}{2!} x^2 + \frac{n(n+1)(n-2)(n+3)}{4!} x^4$$

$$- \frac{n(n+1)(n-2)(n+3)(n-4)(n+5)}{6!} x^6 + \cdots \tag{1.31}$$

and

$$y_1(x) = x - \frac{(n-1)(n+2)}{3!} x^3 + \frac{(n-1)(n+2)(n-3)(n+4)}{5!} x^5$$

$$- \frac{(n-1)(n+2)(n-3)(n+4)(n-5)(n+6)}{7!} x^7 + \cdots \tag{1.32}$$

Legendre's series and polynomial functions will be studied in more detail in Chapter 6.

1.6. FROBENIUS METHOD

Example 1.7. Solve the ODE

$$x^2 y'' + 7xy' + 9y = 0 \tag{1.33}$$

We rewrite (1.33) in the form (1.24) so that the coefficient of y'' is 1

$$y'' + \frac{7}{x} y' + \frac{9}{x^2} y = 0 \tag{1.34}$$

It is apparent that $x = 0$ is a singular point for (1.34). When one attempts to find a series solution of the form (1.17) using the method illustrated by Examples 1.5 and 1.6, one obtains the trivial solution $y = 0$ (see Problem 4 of Exercises 1.3). However, (1.33) is the ODE of Example 1.4, which has a solution

$$y = (C_1 + C_2 \ln x) x^{-3}$$

obtained by the procedure of Section 1.4. We will now show how to modify the series technique to handle regular singular points.

In the ODE (1.24), assume that $x = x_0$ is a singular point. It is a *regular singular point* if $(x - x_0)p(x)$ and $(x - x_0)^2 q(x)$ are each analytic. A singular point which fails to be regular is an *irregular singular point*. We have noted that the power series method fails to give a suitable solution for the ODE (1.33). As a modification to the power series method, let us assume that (1.33) has a solution of the type

$$y = x^r \sum_{k=0}^{\infty} C_k x^k = \sum_{k=0}^{\infty} C_k x^{k+r}, \quad C_0 = 1, \quad x > 0 \qquad (1.35)$$

This is the basic series for the *Frobenius method*. Note that if $x = x_0$ is a regular singular point, substitution of $t = x - x_0$ will change the power series in $(x - x_0)$ into a power series of the form (1.35). Using (1.35) and its derivatives in (1.33), we have

$$x^2 \sum_{k=0}^{\infty} (k + r)(k + r - 1)C_k x^{k+r-2} + 7x \sum_{k=0}^{\infty} (k + r)C_k x^{k+r-1}$$

$$+ 9 \sum_{k=0}^{\infty} C_k x^{k+r} = 0$$

This reduces to

$$\sum_{k=0}^{\infty} [(k + r)(k + r - 1) + 7(k + r) + 9]C_k x^{k+r} = 0 \qquad (1.36)$$

If $k = 0$, then $[r(r - 1) + 7r + 9]C_0 = 0$. But $C_0 \neq 0$, and hence $[r(r - 1) + 7r + 9] = 0$, or $r^2 + 6r + 9 = 0$. This is called the *indicial equation* and its roots, -3 and -3, the *indicial roots* or the *exponents* of the differential equation. If $r = -3$, then from (1.36), $[(k - 3)(k - 4) + 7(k - 3) + 9]C_k = 0$ and $[k^2 - 9 + 9]C_k = 0$. Therefore, $C_k = 0$ for all $k \geq 1$. The solution

$$y_1 = x^{-3} \sum_{k=0}^{\infty} C_k x^k, \quad C_0 = 1, \quad x > 0$$

is then equivalent to

$$y_1 = x^{-3} \qquad (1.37)$$

Since -3 is a double root of the indicial equation only one solution will be obtained by direct substitution in (1.35). A second solution may be obtained by using the method commonly called the *variation of parameters*. The method is based on the assumption that a solution $y_2 = uy_1$ (u a function of x) is a second solution of the differential equation. The assumption is equivalent to $y_2 = ux^{-3}$. Substituting y_2 and its derivatives into (1.33), one obtains

$$x^2(12x^{-5}u - 6x^{-4}u' + x^{-3}u'') + 7x(-3x^{-4}u + x^{-3}u') + 9x^{-3}u = 0$$

This reduces to

$$\frac{u''}{u'} = -\frac{x^{-2}}{x^{-1}} = -\frac{1}{x}$$

Integrating both sides and deferring the constants of integration until the end, we obtain

$$\ln u' = -\ln x = \ln x^{-1}$$

$$u' = \frac{1}{x}$$

and

$$u = \ln x$$

Then

$$y_2 = x^{-3} \ln x \tag{1.38}$$

The linear combination of y_1 and y_2 in (1.37) and (1.38) can be written

$$y = (K_1 + K_2 \ln x)x^{-3}$$

This is equivalent to (1.15a).

Suppose that $x = 0$ is a regular singular point of the ODE (1.24). Assume that r_1 and r_2 are the indicial roots of the equation found from the substitution of $y = x^r \sum_{k=0}^{\infty} C_k x^k$, $C_0 = 1$, in (1.24). Then (1.24) has two linearly independent solutions y_1 and y_2 on the interval $0 < |x| < R$ if the power series for $xp(x)$ and $x^2q(x)$, in powers of x, are valid for $|x| < R$. Three cases follow:

(a) If $r_1 - r_2$ differs from an integer, then

$$y_1 = |x|^{r_1} \sum_{k=0}^{\infty} a_k x^k, \quad a_0 = 1$$

$$y_2 = |x|^{r_2} \sum_{k=0}^{\infty} b_k x^k, \quad b_0 = 1$$

To avoid using $x > 0$, absolute value signs are used.

(b) If $r_1 = r_2 = r$, then

$$y_1 = |x|^r \sum_{k=0}^{\infty} a_k x^k, \quad a_0 = 1$$

$$y_2 = |x|^r \sum_{k=0}^{\infty} b_k x^k + y_1 \ln |x|, \quad b_0 = 1$$

(c) If $r_1 - r_2$ is a positive integer, then

$$y_1 = |x|^{r_1} \sum_{k=0}^{\infty} a_k x^k, \quad a_0 = 1$$

$$y_2 = |x|^{r_2} \sum_{k=0}^{\infty} b_k x^k + B y_1 \ln |x|, \quad b_0 = 1$$

where B is a constant which may be zero.

Treatment of irregular singular points and situations involving complex functions are omitted. For further information see a differential equation text such as Boyce and DiPrima [6, Chapter 4].

We close our discussion of the Frobenius method by obtaining a solution to *Bessel's equation*. This equation, which is very important in applied mathematics, is given by

$$x^2 y'' + xy' + (x^2 - n^2)y = 0 \tag{1.39}$$

Consider the Frobenius series solution for $x > 0$. We let

$$y = x^r \sum_{k=0}^{\infty} c_k x^k = \sum_{k=0}^{\infty} c_k x^{k+r}$$

and substitute this series and its derivatives into (1.39). After simplification we obtain

$$(r^2 - n^2)c_0 x^r + [(1 + r)^2 - n^2]c_1 x^{r+1} + \sum_{k=2}^{\infty} \{[(k + r)^2 - n^2]c_k + c_{k-2}\}x^{k+r} = 0 \tag{1.40}$$

The condition that (1.40) is an identity implies that the coefficient of each x^{k+r} is zero. The coefficient of x^r is $(r^2 - n^2)c_0$, which leads to the indicial equation

$$r^2 - n^2 = 0 \tag{1.41}$$

Thus the indicial roots for Bessel's equation are $r = \pm n$.

First, we consider the case where $r = n$. If $r = n$, the factor $[(1 + n)^2 - n^2] \neq 0$ and this implies that $c_1 = 0$. All remaining coefficients must be zeros, or

$$[(k + n)^2 - n^2]c_k + c_{k-2} = 0$$

Solving for c_k, we have

$$c_k = - \frac{c_{k-2}}{k(k + 2n)} \qquad (1.42)$$

If k is odd, we observe that $c_k = 0$, since $c_1 = 0$. Using (1.42) we can write one solution for the Bessel differential equation (1.39) in the form

$$y = c_0 x^n \left[1 - \frac{x^2}{2^2 1!(n + 1)} + \frac{x^4}{2^4 2!(n + 1)(n + 2)} \right. $$
$$\left. - \frac{x^6}{2^6 3!(n + 1)(n + 2)(n + 3)} + - \cdots \right] \qquad (1.43)$$

If $r = -n$, we replace n with $-n$ in (1.43) and write

$$y = c_0 x^{-n} \left[1 - \frac{x^2}{2^2 1!(1 - n)} + \frac{x^4}{2^4 2!(1 - n)(2 - n)} \right. $$
$$\left. - \frac{x^6}{2^6 3!(1 - n)(2 - n)(3 - n)} + - \cdots \right] \qquad (1.44)$$

Let N represent the natural numbers and let $N_0 = N \cup \{0\}$. If $n = 0$, both (1.43) and (1.44) are the same, but if $n \in N$ (1.44) fails to exist. If $n \notin N_0$ the two solutions (1.43) and (1.44) can be shown to be linearly independent and

$$y = K_0 x^n \left[1 - \frac{x^2}{2^2 1!(n + 1)} + \frac{x^4}{2^4 2!(n + 1)(n + 2)} \right. $$
$$\left. - \frac{x^6}{2^6 3!(n + 1)(n + 2)(n + 3)} + \cdots \right] $$
$$+ K_1 x^{-n} \left[1 - \frac{x^2}{2^2 1!(1 - n)} + \frac{x^4}{2^4 2!(1 - n)(2 - n)} \right. $$
$$\left. - \frac{x^6}{2^6 3!(1 - n)(2 - n)(3 - n)} + - \cdots \right] $$

is a general solution. For the present we investigate the first solution (1.43) and assume that $n \in N_0$.

The solution (1.43) has the arbitrary constant factor c_0. It is customary to assign

$$c_0 = \frac{1}{n!2^n}$$

so that (1.43) becomes

$$J_n(x) = \sum_{k=0}^{\infty} \frac{(-1)^k}{k!(n+k)!} \left(\frac{x}{2}\right)^{n+2k} \tag{1.45}$$

where $J_n(x)$ is a *Bessel function of the first kind of order n*. Naturally, it is a solution of the Bessel differential equation (1.39). According to the ratio test (1.45) is convergent for all real x.

Bessel functions will be discussed more thoroughly in Chapter 5.

Exercises 1.3

1. Find the power series solutions for the following differential equations about a suitable point $x = a$ (which is given for some problems).
   (a) $y' = x - y$.
   (b) $y' = xy$, $y(0) = 3$.
   (c) $xy'' + y = 0$, $a = 1$.
   (d) $(1 - x^2)y'' + y = 0$, $y(0) = 0$, $y'(0) = 1$.
   (e) $y' = x + e^x$ (use a power series for e^x).
   (f) $y'' - xy = 0$ (Airy's equation). Use $a = 0$, then $a = 1$.
   (g) $y'' + xy' - y = 0$, $y(0) = 1$, $y'(0) = 0$.

2. There is a solution obtained by what is known as the *Taylor series method*. A power series solution about an ordinary point is determined by successive differentiation of the differential equation. The resulting coefficients are placed in the Taylor series

$$y(x) = y(a) + \sum_{k=1}^{\infty} \frac{y^{(k)}(a)}{k!} (x - a)^k$$

For example, if $y' = y - x - 1$, let $y(0) = c$. Then

$$y'' = y' - 1, \quad y''' = y'', \quad y^{(4)} = y'', \ldots$$

Evaluation of these expressions at $x = 0$ gives

$$y'(0) = y(0) - 0 - 1 = c - 1, \quad y''(0) = y'(0) - 1 = c - 2, \quad y'''(0) = c - 2$$

Therefore

$$y(x) = c + (c - 1)(x - 0) + \frac{c - 2}{2!} (x - 0)^2 + \frac{c - 2}{3!} (x - 0)^3 + \cdots$$

Solve the following by the Taylor series method:

(a) $y' = y$, $y(0) = 2$;

(b) $y'' + xy = 0$, $y(1) = 2$, $y'(1) = 3$;

(c) $x^2 y'' - y = 0$, $y(2) = 1$, $y'(2) = 0$;

(d) $y'' + xy' + 2y = 0$.

3. For each of the following differential equations, determine the regular singular points (if they exist), the indicial equation and indicial roots. Then for each of the equations (d)–(h) obtain a solution (if one exists) using the Frobenius method.

(a) $4x^2 y'' - 8x^2 y' + (4x^2 + 1)y = 0$.

(b) $xy'' - y' - xy = 0$.

(c) $2x^2 y'' + xy' - (x + 1)y = 0$.

(d) $4xy'' + 2y' + y = 0$.

(e) $2xy'' + y' + xy = 0$.

(f) $x^2 y'' + xy' + (x^2 - \frac{1}{9})y = 0$.

(g) $xy'' - (2x - 1)y' + (x - 1)y = 0$.

(h) $x^2 y'' + xy' - 4y = 0$.

4. Show that the method described in Examples 1.5 and 1.6 only leads to the trivial solution $y \equiv 0$ when applied to equation (1.33).

1.7. NUMERICAL SOLUTIONS

In addition to the series approach described in the previous section there are a number of numerical techniques for approximating solutions to ordinary and partial differential equations. We will describe here the classical fourth order *Runge–Kutta formula*, which represents one of the popular numerical methods for solving differential equations, along with several numerical differentiation formulas which are used to generate approximate solutions for partial differential equations.

Consider an initial value problem of the form

$$y' = f(x, y), \quad y(x_0) = y_0,$$

and suppose a solution is desired on the interval $[a, b]$ with $x_0 = a$. We will subdivide the interval $[a, b]$ with a uniform *step size h*, so that $a = x_0 < x_1 < \cdots < x_n = b$ and $h = x_{i+1} - x_i$. The symbol y_i will represent the approximation to $y(x_i)$, $i = 0, 1, \ldots, n$. The approximations y_i are computed by the recursion formula

$$y_{i+1} = y_i + \frac{h}{6} (m_1 + 2m_2 + 2m_3 + m_4) \tag{1.46}$$

where

$$m_1 = f(x_i, y_i), \quad m_2 = f\left(x_i + \frac{h}{2}, y_i + m_1 \frac{h}{2}\right)$$

$$m_3 = f\left(x_i + \frac{h}{2}, y_i + m_2 \frac{h}{2}\right), \quad m_4 = f(x_{i+1}, y_i + m_3 h)$$

Example 1.8. Consider the initial value problem

$$y' = x + y, \quad y(0) = 1$$

With $x_0 = 0$, $y_0 = 1$, and $h = 0.1$, we obtain

$$m_1 = f(0, 1) = 1, \quad m_2 = f(0.05, 1.05) = 1.10$$

$$m_3 = f(0.05, 1.055) = 1.105, \quad m_4 = f(0.1, 1.1105) = 1.2105$$

Then (1.46) gives

$$y(0.1) = y(x_1) \doteq y_1 = 1.11034167 \quad \text{(actual solution 1.11034184)}$$

In a similar manner we obtain for the next step,

$$m_1 = 1.21034167, \quad m_2 = 1.32085875$$

$$m_3 = 1.32638460, \quad m_4 = 1.44298013$$

and

$$y(0.2) = y(x_2) \doteq y_2 = 1.24280514 \quad \text{(actual solution 1.24280552)}$$

One can easily check that the function y defined by

$$y = 2e^x - x - 1$$

is the solution to this IVP. The actual solution values listed above were obtained by evaluating this function to eight decimal places.

The Runge–Kutta formula (1.46) can be applied to systems of first order ODEs if y, y_0, m_1, m_2, m_3, m_4, and f are interpreted as vectors. Consider the initial value problem

$$y_1' = f_1(x, y_1, y_2, \ldots, y_k), \quad y_1(x_0) = y_{01}$$
$$y_2' = f_2(x, y_1, y_2, \ldots, y_k), \quad y_2(x_0) = y_{02}$$
$$\cdot$$
$$\cdot$$
$$\cdot$$
$$y_k' = f_k(x, y_1, y_2, \ldots, y_k), \quad y_k(x_0) = y_{0k}$$

This can be written in vector form as

$$y' = f(x, y), \quad y(x_0) = y_0$$

where $y = \langle y_1, y_2, \ldots, y_k \rangle$, $y_0 = \langle y_{01}, y_{02}, \ldots, y_{0k} \rangle$, and $f = \langle f_1, f_2, \ldots, f_k \rangle$. The vector Runge–Kutta formula becomes

$$y_{i+1} = y_i + \frac{h}{6}[m_1 + 2m_2 + 2m_3 + m_4] \tag{1.47}$$

where

$$m_1 = f(x_i, y_i), \quad m_2 = f\left(x_i + \frac{h}{2}, y_i + \frac{h}{2}m_1\right)$$

$$m_3 = f\left(x_i + \frac{h}{2}, y_i + \frac{h}{2}m_2\right), \quad m_4 = f(x_{i+1}, y_i + hm_3)$$

Example 1.9. Consider the system of two equations

$$y_1' = y_1 + 5y_2, \quad y_1(0) = 10$$
$$y_2' = 5y_1 + y_2, \quad y_2(0) = 20$$

Let us apply the Runge–Kutta formula with $h = 0.2$, $x_0 = 0$, $y_0 = \langle 10, 20 \rangle$, and $f = \langle f_1, f_2 \rangle$, where

$$f_1(x, y_1, y_2) = y_1 + 5y_2$$
$$f_2(x, y_1, y_2) = 5y_1 + y_2$$

From the vector formulas given above we obtain

$$m_1 = f(x_0, y_0) = \langle f_1(x_0, y_{01}, y_{02}), f_2(x_0, y_{01}, y_{02}) \rangle = \langle 110, 70 \rangle$$

$$y_0 + \frac{h}{2}m_1 = \langle 10, 20 \rangle + 0.1\langle 110, 70 \rangle = \langle 21, 27 \rangle$$

$$m_2 = f(0.1, 21, 27) = \langle 156, 132 \rangle$$

$$y_0 + \frac{h}{2}m_2 = \langle 10, 20 \rangle + 0.1\langle 156, 132 \rangle = \langle 25.6, 33.2 \rangle$$

$$m_3 = f(0.1, 25.6, 33.2) = \langle 191.6, 161.2 \rangle$$

$$y_0 + hm_3 = \langle 10, 20 \rangle + 0.2\langle 191.6, 161.2 \rangle = \langle 48.32, 52.24 \rangle$$

$$m_4 = f(0.2, 48.32, 52.24) = \langle 309.52, 293.84 \rangle$$

Substituting these in (1.47) we have

$$y_1 = \langle 10, 20 \rangle + 0.033333\langle 1114.72, 950.24 \rangle = \langle 47.157333, 51.674667 \rangle$$

One can easily check that the solution to this initial value problem is given by

$$y_1 = 15e^{6x} - 5e^{-4x}, \quad y_2 = 15e^{6x} + 5e^{-4x}$$

For $x = 0.2$ this gives $\langle 47.555109, 52.048399 \rangle$ for the actual solution to six decimals.

In theory, one can improve the approximations by decreasing the step size; in practice there will be a point of diminishing returns where the round-off error begins to dominate the approximation error. Further discussion of this and other numerical methods for solving ODEs can be found in Conte and de Boor [16] or Szidarovszky and Yakowitz [46].

One technique used to obtain numerical solutions to partial differential equations is to approximate some or all of the derivatives by so-called finite difference formulas. This leads in some cases to a linear system of algebraic equations and in other cases to a system of first order ODEs involving the values of the solution at grid points in the domain. We list without derivation the following numerical differentiation formulas:

$$f'(x) \doteq \frac{f(x+h) - f(x)}{h} \tag{1.48}$$

$$f'(x) \doteq \frac{f(x+h) - f(x-h)}{2h} \tag{1.49}$$

$$f''(x) \doteq \frac{f(x+h) - 2f(x) + f(x-h)}{h^2} \tag{1.50}$$

In these formulas h represents a small positive increment. The formula (1.48) is called a *forward difference approximation*, while (1.49) and (1.50) are referred to as *centered difference approximations*. Derivations and further discussion of these formulas are given in Conte and de Boor [16] or Szidarovszky and Yakowitz [46].

Exercises 1.4

1. Approximate $y(1.4)$ and $y(1.8)$ using the Runge–Kutta formula with step size $h = 0.4$ if

$$y' = 2xy, \quad y(1) = 10$$

2. Consider the system of equations

$$y_1' = (x+2)y_1 + y_1 y_2, \quad y_1(0) = 1$$
$$y_2' = y_1 y_2 + 3xy_2, \quad y_2(0) = 2$$

Approximate $y_1(0.1)$ and $y_2(0.1)$ using the Runge–Kutta formula for systems with step size $h = 0.1$.

3. Let $f(x) = \ln x$ and $h = 0.1$. Approximate
   (a) $f'(3)$ using (i) (1.48) and (ii) (1.49),
   (b) $f''(3)$ using (1.50),
   (c) repeat (a) and (b) using $h = 0.01$.
   In each case compare your approximation to the actual value.

1.8. LINEAR PDEs

A PDE is called *linear* if L is a linear partial differential operator so that

$$Lu = f \tag{1.51}$$

The variable u is dependent and f is a function of the independent variables alone. If the equation is not linear it is described as *nonlinear*. Equation (1.51) is *homogeneous* if $f \equiv 0$; otherwise it is referred to as *nonhomogeneous*. A *solution* for the equation is a function of independent variables which satisfies (1.51). The order of a PDE is the order of its highest order derivative. The following are examples of PDEs.

$$Lu = u_x + u_y = x(x + 2y) \tag{1.52}$$

$$Lu = u_{xy} + u_{yy} = 0 \tag{1.53}$$

$$Lu = u_y u_{yy} + uu_x = 0 \tag{1.54}$$

Equation (1.52) is linear, nonhomogeneous of order 1 with a solution $u = x^2 y$. The second equation (1.53) is linear, homogeneous of order 2. One can verify that $u = \sin x$, $u = e^{y-x}$, $u = g(x)$ and $u = h(y - x)$ are all solutions of (1.53). The functions g and h are arbitrary. The last equation (1.54) is nonlinear, homogeneous of order 2. It has a solution $u = \sin(x + y)$.

For ODEs of nth order, general solutions are families of functions with n arbitrary constants. Instead of arbitrary constants, general solutions for PDEs are arbitrary functions of definite functions. The last two solutions mentioned for (1.53) were arbitrary functions $g(x)$ and $h(y - x)$. This implies that functions e^x, $\cos x$, $\sin(y - x)$, $(y - x)^2$, $\ln(y - x)$, and all others that are appropriately differentiable functions of x alone or $y - x$ are solutions of (1.53). Finding a particular solution from a general solution satisfying a constraint may be a difficult task. It may be preferable to find a particular solution satisfying specified conditions directly.

1.9. CLASSIFICATION OF A LINEAR PDE OF SECOND ORDER

A second order linear PDE with two independent variables has the form

$$Au_{xx} + Bu_{xy} + Cu_{yy} + Du_x + Eu_y + Fu = G \tag{1.55}$$

where coefficients $A, \ldots, G$ are functions of x and y alone. The equation is *hyperbolic*, *elliptic*, or *parabolic* at a specific point in a domain as

$$B^2 - 4AC \qquad\qquad (1.56)$$

is positive, negative, or zero. The classification is analogous to the analytic geometry classification of conic sections. It can be shown by proper coordinate transformation that the nature of (1.55) is invariant and the sign of (1.56) is unaltered. Equation (1.55) can be classified different at different points. Should the coefficients $A, \ldots, G$ be constants, then the equation is a single type for all points of the domain. For details of the classification, and information on canonical forms and characteristic equations, the reader may refer to Sommerfeld [44, pp. 36–43]. Illustrations of the classification follow:

(a) $u_{xx} - u_{yy} = 0$ is hyperbolic with $B^2 - 4AC = 4$.
(b) $u_{xx} + u_{yy} + u = xy$ is elliptic with $B^2 - 4AC = -4$.
(c) $u_{xx} + u_x - u_y + u = 0$ is parabolic with $B^2 - 4AC = 0$.
(d) $u_{xx} + xu_{yy} = 0$ is elliptic, parabolic, or hyperbolic as $x > 0$, $x = 0$, or $x < 0$ since $B^2 - 4AC = -4x$.

1.10. BOUNDARY VALUE PROBLEMS WITH PDEs

A mathematical problem composed of a PDE and certain constraints on the boundary of the domain is called a *boundary value problem*. If u is the dependent variable of the PDE it must satisfy the PDE in a domain of its independent variables and also constraint equations involving u and appropriate partial derivatives of u.

Problems involving time t as one of the independent variables of the PDE may have a condition given at one specified time, frequently when $t = 0$. Such a constraint is referred to as an initial condition. If all the supplementary conditions are initial conditions then the problem is an *initial value problem*. A problem that has both initial and boundary conditions is properly called an *initial-boundary value problem*. In the literature one often finds the use of the terminology *boundary value problem* to include the initial-boundary value problem or mixed problem. In the problem

$$u_t(x, t) = a^2 u_{xx}(x, t), \quad (0 < x < 1, t > 0) \qquad (1.57)$$

$$u(0, t) = u(1, t) = 0, \quad (t \geq 0) \qquad (1.58)$$

$$u(x, 0) = f(x), \quad (0 \leq x \leq 1) \qquad (1.59)$$

the condition (1.59) is an initial condition, while (1.58) are boundary conditions. The problem (1.57)–(1.59) is an initial-boundary value problem or simply a boundary value problem depending on one's preference.

Existence and uniqueness are important topics for boundary or initial value problems of PDEs. At this time we indicate only a Cauchy–Kowalewsky theorem for the second order PDE with initial conditions. For details see Zachmanoglou and Thoe [55, pp. 100–109].

Theorem.* Let

$$u_{tt} = F(t, x, u_t, u_x, u_{tx}, u_{xx}) \tag{1.60}$$

be the PDE with initial conditions

$$u(0, x) = f(x), \quad u_t(0, x) = g(x) \tag{1.61}$$

Functions $f(x)$ and $g(x)$ are defined on an interval of the x axis containing the origin. Assume that $f(x)$ and $g(x)$ are analytic in a neighborhood of the origin and F is analytic in a neighborhood of the point $(0, 0, f(0), g(0), f'(0), g'(0), f''(0))$. Then the problem (1.60), (1.61) has a unique analytic solution $u(x, t)$ in a neighborhood of the origin.

The Cauchy–Kowalewsky theorem serves as an example of an existence-uniqueness theorem for an IVP with a PDE. At a later time we will investigate properties of existence and uniqueness for a few problems of mathematical physics.

A mathematical problem is *well posed* if it has a unique solution that depends continuously on initial or boundary data. The last requirement implied above is sometimes referred to as *stability*. For a mathematical model to describe a specified phenomenon, a small modification in the original data should result only in a small variation of the solution. Even though most of our problems are well posed, it is important to know that there are problems that fail to meet these conditions. From a family of examples attributed to Hadamard [23, pp. 33–34] the elliptic equation

$$u_{xx} + u_{yy} = 0, \quad -\infty < x < \infty, \quad y > 0$$

with the initial conditions on the x axis

$$u(x, 0) = 0, \quad -\infty < x < \infty$$
$$u_y(x, 0) = e^{-\sqrt{n}} \sin nx, \quad -\infty < x < \infty$$

has the solution

*From Zachmanoglou and Thoe [55], by permission of Williams & Wilkins Co.

$$u(x, y) = \frac{e^{-\sqrt{n}}}{n} \sin nx \sinh ny \qquad (1.62)$$

As $n \to \infty$, $e^{-\sqrt{n}} \sin nx \to 0$, but for $x \neq 0$ the solution $(e^{-\sqrt{n}}/n) \sin nx$ $\sinh ny \to \infty$ for any $y \neq 0$. The solution (1.62) fails to depend continuously on the initial data, and therefore is unstable.

1.11. SECOND ORDER LINEAR PDEs WITH CONSTANT COEFFICIENTS

One of the simplest equations in this category is a second order partial derivative equal to a function of the independent variables. Illustrations of this type follow.

Example 1.10. Find a solution for the PDE

$$u_{xy} = xy^2$$

First integrate relative to y with x fixed. Then

$$u_x = \frac{xy^3}{3} + f'(x)$$

where $f'(x)$ is an arbitrary function of x only. A second integration relative to x with y fixed produces the solution

$$u = \frac{x^2 y^3}{6} + f(x) + g(y)$$

where $g(y)$ is an arbitrary function of y alone. Anticipating an integration relative to x, we select an arbitrary function $f'(x)$ in derivative form in the first step.

Example 1.11. Solve the PDE

$$u_{yy} = e^y$$

with the supplementary conditions

$$u_y(x, 0) = x^3 \quad \text{and} \quad u(x, 0) = e^x$$

Integrating the PDE relative to y, one obtains

$$u_y = e^y + f(x)$$

Due to the nature of the first supplementary condition we determine $f(x)$ before finding u.

$$u_y(x, 0) = x^3 = 1 + f(x)$$

This implies that

$$f(x) = x^3 - 1$$

Therefore,

$$u_y = e^y + x^3 - 1$$

Integrating a second time relative to y, one finds

$$u = e^y + x^3 y - y + g(x)$$

To determine $g(x)$ we use the second condition,

$$u(x, 0) = e^x = 1 + g(x)$$

It follows that

$$g(x) = e^x - 1$$

The solution for the problem is

$$u = e^y + x^3 y - y + e^x - 1$$

For a second type, we consider the equation with second partial derivatives only

$$Au_{xx} + Bu_{xy} + Cu_{yy} = 0 \tag{1.63}$$

where A, B, and C are real constants. Let

$$u = f(y + mx) \tag{1.64}$$

be a proposed solution. We attempt to find m so that (1.64) satisfies (1.63). If f is a solution of (1.63) it must be twice differentiable. Substituting (1.64) into (1.63), we obtain

$$Am^2 f''(y + mx) + Bmf''(y + mx) + Cf''(y + mx) = 0$$

If $f''(y + mx) \neq 0$,

$$Am^2 + Bm + C = 0 \tag{1.65}$$

The polynomial equation (1.65) is a characteristic equation. If it has distinct roots $m = m_1$ and $m = m_2$ then $u = f(y + m_1 x)$ and $u = g(y + m_2 x)$ are solutions of (1.63). The linear combination

$$u = f(y + m_1 x) + g(y + m_2 x) \qquad (1.66)$$

is a *general solution* of (1.63).

If m_1 and m_2 are distinct and new variables

$$r = y + m_1 x \quad \text{and} \quad s = y + m_2 x \qquad (1.67)$$

are introduced in (1.63), the new equation is (see Hildebrand [25, Chapter 8])

$$A[m_1^2 u_{rr} + 2m_1 m_2 u_{rs} + m_2^2 u_{ss}] + B[m_1 u_{rr} + (m_1 + m_2) u_{rs} + m_2 u_{ss}]$$
$$+ C[u_{rr} + 2u_{rs} + u_{ss}] = 0 \qquad (1.68)$$

assuming $u_{rs} = u_{sr}$. Equation (1.68) can be simplified so that the coefficients of u_{rr} and u_{ss} are both zero, and

$$u_{rs} = 0 \qquad (1.69)$$

Equation (1.69) is a special type solvable by integration. It has the solution

$$u = f(r) + g(s)$$

Replacing r and s as given in (1.67) one obtains the solution (1.66).

The d'Alembert solution of the wave equation

$$u_{tt} = c^2 u_{xx}, \quad c > 0 \qquad (1.70)$$

is a good illustration of the transformation described in (1.67). Equation (1.70) is hyperbolic. The auxiliary equation is

$$m^2 - c^2 = 0 \qquad (1.71)$$

The transformation (1.67) becomes

$$r = x + ct \quad \text{and} \quad s = x - ct \qquad (1.72)$$

Using (1.72) as described above, we obtain

$$u = f(x + ct) + g(x - ct)$$

for the solution of the wave equation.

The solutions of the characteristic equation (1.65) may be (a) real and distinct, (b) double, or (c) conjugate (imaginary part nonzero) complex numbers. The discriminant for the quadratic equation (1.65), is the same as the discriminant for (1.63). Therefore, a hyperbolic PDE (1.63) is matched by real and distinct roots in (1.65); an elliptic equation (1.63) is paired with conjugate complex roots in (1.65); and a parabolic equation (1.63) is associated with a double root in (1.65).

If $m_1 = m_2$ in (1.65), then $B^2 - 4AC = 0$. The two roots are $m_1 = -B/2A$. A second solution for (1.63) is

$$u = xg(y + m_1 x)$$

This result can be verified if $m_1 = m_2 = -B/2A$ is employed. In this case

$$u = f(y + m_1 x) + xg(y + m_1 x) \tag{1.73}$$

is a general solution for (1.63). One can show that

$$u = f(y + m_1 x) + yg(y + m_1 x) \tag{1.74}$$

is a general solution of (1.63) also.

Example 1.12. Find a general solution for $u_{xx} + 4u_{xy} + 4u_{yy} = 0$.

This equation is parabolic. The characteristic equation has a double root -2. A general solution using (1.73) is

$$u = f(y - 2x) + xg(y - 2x)$$

If (1.74) is used

$$u = f(y - 2x) + yg(y - 2x)$$

is a general solution.

Example 1.13. Determine a solution for $u_{xx} + 4u_{yy} = 0$.

The discriminant $B^2 - 4AC < 0$. Therefore, the equation is elliptic. The characteristic equation has roots $\pm 2i$. The general solution is written in the same form as (1.66). For this PDE

$$u = f(y - 2ix) + g(y + 2ix)$$

is a general solution.

By comparison with an ODE one may suspect the existence of an exponential solution for the homogeneous PDE

$$Au_{xx} + Bu_{xy} + Cu_{yy} + Du_x + Eu_y + Fu = 0 \qquad (1.75)$$

where the coefficients $A, \ldots, F$ are real constants. Let

$$u = e^{\alpha x + \beta y} \qquad (1.76)$$

where α and β are real, be a proposed solution. Substituting (1.76) in (1.75), one obtains the condition

$$A\alpha^2 + B\alpha\beta + C\beta^2 + D\alpha + E\beta + F = 0 \qquad (1.77)$$

In the quadratic equation (1.77), one may solve for β as a function of α or α as a function of β. Assume that we solve for β and obtain $\beta_1(\alpha)$ and $\beta_2(\alpha)$. A particular solution

$$u = K_1 e^{\alpha x + \beta_1(\alpha)y} + K_2 e^{\alpha x + \beta_2(\alpha)y}$$

is the result.

Example 1.14. Determine a solution for the PDE

$$u_{xx} - u_{yy} - 2u_x + u = 0 \qquad (1.78)$$

Substitute the exponential function

$$u = e^{\alpha x + \beta y}$$

in (1.78). The characteristic equation

$$\alpha^2 - \beta^2 - 2\alpha + 1 = 0$$

has solutions

$$\beta = \alpha - 1 \quad \text{and} \quad \beta = -\alpha + 1$$

Using superposition of the two solutions one finds the particular solution

$$u = K_1 e^{\alpha x + (\alpha - 1)y} + K_2 e^{\alpha x + (-\alpha + 1)y}$$

This solution may be written

$$u = K_1 e^{-y} e^{\alpha(x+y)} + K_2 e^{y} e^{\alpha(x-y)}$$

We may conjecture that a general solution has the form

$$u = e^{-y}f(x + y) + e^{y}g(x - y) \qquad (1.79)$$

where f and g are twice differentiable arbitrary functions. By substituting (1.79) into (1.78), we confirm that (1.79) is a solution.

When the left member of (1.77) has distinct linear factors, the type of simplification discussed is possible. The case of a repeated linear factor may be considered by using a result comparable to (1.73) or (1.74).

Example 1.15. Examine

$$u_{xx} - 2u_{xy} + u_{yy} - 2u_y + 2u_x + u = 0$$

for a general solution.

Let $u = e^{\alpha x + \beta y}$ and obtain a characteristic equation

$$\alpha^2 - 2\alpha\beta + \beta^2 - 2\beta + 2\alpha + 1 = 0$$

The double root is

$$\beta = \alpha + 1$$

An exponential form of a solution is

$$u = e^y [K_1 e^{\alpha(x+y)} + K_2 x e^{\alpha(x+y)}]$$

A general solution

$$u = e^y [f(x + y) + xg(x + y)]$$

can be verified.

Certain cases may arise in (1.77) where linear factors with imaginary elements appear.

Example 1.16. Investigate a solution for the equation

$$u_{xx} + u_{yy} - 2u_y + u = 0 \tag{1.80}$$

Let

$$u = e^{\alpha x + \beta y}$$

be a proposed solution. The characteristic equation

$$\alpha^2 + \beta^2 - 2\beta + 1 = 0$$

has two linear factors with imaginary elements for which

$$\beta = 1 \pm i\alpha$$

An exponential solution is

$$u = e^y[e^{\alpha(x+iy)} + e^{\alpha(x-iy)}] \tag{1.81}$$

A general solution for (1.80) is suggested by (1.81)

$$u = e^y[f(x + iy) + g(x - iy)] \tag{1.82}$$

It is easy to verify that (1.82) is a solution of (1.80).

In some situations the exponential procedure may produce a set of useful particular solutions, but fail to suggest a general solution.

Example 1.17. Determine a solution for the equation

$$u_{xx} + u_{yy} + 4u = 0$$

One obtains a characteristic equation

$$\alpha^2 + \beta^2 + 4 = 0$$

with

$$\beta = \pm i\sqrt{\alpha^2 + 4}$$

If the exponential substitution is followed then

$$u = e^{\alpha x}[K_1 e^{i\sqrt{\alpha^2+4}\, y} + K_2 e^{-i\sqrt{\alpha^2+4}\, y}]$$

This solution can be expressed

$$u = e^{\alpha x}[M_1 \cos \sqrt{\alpha^2 + 4}\, y + M_2 \sin \sqrt{\alpha^2 + 4}\, y]$$

if K_1 and K_2 are properly related to M_1 and M_2 using Euler's identity.

Equation (1.75) can be solved almost like an ODE if only partial derivatives with respect to one variable appear. Arbitrary constants of the ODE solution become arbitrary functions of the remaining variable.

Example 1.18. Solve the PDE

$$u_{yy} - 4u_y + 3u = 0$$

The dependent variable u is a function of x and y, but the only derivatives involved are relative to y alone. The corresponding ODE, with u as a function of y,

$$\frac{d^2u}{dy^2} - 4\frac{du}{dy} + 3u = 0$$

has a solution

$$u = c_1 e^{3y} + c_2 e^y$$

Arbitrary constants c_1 and c_2 are replaced by arbitrary functions of x alone. The general solution becomes

$$u = e^{3y}f(x) + e^y g(x)$$

Other PDEs may be solved by using comparable solutions of ODEs.

Example 1.19. Find a solution for the PDE

$$xu_{xy} + 2u_y = y^2$$

We observe that the equation may be written

$$\frac{\partial}{\partial y}[xu_x + 2u] = y^2$$

By integrating, we obtain

$$xu_x + 2u = \frac{y^3}{3} + f(x)$$

Dividing by x, with y fixed, one recognizes a linear differential equation of first order

$$u_x + \frac{2}{x}u = \frac{y^3}{3x} + \frac{f(x)}{x}$$

The integrating factor is x^2. This equation may be displayed

$$\frac{\partial}{\partial x}(x^2 u) = \frac{xy^3}{3} + xf(x)$$

Integrating the most recent equation, we obtain

$$x^2 u = \frac{x^2 y^3}{6} + f^*(x) + G(y)$$

An explicit form of the solution is

$$u = \frac{y^3}{6} + F(x) + \frac{1}{x^2}G(y)$$

For more information regarding Section 1.11 the reader may consult Hildebrand [25, Chapter 8].

Exercises 1.5

1. Solve the boundary value problem

$$u_{xy} = 0, \quad u_x(x, 0) = \cos x, \quad u\left(\frac{\pi}{2}, y\right) = \sin y$$

2. Find the solution for

$$u_{yx} = x^2 y, \quad u_y(0, y) = y^2, \quad u(x, 1) = \cos x$$

3. Determine a solution for $u_{xx} = \cos x$ if

$$u(0, y) = y^2 \quad \text{and} \quad u(\pi, y) = \pi \sin y$$

4. Classify the following PDEs as hyperbolic, parabolic or elliptic:
   (a) $yu_{xx} + xu_{yy} = 0$;
   (b) $x^2 u_{xx} + 2xyu_{xy} + y^2 u_{yy} + u_x + u_y = 0$;
   (c) $u_{xx} + 2u_{xy} - 3u_{yy} = 0$;
   (d) $u_{xx} - 2u_{xy} + u_{yy} = 0$;
   (e) $u_{xx} + a^2 u_{yy} = 0, \ a > 0$;
   (f) $u_{xx} - 2u_{xy} + 2u_{yy} = 0$.
   Solve the equations (c)–(f).
5. The d'Alembert solution of the wave equation (1.70) is

$$u = f(x + ct) + g(x - ct)$$

   Solve the wave equation if $u(x, 0) = 0$ and $u_t(x, 0) = \phi(x)$.
6. (a) Determine a general solution for (c) in Exercise 4 by using the transformations $s = y - 3x$, $r = y + x$.
   (b) If $u(0, y) = 0$ and $u_x(0, y) = \phi(y)$ in (a), show that

$$u = \frac{1}{4} \int_{y-3x}^{y+x} \phi(\alpha) \, d\alpha$$

7. Determine a solution for $u_{xx} + 2u_{xy} + u_{yy} + u_x + u_y = 0$ by letting $u = e^{\alpha x + \beta y}$. After finding β as a function of α, propose a general solution. Verify the general solution.
8. Using the substitution $u = e^{\alpha x + \beta y}$
   (a) find an exponential solution for $4u_{xx} - u_{yy} - 4u_x + 2u_y = 0$;
   (b) propose and verify a general solution for the equation.

9. Solve the PDE $xu_{xy} + 3u_y = y^3$.

10. If $Au_{xx} + Bu_{xy} + Cu_{yy} = F(x, y)$, A, B, and C are constants, then the equation has a general solution

$$u = u_c(x, y) + u_p(x, y)$$

where $u_c(x, y)$ is a general solution of $Au_{xx} + Bu_{xy} + Cu_{yy} = 0$ and $u_p(x, y)$ is a particular solution of the original equation. Find a general solution for the following equations:

(a) $u_{xx} - 2u_{xy} - 3u_{yy} = e^x$;

(b) $u_{xx} - u_{xy} - 2u_{yy} = \sin y$.

1.12. SEPARATION OF VARIABLES

It is assumed in this method that the solution of a PDE can be expressed in the form of a product of functions of single independent variables. Using this procedure we produce an equation with one member a function of a single variable and the other member a function of the remaining variables. Each member can be a constant but not a function of all the original independent variables. This process is illustrated in the following examples.

Example 1.20. Find a solution for the PDE

$$u_t = 4u_{xx} \tag{1.83}$$

using the separation of variables.

We assume that the solution of (1.83) has the form

$$u(x, t) = X(x)T(t) \tag{1.84}$$

where X is a function of x alone and T is a function of t alone. Inserting (1.84) into (1.83) we obtain

$$XT' = 4X''T$$

After dividing by $4XT$, one has the variables separated in the form

$$\frac{T'}{4T} = \frac{X''}{X} \tag{1.85}$$

If (1.85) is differentiated partially relative to t, one attains the result

$$\frac{\partial}{\partial t}\left(\frac{T'}{4T}\right) = 0 \tag{1.86}$$

Assuming ϕ is an arbitrary function of x alone, the solution of (1.86) is

$$\frac{T'}{4T} = \phi(x)$$

This violates the condition that T is a function of t alone unless $\phi(x)$ is a constant. A similar partial differentiation of (1.85) relative to x leads to a PDE which has a solution

$$\frac{X''}{X} = \psi(t)$$

valid only if $\psi(t)$ is constant. Therefore both members of (1.85) must be equal to the same constant, say α^2 or $-\alpha^2$.

If α^2 is used (1.85) becomes

$$\frac{T'}{4T} = \frac{X''}{X} = \alpha^2 \qquad (1.87)$$

Result (1.87) is equivalent to two ODEs

$$T' - 4\alpha^2 T = 0, \quad X'' - \alpha^2 X = 0 \qquad (1.88)$$

The solutions of the two ODEs of (1.88) are respectively,

$$T = A e^{4\alpha^2 t}, \quad X = B_1 e^{\alpha x} + B_2 e^{-\alpha x} \qquad (1.89)$$

Inserting the solutions of (1.89) in (1.84) we find a solution

$$u(x, y) = e^{4\alpha^2 t}[C_1 e^{\alpha x} + C_2 e^{-\alpha x}]$$

where $C_1 = AB_1$ and $C_2 = AB_2$.

If $-\alpha^2$ is used instead of α^2 in (1.87) the two ODEs are

$$T' + 4\alpha^2 T = 0, \quad X'' + \alpha^2 X = 0 \qquad (1.90)$$

The solutions of (1.90) are

$$T = A^* e^{-4\alpha^2 t}, \quad X = B_1^* \cos \alpha x + B_2^* \sin \alpha x \qquad (1.91)$$

Using the solutions of (1.91) in (1.84) we have

$$u = e^{-4\alpha^2 t}[C_1^* \cos \alpha x + C_2^* \sin \alpha x]$$

In most of our BVPs a bounded solution will be necessary. The constants α^2 or $-\alpha^2$ must be selected to satisfy this requirement.

Example 1.21. Determine a solution for

$$u_t = a^2(u_{xx} + u_{yy}) \tag{1.92}$$

Since three independent variables appear in (1.92), we let

$$u(x, y, t) = T(t)X(x)Y(y) \tag{1.93}$$

Equation (1.92) has the form

$$T'XY = a^2(TX''Y + TXY'') \tag{1.94}$$

after substituting (1.93) in the PDE. Equation (1.94) has another form

$$\frac{T'}{a^2 T} = \frac{X''}{X} + \frac{Y''}{Y} \tag{1.95}$$

Partially differentiating (1.95) relative to x, then y, and finally t, we have respectively

$$\frac{\partial}{\partial x}\left(\frac{X''}{X}\right) = 0, \quad \frac{\partial}{\partial y}\left(\frac{Y''}{Y}\right) = 0, \quad \frac{\partial}{\partial t}\left(\frac{T'}{a^2 T}\right) = 0 \tag{1.96}$$

Solutions of the three PDEs of (1.96) are

$$\frac{X''}{X} = -\alpha^2, \quad \frac{Y''}{Y} = -\beta^2, \quad \frac{T'}{a^2 T} = -(\alpha^2 + \beta^2) \tag{1.97}$$

In order that (1.95) be satisfied we select $-(\alpha^2 + \beta^2)$ as the constant in the solution of the T equation.

The three associated ODEs

$$X'' + \alpha^2 X = 0$$
$$Y'' + \beta^2 Y = 0$$
$$T' + (\alpha^2 + \beta^2)a^2 T = 0$$

have solutions

$$X = B_1 \cos \alpha x + B_2 \sin \alpha x$$
$$Y = C_1 \cos \beta y + C_2 \sin \beta y$$
$$T = A \exp[-(\alpha^2 + \beta^2)a^2 t]$$

Therefore,

$$u = \exp[-(\alpha^2 + \beta^2)a^2 t][B_1^* \cos \alpha x + B_2^* \sin \alpha x][C_1 \cos \beta y + C_2 \sin \beta y]$$

is a solution of (1.92). Other forms for the solution are available. The one displayed is a bounded solution.

The method of separation of variables is valuable for solving a number of important problems of mathematical physics, yet it fails for many PDEs and BVPs. Myint-U [35, pp. 128–129] shows that the second order PDE* with variable coefficients in x and y

$$A(x, y)u_{xx} + C(x, y)u_{yy} + D(x, y)u_x + E(x, y)u_y + F(x, y)u = 0 \tag{1.98}$$

is separable when a functional multiplier $1/[\phi(x, y)]$ converts the new equation

$$A(x, y)X''Y + C(x, y)XY'' + D(x, y)X'Y + E(x, y)XY' + F(x, y)XY = 0$$

into the form

$$A_1(x)X''Y + B_1(y)XY'' + A_2(x)X'Y + B_2(y)XY' + [A_3(x)$$
$$+ B_3(y)]XY = 0$$

Explicit rules for the workability of this method are a bit elusive. Types of differential equations, kinds of coordinate systems, and forms of boundary conditions are all important items for the success of the procedure.

Exercises 1.6

1. Test the following PDEs for the method of separation of variables. If the method is successful, solve the PDE.
   (a) $u_{xy} - u = 0$.
   (b) $u_{tt} - u_{xx} = 0$.
   (c) $u_{xx} - u_{yy} - 2u_y = 0$.
   (d) $u_{xx} - u_{yy} + 2u_x - 2u_y + u = 0$.
   (e) $t^2 u_{tt} - x^2 u_{xx} = 0$.
   (f) $(t^2 + x^2)u_{tt} + u_{xx} = 0$.
   (g) $u_{xx} - y^2 u_{yy} - yu_y = 0$.
   (h) $u_{xy} = 0$.
   (i) $u_{xx} - u_{xy} + u_{yy} = 2x$.
   (j) $u_{xx} - u_{yy} - u_y = 0$.
   (k) $u_t = u_{xx}$.

*The example that follows is from Myint-U [35], by permission of Elsevier/North-Holland, Inc.

2. Find a solution for the boundary (or initial) value problems:

(a) $u_{tt} - u_{xx} = 0$, $\quad u(x, 0) = u(0, t) = 0$;

(b) $u_{xx} - u_{yy} - 2u_y = 0$, $\quad u_x(0, y) = u(x, 0) = 0$;

(c) $u_t = u_{xx}$, $\quad u_x(0, t) = 0$.

3. (a) Show that the equation with constant coefficients

$$Au_{xx} + Bu_{xy} + Cu_{yy} = 0$$

is separable if the coefficients meet proper conditions. Determine appropriate conditions. *Note:* Let $u(x, y) = X(x)Y(y)$ and show that a result

$$\left(\frac{X''}{X}\right)' + \frac{B}{A}\left(\frac{X'}{X}\right)'\left(\frac{Y'}{Y}\right) = 0$$

is obtained from

$$\frac{X''}{X} + \frac{B}{A}\frac{X'}{X}\frac{Y'}{Y} + \frac{C}{A}\frac{Y''}{Y} = 0$$

Finally, show that

$$Y'' + \lambda Y = 0 \quad \text{and} \quad X'' - \lambda \frac{B}{A}X' + \lambda^2 \frac{C}{A}X = 0$$

are related ODEs.

(b) Find a solution for $u_{xx} - u_{xy} + u_{yy} = 0$ by separating variables.

2

ORTHOGONAL SETS OF FUNCTIONS

The first concept of orthogonality for many of us is associated with perpendicularity in geometry. Here our first reference to orthogonality and orthonormality pertains to right angle relations with vectors. Later we define orthogonal and orthonormal functions. We describe several types of orthogonality. Finally we consider a BVP that has a solution set of special orthogonal functions. The formation of a series based upon a set of orthogonal functions is fundamental for our development of Fourier series.

2.1. ORTHOGONALITY AND VECTORS

Vectors furnish good examples of orthogonal sets. Using the component form, $\mathbf{i} = \langle 1, 0, 0 \rangle$, $\mathbf{j} = \langle 0, 1, 0 \rangle$, and $\mathbf{k} = \langle 0, 0, 1 \rangle$. Then the set $\{\mathbf{i}, \mathbf{j}, \mathbf{k}\}$ is an orthogonal set of unit position vectors. This means that $\mathbf{i}$, $\mathbf{j}$, and $\mathbf{k}$ are mutually perpendicular. If vector $\mathbf{A} = \langle a_1, a_2, a_3 \rangle$, then its length or norm $\|\mathbf{A}\| = (a_1^2 + a_2^2 + a_3^2)^{1/2}$. If a second vector $\mathbf{B} = \langle b_1, b_2, b_3 \rangle$, then the inner product of $\mathbf{A}$ and $\mathbf{B}$ is defined by

$$\mathbf{A} \cdot \mathbf{B} = (\mathbf{A}, \mathbf{B}) = \|\mathbf{A}\| \, \|\mathbf{B}\| \cos \theta = a_1 b_1 + a_2 b_2 + a_3 b_3 \qquad (2.1)$$

where θ is the angle between the two vectors. See Figure 2.1.

Consider a set of orthogonal vectors $\{\mathbf{e}_1, \mathbf{e}_2, \mathbf{e}_3\}$. Since the set is orthogonal, the definition (2.1) implies that

$$(\mathbf{e}_1, \mathbf{e}_2) = (\mathbf{e}_1, \mathbf{e}_3) = (\mathbf{e}_2, \mathbf{e}_3) = 0 \qquad (2.2)$$

Let vector $\mathbf{V} = \langle v_1, v_2, v_3 \rangle$ be related to $\{\mathbf{e}_r\}$, $r = 1, 2, 3$ by

$$\mathbf{V} = v_1 \mathbf{e}_1 + v_2 \mathbf{e}_2 + v_3 \mathbf{e}_3 = \sum_{r=1}^{3} v_r \mathbf{e}_r \qquad (2.3)$$

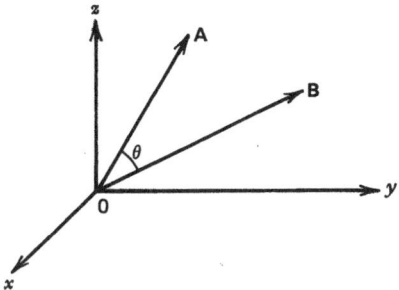

Figure 2.1. Two position vectors in a three dimensional rectangular coordinate system.

This implies that $\mathbf{V}$ is referenced to a coordinate system having axes along which vectors $\mathbf{e}_1, \mathbf{e}_2, \mathbf{e}_3$ lie as position vectors. If $\mathbf{V} = u_1\mathbf{i} + u_2\mathbf{j} + u_3\mathbf{k} = \langle u_1, u_2, u_3 \rangle_{\mathbf{ijk}}$ then $\mathbf{V}$ is related to $\mathbf{i}, \mathbf{j}, \mathbf{k}$ referenced to the x, y, z axes. Therefore,

$$u_1\mathbf{i} + u_2\mathbf{j} + u_3\mathbf{k} = v_1\mathbf{e}_1 + v_2\mathbf{e}_2 + v_3\mathbf{e}_3$$

Using (2.1), and assuming that $\{\mathbf{e}_1, \mathbf{e}_2, \mathbf{e}_3\}$ is related to $\{\mathbf{i}, \mathbf{j}, \mathbf{k}\}$,

$$\langle u_1, u_2, u_3 \rangle \cdot \mathbf{e}_1 = \|\mathbf{e}_1\|^2 v_1$$
$$\langle u_1, u_2, u_3 \rangle \cdot \mathbf{e}_2 = \|\mathbf{e}_2\|^2 v_2$$
$$\langle u_1, u_2, u_3 \rangle \cdot \mathbf{e}_3 = \|\mathbf{e}_3\|^2 v_3$$

Then

$$v_1 = \|\mathbf{e}_1\|^{-2} \langle u_1, u_2, u_3 \rangle \cdot \mathbf{e}_1$$
$$v_2 = \|\mathbf{e}_2\|^{-2} \langle u_1, u_2, u_3 \rangle \cdot \mathbf{e}_2 \qquad (2.4)$$
$$v_3 = \|\mathbf{e}_3\|^{-2} \langle u_1, u_2, u_3 \rangle \cdot \mathbf{e}_3$$

Example 2.1. If the reference set of vectors $\{\mathbf{e}_1, \mathbf{e}_2, \mathbf{e}_3\}$ is given by $\mathbf{e}_1 = \langle 1, 1, 0 \rangle$, $\mathbf{e}_2 = \langle -1, 1, 0 \rangle$, and $\mathbf{e}_3 = \langle 0, 0, 1 \rangle$, find $\mathbf{V}_{\mathbf{ijk}} = \langle 1, 2, 3 \rangle$ related to $\{\mathbf{e}_1, \mathbf{e}_2, \mathbf{e}_3\}$ or $\mathbf{V}_{\mathbf{e}_1\mathbf{e}_2\mathbf{e}_3}$.

Using (2.1), we observe that

$$(\mathbf{e}_1, \mathbf{e}_2) = (\mathbf{e}_1, \mathbf{e}_3) = (\mathbf{e}_2, \mathbf{e}_3) = 0$$

and the set $\{\mathbf{e}_1, \mathbf{e}_2, \mathbf{e}_3\}$ is orthogonal. If we let

$$\mathbf{i} + 2\mathbf{j} + 3\mathbf{k} = v_1\mathbf{e}_1 + v_2\mathbf{e}_2 + v_3\mathbf{e}_3$$

then

$$\langle 1, 2, 3 \rangle \cdot \langle 1, 1, 0 \rangle = \| \langle 1, 1, 0 \rangle \|^2 v_1$$
$$\langle 1, 2, 3 \rangle \cdot \langle -1, 1, 0 \rangle = \| \langle -1, 1, 0 \rangle \|^2 v_2$$
$$\langle 1, 2, 3 \rangle \cdot \langle 0, 0, 1 \rangle = \| \langle 0, 0, 1 \rangle \|^2 v_3$$

Completing the computation, we have $v_1 = \frac{3}{2}$, $v_2 = \frac{1}{2}$, $v_3 = 3$. The vector $V_{e_1 e_2 e_3} = \frac{3}{2} e_1 + \frac{1}{2} e_2 + 3 e_3$.

If in addition to the conditions (2.2) we have

$$\| e_1 \| = \| e_2 \| = \| e_3 \| = 1$$

then the set $\{e_r\}$, $r = 1, 2, 3$, is composed of vectors having norms of 1. In this case

$$(e_r, e_s) = \begin{cases} 0 & \text{when } r \neq s \\ 1 & \text{when } r = s \end{cases}$$

and the set $\{e_r\}$, $r = 1, 2, 3$, is referred to as an *orthonormal set of vectors*. The set $\{i, j, k\}$ is an orthonormal set of vectors. The set in Example 2.1 is orthogonal, but not orthonormal. If the set of orthonormal vectors $\{e_r\}$, $r = 1, 2, 3$, is used as a *basis* or reference set for (2.3) then (2.4) becomes

$$v_1 = \langle u_1, u_2, u_3 \rangle \cdot e_1$$
$$v_2 = \langle u_1, u_2, u_3 \rangle \cdot e_2$$
$$v_3 = \langle u_1, u_2, u_3 \rangle \cdot e_3$$

The idea we have expressed can be generalized so that the vectors have n components and an orthogonal basis $\{e_r\}$, $r = 1, 2, \ldots, n$. Assume a vector $V = \langle u_1, u_2, \ldots, u_n \rangle_{a_1 a_2 \ldots a_n}$ where $a_1 = \langle 1, 0, \ldots, 0 \rangle$, $a_2 = \langle 0, 1, \ldots, 0 \rangle, \ldots, a_n = \langle 0, 0, \ldots, 1 \rangle$. The set $\{a_r\}$, $r = 1, 2, \ldots, n$, is orthonormal. If the set $\{e_r\}$, $r = 1, 2, \ldots, n$ is related to $\{a_r\}$, $r = 1, 2, \ldots, n$, then

$$u_1 a_1 + u_2 a_2 + \cdots + u_n a_n = v_1 e_1 + v_2 e_2 + \cdots + v_n e_n$$

and one obtains

$$v_r = \| e_r \|^{-2} \langle u_1, u_2, \ldots, u_n \rangle \cdot e_r, \quad r = 1, 2, \ldots, n$$

2.2. ORTHOGONAL FUNCTIONS

It is possible to consider a function $A(x)$, $a \leq x \leq b$, analogous to a vector having an infinity of components, each component specified by the value of

$A(x)$ at a particular value of $x \in (a, b)$. Instead of using a sum in this case we use a limit of a sum or an integral.

The norm of $A(x)$ is defined by

$$\|A(x)\|^2 = \int_a^b [A(x)]^2 \, dx$$

and the inner product of two functions $A(x)$ and $B(x)$, $a \le x \le b$, by

$$(A, B) = \int_a^b A(x)B(x) \, dx$$

For the analogy to be extended, the condition that $A(x)$ and $B(x)$ are orthogonal is defined by

$$(A, B) = \int_a^b A(x)B(x) \, dx = 0 \tag{2.5}$$

As a special case, the inner product

$$(A, A) = \|A\|^2$$

Although we have suggested some analogies with functions and vectors, we hasten to add that our geometrical significance is gone. The concept of orthogonality as defined by (2.5) bears fruit when a study of Fourier series is undertaken.

If we consider an orthogonal set of functions $\{f_n(x)\}$, $n \in \mathbf{N}$ (N the set of natural numbers), $a \le x \le b$, then

$$(f_n, f_m) = \int_a^b f_n(x)f_m(x) \, dx = 0 \quad \text{when } n \ne m$$

If the set $\{g_n(x)\}$, $n \in \mathbf{N}$, $a \le x \le b$, is orthonormal, then

$$(g_n, g_m) = \begin{cases} 0 & \text{when } n \ne m \\ 1 & \text{when } n = m \end{cases}$$

A set $\{f_n(x)\}$ which is orthogonal, but not orthonormal, can be transformed into an orthonormal set by dividing each function of the set by its norm $\|f_n\|$. Naturally this process of *normalization* is possible only if all norms are nonzero.

Example 2.2. Show that the set of functions $\{\sin nx\}$, $0 \le x \le \pi$, $n \in \mathbf{N}$, is orthogonal. Find a normalizing factor and display the corresponding orthonormal set.

$$(\sin nx, \sin mx) = \int_0^{\pi} \sin nx \sin mx \, dx = 0 \quad \text{if } n \neq m$$

$$\|\sin nx\|^2 = (\sin nx, \sin nx) = \int_0^{\pi} \sin^2 nx \, dx = \frac{\pi}{2}$$

The norm of $\sin nx$ is $\sqrt{\pi/2}$, and the orthonormal set corresponding to $\{\sin nx\}$ is $\{\sqrt{2/\pi} \sin nx\}$, $n \in N$, $0 \leq x \leq \pi$.

Exercises 2.1

1. The vector $V = 2i + 3j - k$. Find its representation for the basis $e_1 = i + j$, $e_2 = -i + j + k$, $e_3 = i - j + 2k$.
2. (a) Given the set of vectors $e_1 = i + 2k$, $e_2 = -2i + \alpha j + k$, $e_3 = 2i + j + \beta k$, determine α and β so that $\{e_1, e_2, e_3\}$ is an orthogonal set of vectors.
   (b) If $V = 2e_1 + 3e_2 - e_3$, determine V_{ijk}.
   (c) Check your transformation by assuming V_{ijk} obtained in (b) and use (2.4) to verify that $V = 2e_1 + 3e_2 - e_3$.
3. (a) If $\{H_1, H_2, H_3\}$ is a linearly independent set of vectors ($\alpha_1 H_1 + \alpha_2 H_2 + \alpha_3 H_3 = 0$ implies that $\alpha_1 = \alpha_2 = \alpha_3 = 0$ only) then $\{H_1, H_2, H_3\}$ is a basis. We let

$$K_1 = K_1$$
$$K_2 = H_2 + A_{22}K_1$$
$$K_3 = H_3 + A_{32}K_2 + A_{33}K_1$$

with A_{22}, A_{32}, and A_{33} scalars. If the set $\{K_1, K_2, K_3\}$ is designed as orthogonal, show that

$$A_{22} = -\frac{(K_1, H_2)}{(K_1, K_1)}, \quad A_{32} = -\frac{(K_2, H_3)}{(K_2, K_2)}, \quad A_{33} = -\frac{(K_1, H_3)}{(K_1, K_1)}$$

This is the well-known *Gram–Schmidt Orthogonalization Process* for 3-space vectors. For the basis $\{H_1, H_2, \ldots, H_n\}$ and the orthogonal set $\{K_1, K_2, \ldots, K_n\}$, the relationships of the vectors follow the pattern above with

$$K_n = H_n + A_{n2}K_{n-1} + A_{n3}K_{n-2} + \cdots + A_{nn}K_1$$

   (b) If $H_1 = i + 2j + k$, $H_2 = i - j + 2k$, $H_3 = 2i - j + k$, find an orthogonal set $\{K_1, K_2, K_3\}$ by the Gram–Schmidt process.

4. (a) Show that the set of functions $\{\sin n\pi x\}$, $-1 < x < 1$, $n \in \mathbb{N}$, is orthogonal.

   (b) Is the set in part (a) orthonormal?

5. (a) Find α so that $\{1, x, 1 + \alpha x^2\}$ on $(-1, 1)$ is orthogonal.

   (b) Normalize the set obtained in (a).

6. (a) Show that the set $\{1, \cos(n\pi x/L), \sin(m\pi x/L)\}$, $n, m \in \mathbb{N}$, $-L < x < L$, is orthogonal but not orthonormal.

   (b) Normalize the set of (a).

7. (a) Is the set $\{\cos(n\pi x/2)\}$, $n \in \mathbb{N}_0$ ($\mathbb{N}_0$ is the set of natural numbers plus 0), $0 < x < 2$, orthonormal?

   (b) If it fails to be orthonormal, write the corresponding orthonormal set.

8. Given the set of nontrivial orthogonal functions $\{f_n(x)\}$, $a \le x \le b$, $n \in \mathbb{N}_0$, show that the set is linearly independent.

9. The set of continuous functions $\{g_n(x)\}$, $a \le x \le b$, $n \in \mathbb{N}$, is linearly independent. A new set $\{f_n(x)\}$, $a \le x \le b$, $n \in \mathbb{N}$, designed to be orthogonal, has the relationships with the first set:

$$f_1 = g_1$$
$$f_2 = g_2 + A_{22}f_1$$
$$f_3 = g_3 + A_{32}f_2 + A_{33}f_1$$
$$\cdot$$
$$\cdot$$
$$\cdot$$
$$f_n = g_n + A_{n2}f_{n-1} + A_{n3}f_{n-2} + \cdots + A_{nn}f_1$$

where $A_{22}, A_{32}, A_{33}, \ldots, A_{n2}, A_{n3}, \ldots, A_{nn}$ are constants. As with vectors, this is the *Gram–Schmidt orthogonalization process* for functions. Show that for the set $\{g_1, g_2, g_3\}$ on $[a, b]$, the orthogonal set given by the Gram–Schmidt process is

$$f_1 = g_1$$

$$f_2 = g_2 - \frac{(f_1, g_2)}{(f_1, f_1)} f_1$$

$$f_3 = g_3 - \frac{(f_2, g_3)}{(f_2, f_2)} f_2 - \frac{(f_1, g_3)}{(f_1, f_1)} f_1$$

10. The set $\{1, x, x^2, x^3\}$, $-1 \le x \le 1$, is linearly independent and continuous. By the Gram–Schmidt process of Exercise 9, find the corresponding orthogonal set.

11. The integral $\int_a^b \int_a^b [f(x)g(y) - f(y)g(x)]^2 \, dx \, dy \geq 0$. Show that $\frac{1}{2} \int_a^b \int_a^b [f(x)g(y) - f(y)g(x)]^2 \, dx \, dy = \|f\|^2 \|g\|^2 - (f, g)^2$ and prove the *Schwarz inequality*: $(f, g) \leq \|f\| \|g\|$.

12. If f and g are continuous functions on (a, b) and one function has a zero norm, then the inner product $(f, g) = 0$. Show this statement is true. Use Exercise 11.

2.3. COMPLEX FUNCTIONS

Before discussing an orthogonality concept involving complex functions, we state a few definitions and operations on these functions. A complex function of a real variable is defined by

$$w(x) = u(x) + iv(x)$$

where u (real part of w) and v (imaginary part of w) are both real functions of the real variable x. If $v(x)$ is identically zero, then $w(x)$ is a real function. The complex conjugate of w, denoted by $\bar{w}$, is defined by

$$\overline{w(x)} = u(x) - iv(x)$$

The absolute value of w, written $|w|$, is

$$|w| = [u^2 + v^2]^{1/2}$$

For $w(x)$

$$w'(x) = u'(x) + iv'(x)$$

and

$$\int_a^b w(x) \, dx = \int_a^b u(x) \, dx + i \int_a^b v(x) \, dx$$

We define a complex function of a complex variable z by

$$w(z) = u(x, y) + iv(x, y)$$

where $z = x + iy$ with u and v real functions of the real variables x and y. The exponential function $\exp z$ or e^z is expressed by

$$\exp z = e^x(\cos y + i \sin y)$$

Other definitions of elementary functions follow:

$$\sin z = \frac{1}{2i} (e^{iz} - e^{-iz})$$

$$\cos z = \frac{1}{2} (e^{iz} + e^{-iz})$$

$$\sinh z = \frac{1}{2} (e^z - e^{-z})$$

$$\cosh z = \frac{1}{2} (e^z + e^{-z})$$

The derivatives of these elementary complex functions follow forms of the derivatives of the corresponding real elementary functions. For example,

$$\frac{d}{dz} (e^z) = e^z$$

$$\frac{d}{dz} (\sin z) = \cos z$$

$$\frac{d}{dz} (\cos z) = -\sin z$$

If $z(s) = x(s) + iy(s)$, then

$$\frac{d}{ds} (e^z) = e^z \frac{dz}{ds}$$

$$\frac{d}{ds} (\sin z) = \cos z \frac{dz}{ds}$$

$$\frac{d}{ds} (\cos z) = -\sin z \frac{dz}{ds}$$

The integral

$$\int_a^b e^{z(s)} z'(s) \, ds = [e^{z(s)}]_a^b = e^{z(b)} - e^{z(a)}$$

2.4. ADDITIONAL CONCEPTS OF ORTHOGONALITY

If $f_n(x) = u_n(x) + iv_n(x)$, then the set of complex functions $\{f_n(x)\}$, $a \leqslant x \leqslant b$, is *orthogonal in the Hermitian sense* if

$$\int_a^b f_n(x)\overline{f_m(x)} \, dx = 0 \quad \text{when } m \neq n$$

The square of the norm in this case is

$$\|f_n(x)\|^2 = \int_a^b f_n(x)\overline{f_n(x)} \, dx = \int_a^b [u_n^2(x) + v_n^2(x)] \, dx$$

Example 2.3. Show that the set of functions $\{\exp[2n\pi ix/(b-a)]\}$, $a \le x \le b$, $n \in \mathbf{Z}$ ($\mathbf{Z}$ is the set of all integers), is orthogonal in the Hermitian sense.

To show Hermitian orthogonality we must demonstrate that the integral

$$\int_a^b \exp\left(\frac{2n\pi ix}{b-a}\right) \overline{\exp\left(\frac{2m\pi ix}{b-a}\right)} \, dx \tag{2.6}$$

is zero when $n \ne m$. It can be shown that

$$\overline{\exp\left(\frac{2m\pi ix}{b-a}\right)} = \exp\left(\frac{-2m\pi ix}{b-a}\right)$$

Integral (2.6) may be rewritten

$$\int_a^b \exp\left(\frac{2(n-m)\pi ix}{b-a}\right) dx = \frac{b-a}{2(n-m)\pi i}\left[\exp\left(\frac{2(n-m)\pi ix}{b-a}\right)\right]_a^b$$

$$= \frac{b-a}{2(n-m)\pi i}\exp\left(\frac{2(n-m)\pi ia}{b-a}\right)\left[\exp\left(\frac{2(n-m)\pi i(b-a)}{b-a}\right)-1\right]$$

$$= 0 \quad \text{if } n \ne m$$

The set satisfies the definition for Hermitian orthogonality.

The definition given in Section 2.2 for orthogonality of a set of functions $\{f_n(x)\}$, $n \in \mathbf{N}$, $a \le x \le b$, is a special case of a concept we consider at this time. We say that the set is *orthogonal with respect to a weight function* $r(x) \ge 0$ if

$$\int_a^b r(x)f_n(x)f_m(x) \, dx = 0 \quad \text{when } m \ne n$$

The square of the norm is written

$$\|f_n(x)\|^2 = \int_a^b r(x)f_n^2(x) \, dx$$

One observes that the *ordinary* type of orthogonality occurs when $r(x) = 1$. Orthogonality with respect to a weight function $r(x)$ reduces to ordinary orthogonality if the set $\{\sqrt{r(x)}f_n(x)\}$ replaces the set $\{f_n(x)\}$ in our new definition. The importance of orthogonality with respect to weight functions will become apparent when we consider orthogonality of eigenfunctions of Strum–Liouville problems.

Example 2.4. Functions $T_n(x) = \cos(n \arccos x)$ are called Tchebycheff polynomials of the first kind. Show that the set $\{T_n(x)\}$, $-1 \le x \le 1$, $n \in \mathbf{N}_0$, is orthogonal with respect to the weight function $(1 - x^2)^{-1/2}$.

In compliance with the definition, we must test the integral

$$\int_{-1}^{1} (1 - x^2)^{-1/2} T_n(x) T_m(x) \, dx \tag{2.7}$$

To evaluate the integral, let $\theta = \arccos x$ or $x = \cos \theta$. Then (2.7) may be written

$$\int_{0}^{\pi} \cos n\theta \cos m\theta \, d\theta$$

$$= \frac{1}{2} \left[\frac{1}{n+m} \sin (n+m)\theta + \frac{1}{n-m} \sin (n-m)\theta \right]_{0}^{\pi} = 0 \quad \text{if } n \neq m$$

Thus the set $\{T_n(x)\}$ is orthogonal with respect to the weight function given.

Exercises 2.2

1. Show that $e^{2n\pi i} = 1$ and $e^{(2n+1)\pi i} = -1$, $n \in N_0$.
2. (a) Show that the set $\{\exp(inx)\}$, $-\pi \leqslant x \leqslant \pi$, $n \in Z$, is orthogonal in the Hermitian sense.
   (b) Determine the norm for the set in (a).
3. (a) Is the set $F_n(x) = \sin[(n+1)\arccos x]$ on $(-1, 1)$, $n \in N_0$, orthogonal with respect to the weight function $(1 - x^2)^{1/2}$?
   (b) Tchebycheff polynomials of the second kind are defined by $U_n(x) = (1 - x^2)^{-1/2} \sin[(n+1)\arccos x]$. Show that the set $\{U_n(x)\}$, $-1 \leqslant x \leqslant 1$ is orthogonal relative to the weight function $(1 - x^2)^{1/2}$.
4. (a) The set of functions $\{L_n(x)\}$ satisfies the Laguerre differential equation $xy'' + (1 - x)y' + ny = 0$. $L_n(x)$ is generated so that $L_0(x) = 1$ and $L_n(x) = (e^x/n!)d^n/dx^n(x^n e^{-x})$, $n \in N$. Compute $L_n(x)$ for $n = 1, 2$, and show that $\{L_0, L_1, L_2\}$ satisfy the differential equation.
   (b) By direct integration show that $\{L_0, L_1, L_2\}$ form an orthogonal set on the real axis $(0, \infty)$ with respect to the weight function e^{-x}.
   (c) The set $\{H_n(x)\}$ satisfies the Hermite differential equation $y'' - 2xy' + 2ny = 0$. $H_n(x)$ has properties $H_0(x) = 1$ and $H_n(x) = (-1)^n e^{x^2}(d^n/dx^n)(e^{-x^2})$, $n \in N$. Determine $H_n(x)$ for $n = 1, 2$ and show that $\{H_0, H_1, H_2\}$ satisfy the differential equation.
   (d) Using the appropriate definition show that $\{H_0, H_1, H_2\}$ is an orthogonal set with respect to the weight function e^{-x^2} for $-\infty < x < \infty$.
      Definitions for $L_n(x)$ and $H_n(x)$ appear in Brand [8, pp. 475 and 478]. Appropriate weight functions and definitions for $T_n(x)$ and $U_n(x)$ are found in Abramowitz and Stegun [1, pp. 774–776].

2.5. THE STURM–LIOUVILLE BOUNDARY VALUE PROBLEM

In the discussion of the method of separation of variables in Chapter 1, equations of the type $y'' + \alpha^2 y = 0$ were discovered. These equations are special forms of

$$c_0(x)y'' + c_1(x)y' + [c_2(x) + \lambda]y = 0 \tag{2.8}$$

For the first two terms of (2.8) to be written $[p(x)y']'$ we multiply the equation by an integrating factor $p(x) = \exp[\int (c_1(x)/c_0(x))\, dx]$. If in addition we let $q(x) = [c_2(x)/c_0(x)]p(x)$ and $r(x) = p(x)/c_0(x)$ then (2.8) becomes

$$[p(x)y']' + [q(x) + \lambda r(x)]y = 0 \tag{2.9}$$

which is called a *Sturm–Liouville differential equation* (SLDE). It is *regular* in $[a, b]$ if $p(x)$ and $r(x)$ are positive in the interval. For a given λ two linearly independent solutions of a regular SLDE exist in $[a, b]$.

The boundary value problem containing the SLDE, $a \leqslant x \leqslant b$, along with the separated end conditions,

$$a_1 y(a) + a_2 y'(a) = 0, \quad b_1 y(b) + b_2 y'(b) = 0 \tag{2.10}$$

forms a *Sturm–Liouville problem* (SLP). If the coefficients a_1, a_2 and b_1, b_2 are real constants such that $a_1^2 + a_2^2 \neq 0$ and $b_1^2 + b_2^2 \neq 0$ and the SLDE is regular, then the problem is a *regular* SLP. The trivial solution $y = 0$ satisfies the SLP for any value of the parameter λ. Nontrivial solutions are called *eigenfunctions* or *characteristic functions* of the SLP. The corresponding values of λ for which these nontrivial solutions exist are known as *eigenvalues* or *characteristic values*.

Theorem 2.1. Assume that the functions $p(x)$, $q(x)$, and $r(x)$ of the regular SLP (2.9) and (2.10) are real and continuous in $[a, b]$. If $y_n(x)$ and $y_m(x)$ are continuously differentiable eigenfunctions of the matching distinct eigenvalues λ_n and λ_m, respectively, then $y_n(x)$ and $y_m(x)$ are orthogonal in the interval with respect to the weight function $r(x)$.

Since y_n and y_m are solutions corresponding to λ_n and λ_m

$$[py_n']' + [q + \lambda_n r]y_n = 0 \tag{2.11}$$

$$[py_m']' + [q + \lambda_m r]y_m = 0 \tag{2.12}$$

By multiplying (2.11) by y_m and (2.12) by y_n and then subtracting (2.12)–(2.11) we obtain

$$[py_m']'y_n - [py_n']'y_m + (\lambda_m - \lambda_n)ry_ny_m = 0 \tag{2.13}$$

However,

$$\frac{d}{dx}\{[py_m']y_n - [py_n']y_m\} = [py_m']'y_n - [py_n']'y_m \tag{2.14}$$

Using (2.14), equation (2.13) may be written

$$\frac{d}{dx}\{p[y_m'y_n - y_n'y_m]\} = (\lambda_n - \lambda_m)ry_ny_m \tag{2.15}$$

Integrating (2.15) over $[a, b]$

$$[p(y_m'y_n - y_n'y_m)]_a^b = (\lambda_n - \lambda_m)\int_a^b ry_ny_m \, dx \tag{2.16}$$

We observe that y_n and y_m must satisfy the conditions of (2.10). Therefore

$$a_1y_n(a) + a_2y_n'(a) = 0$$
$$a_1y_m(a) + a_2y_m'(a) = 0 \tag{2.10a}$$

and

$$b_1y_n(b) + b_2y_n'(b) = 0$$
$$b_1y_m(b) + b_2y_m'(b) = 0 \tag{2.10b}$$

The condition that (2.10a) has a solution other than $a_1 = a_2 = 0$ is

$$\begin{vmatrix} y_n(a) & y_n'(a) \\ y_m(a) & y_m'(a) \end{vmatrix} = y_m'(a)y_n(a) - y_n'(a)y_m(a) = 0 \tag{2.17}$$

Similarly if (2.10b) has a nontrivial solution then

$$\begin{vmatrix} y_n(b) & y_n'(b) \\ y_m(b) & y_m'(b) \end{vmatrix} = y_m'(b)y_n(b) - y_n'(b)y_m(b) = 0 \tag{2.18}$$

Conditions (2.17) and (2.18) permit us to restate (2.16) as

$$(\lambda_n - \lambda_m)\int_a^b ry_ny_m \, dx = 0 \tag{2.19}$$

Since $\lambda_n - \lambda_m \neq 0$, then

$$\int_a^b ry_ny_m \, dx = 0 \quad \text{if } n \neq m$$

This concludes the proof.

Example 2.5. Find all the eigenvalues and eigenfunctions for the SLP: $y'' + \lambda y = 0$; $y'(0) = y(1) = 0$.

In this problem $p(x) = 1$, $q(x) = 0$, $r(x) = 1$, and $\lambda = \lambda$ when the ODE is compared to the SLDE (2.9). We check the cases when $\lambda < 0$, $\lambda = 0$, and $\lambda > 0$. First, if $\lambda = -\alpha^2$, α a real constant, the ODE has the form

$$y'' - \alpha^2 y = 0$$

and the general solution is

$$y = K_1 e^{\alpha x} + K_2 e^{-\alpha x}$$
$$y' = \alpha K_1 e^{\alpha x} - \alpha K_2 e^{-\alpha x}$$

Applying the two boundary conditions, we have

$$y'(0) = \alpha K_1 - \alpha K_2 = 0$$
$$y(1) = K_1 e^{\alpha} + K_2 e^{-\alpha} = 0 \qquad (2.20)$$

The only solution for (2.20) is $K_1 = K_2 = 0$. Therefore $y = 0$, the forbidden trivial solution, is the only solution if $\lambda < 0$. If $\lambda = 0$, then

$$y'' = 0, \quad y = K_1 x + K_2$$

After using the two boundary conditions, $K_1 = K_2 = 0$, and $y = 0$ is the only solution. If $\lambda = \alpha^2$, α a real constant, then

$$y'' + \alpha^2 y = 0$$

and

$$y = c_1 \cos \alpha x + c_2 \sin \alpha x$$
$$y' = -\alpha c_1 \sin \alpha x + \alpha c_2 \cos \alpha x$$

In this case

$$y'(0) = \alpha c_2 = 0$$

and $c_2 = 0$, since $\alpha \neq 0$.

$$y(1) = c_1 \cos \alpha = 0$$

If $c_1 \neq 0$, then $\cos \alpha = 0$. If $\cos \alpha = 0$, then $\alpha = (2n - 1)\pi/2$.

$$\lambda_n = \alpha_n^2 = \frac{(2n-1)^2\pi^2}{4}, \quad n \in \mathbf{N}$$

is the set of *eigenvalues*.

$$y_n(x) = \cos\left[\frac{(2n-1)\pi x}{2}\right], \quad n \in \mathbf{N}$$

is the set of *eigenfunctions*. We have stated the set using $c_1 = 1$. The set $\{\cos[(2n-1)\pi x/2]\}$, $n \in \mathbf{N}$, $0 \le x \le 1$, is orthogonal with weight function $r(x) = 1$.

The SLP composed of the SLDE (2.9) with $p(a) = p(b) > 0$ and the periodic end constraints

$$y(a) = y(b), \quad y'(a) = y'(b) \tag{2.21}$$

is a *periodic SLP*. If $f(x + P) = f(x)$, $f(x)$ is periodic with a period P.

Theorem 2.2. Let $y_n(x)$ and $y_m(x)$ be continuously differentiable eigenfunctions matching distinct λ_n and λ_m, respectively, for a periodic SLP. Then $y_n(x)$ and $y_m(x)$ are orthogonal relative to the weight function $r(x)$ in $[a, b]$.

Since the solutions y_n and y_m must satisfy (2.21), it follows that

$$y_n(a) = y_n(b) \quad \text{and} \quad y_n'(a) = y_n'(b)$$
$$y_m(a) = y_m(b) \quad \text{and} \quad y_m'(a) = y_m'(b) \tag{2.21a}$$

If the end constraints (2.21a) are used in (2.16), then

$$[p(b) - p(a)][y_m'(a)y_n(a) - y_n'(a)y_m(a)] = (\lambda_n - \lambda_m)\int_a^b r y_n y_m \, dx \tag{2.22}$$

From the hypothesis of the theorem it is implied that $p(b) - p(a) = 0$. Therefore (2.22) becomes

$$(\lambda_n - \lambda_m)\int_a^b r y_n y_m \, dx = 0$$

Since λ_n and λ_m are distinct, $\lambda_n \neq \lambda_m$ and

$$\int_a^b r y_n y_m \, dx = 0 \quad \text{if } m \neq n$$

The final statement implies orthogonality.

Example 2.6. Determine the eigenvalues and eigenfunctions for the periodic SLP: $y'' + \lambda y = 0$; $y(-L) = y(L)$; $y'(-L) = y'(L)$.

If $\lambda < 0$, the only solution is trivial. If $\lambda = 0$, then

$$y'' = 0, \quad y = K_1 x + K_2$$

When $y(-L) = y(L)$, $K_1 = 0$. If $y'(-L) = y'(L)$, there is no restriction on K_2. Therefore $y_0 = 1$ is a solution matching $\lambda = 0$. When $\lambda = \alpha^2$

$$y'' + \alpha^2 y = 0$$

and

$$y = c_1 \cos \alpha x + c_2 \sin \alpha x$$
$$y' = -\alpha c_1 \sin \alpha x + \alpha c_2 \cos \alpha x$$

The two boundary conditions in this case permit us to write

$$c_1 \sin \alpha L = 0 \quad \text{and} \quad c_2 \sin \alpha L = 0$$

If $c_1 \neq 0$ and $c_2 \neq 0$, then $\sin \alpha L = 0$ so that $\alpha L = n\pi$. Therefore, $\alpha = n\pi/L$ and

$$\lambda_n = \alpha_n^2 = \frac{n^2 \pi^2}{L^2}$$

for the set of *eigenvalues*. The set of *eigenfunctions* is displayed by

$$\left\{ 1, \sin \frac{n\pi x}{L}, \cos \frac{m\pi x}{L} \right\}, \quad n, m \in \mathbb{N}, \quad -L < x < L \qquad (2.23)$$

The set (2.23) is orthogonal in the ordinary sense. That is $r(x) = 1$.

A SLDE can be associated with unbounded intervals such as $(0, \infty)$ or $(-\infty, \infty)$ as well as finite intervals. When the interval is unbounded, when the interval is finite with $p(x)$ or $r(x)$ zero at one or both endpoints, or when $q(x)$ is discontinuous at these points, the SLDE is described as *singular*. We avoid discussing the corresponding *singular SLP* here, but refer the reader to Birkhoff and Rota [2, pp. 263–265].

In addition to orthogonality other properties of the eigenvalues and eigenfunctions are important in the study of the SLP.

Theorem 2.3. For a regular SLP with $p(x) > 0$ all the eigenvalues are real if $p(x)$, $q(x)$, and $r(x)$ are real functions and the eigenfunctions are differentiable and continuous.

Let $y(x)$ be an eigenfunction corresponding to an eigenvalue $\gamma + i\delta$ where γ and δ are real numbers. We assume y has the form $u + iv$ where u and v are real functions of x. Substitution of the complex form of y and λ in the SLDE (2.9) results in

$$[p(u' + iv')]' + [q + (\gamma + i\delta)r][u + iv] = 0 \qquad (2.24)$$

Equating the real and imaginary parts to zero in (2.24) one writes

$$(pu')' + (q + \gamma r)u - \delta rv = 0 \qquad (2.25)$$

$$(pv')' + (q + \gamma r)v + \delta ru = 0 \qquad (2.26)$$

If we multiply (2.25) by v and (2.26) by u and subtract the results, we find that

$$u(pv')' - v(pu')' + \delta r(u^2 + v^2) = 0 \qquad (2.27)$$

Using a procedure similar to (2.14) we obtain for (2.27)

$$\frac{d}{dx}[(pv')u - (pu')v] + \delta r(u^2 + v^2) = 0$$

Integrating over $[a, b]$ one finds that

$$-\delta \int_a^b r(u^2 + v^2)\, dx = [p(uv' - vu')]_a^b \qquad (2.28)$$

If the complex form is inserted in (2.10) we have

$$a_1[u(a) + iv(a)] + a_2[u'(a) + iv'(a)] = 0$$
$$b_1[u(b) + iv(b)] + b_2[u'(b) + iv'(b)] = 0 \qquad (2.29)$$

The two zero complex constraints of (2.29) imply that

$$a_1 u(a) + a_2 u'(a) = 0$$
$$a_1 v(a) + a_2 v'(a) = 0 \qquad (2.30)$$

and

$$b_1 u(b) + b_2 u'(b) = 0$$
$$b_1 v(b) + b_2 v'(b) = 0 \qquad (2.31)$$

The conditions $a_1^2 + a_2^2 \neq 0$ and $b_1^2 + b_2^2 \neq 0$ imply that not both a_1 and a_2 are zero, and b_1 and b_2 are not both zero. Therefore

$$\begin{vmatrix} u(a) & u'(a) \\ v(a) & v'(a) \end{vmatrix} = u(a)v'(a) - u'(a)v(a) = 0 \tag{2.32}$$

and

$$\begin{vmatrix} u(b) & u'(b) \\ v(b) & v'(b) \end{vmatrix} = u(b)v'(b) - u'(b)v(b) = 0 \tag{2.33}$$

Using (2.32) and (2.33) in the evaluation of the right member of (2.28) we obtain

$$-\delta \int_a^b r(u^2 + v^2) \, dx = 0 \tag{2.34}$$

If $r(x) > 0$, the integral of (2.34) is positive. Therefore $\delta = 0$. Since $\delta = 0$, then $\lambda = \gamma$. This implies that the eigenvalues are real.

Theorem 2.4. Let $y_n(x)$ and $y_m(x)$ be any two solutions of the SLDE (2.9) on $[a, b]$ for a given λ. We assume that $y_n(x)$ and $y_m(x)$ are continuous and differentiable. If $W(y_n, y_m)$ is the Wronskian of the solutions, then $p(x)W(y_n, y_m)$ is a constant.

Since y_n and y_m are solutions of (2.9) then

$$[py_n']' + [q + \lambda r]y_n = 0 \tag{2.35}$$

$$[py_m']' + [q + \lambda r]y_m = 0 \tag{2.36}$$

Multiplying (2.35) by y_m and (2.36) by y_n and then subtracting, one obtains

$$[py_m']'y_n - [py_n']'y_m = 0 \tag{2.37}$$

If we use the relation (2.14), then (2.37) becomes

$$\frac{d}{dx}\{p(y_m'y_n - y_n'y_m)\} = 0 \tag{2.38}$$

By integrating (2.38) over the interval a to x, $a \leqslant x \leqslant b$, we find that

$$p(x)W(y_n(x), y_m(x)) = p(a)[y_m'(a)y_n(a) - y_n'(a)y_m(a)] \tag{2.39}$$

The left member of (2.39) is a constant C. Therefore

$$p(x)W(y_n(x), y_m(x)) = C$$

Theorem 2.5. Two eigenfunctions $y_n(x)$ and $y_m(x)$ matching a single λ of a regular SLP are linearly dependent. We assume the eigenfunctions are differentiable and continuous.

We let $y_n(x)$ and $y_m(x)$ be eigenfunctions for a single eigenvalue λ. By Theorem 2.4 we know that

$$p(x)W(y_n(x), y_m(x)) = C$$

with $p(x) > 0$. According to the discussion of the Wronskian, Section 1.2, if $W(y_n(x), y_m(x))$ vanishes at a point in $[a, b]$ it must vanish at every point in the interval.

The eigenfunctions $y_n(x)$ and $y_m(x)$ satisfy the end constraints at $x = a$, so that

$$a_1 y_n(a) + a_2 y_n'(a) = 0$$
$$a_1 y_m(a) + a_2 y_m'(a) = 0$$

Recall that $a_1^2 + a_2^2 \neq 0$, so that

$$\begin{vmatrix} y_n(a) & y_n'(a) \\ y_m(a) & y_m'(a) \end{vmatrix} = 0$$

This determinant is $W(y_n(a), y_m(a))$ after a row-column interchange. Therefore $W(y_n(a), y_m(a)) = 0$. It is a sufficient condition that $y_n(x)$ and $y_m(x)$ matching a single λ are dependent. This implies that one eigenfunction $y_n(x)$ is a constant times $y_m(x)$. Another way of stating the result is to say that for a *single λ in the regular SLP the matching eigenfunction is unique except for a constant factor*.

Theorem 2.6.* A regular SLP has an infinite sequence of real eigenvalues $\lambda_0 < \lambda_1 < \lambda_2 < \cdots$, with $\lim_{n \to \infty} \lambda_n = \infty$. Every eigenfunction $y_n(x)$ matching the eigenvalue λ_n has exactly n zeros in (a, b). The eigenfunction is unique except for a constant factor.

The last sentence of the theorem comes as a result of Theorem 2.5. The remainder of Theorem 2.6 is established by a sequence of ideas terminating with the statement in Birkhoff and Rota [2, p. 273].

Exercises 2.3

1. Find all the eigenvalues and eigenfunctions for the following regular SLPs:

*Adapted from Birkhoff and Rota [2], by permission of John Wiley & Sons, Inc.

    (a) $y'' + \lambda y = 0$, $y(0) = y(1) = 0$;

    (b) $y'' + \lambda y = 0$, $y(0) = y'(2) = 0$;

    (c) $y'' + \lambda y = 0$, $y'(0) = y'(\pi/2) = 0$;

    (d) $y'' + \lambda y = 0$, $y(0) = 0$, $y(1) + y'(1) = 0$.

2. For the periodic SLPs, determine the set of eigenvalues and the set of eigenfunctions:

    (a) $y'' + \lambda y = 0$, $y(-\pi) = y(\pi)$, $y'(-\pi) = y'(\pi)$;

    (b) $y'' + \lambda y = 0$, $y(0) = y(1)$, $y'(0) = y'(1)$;

    (c) $y'' + \lambda y = 0$, $y(-1) = y(1)$, $y'(-1) = y'(1)$.

3. Determine the eigenvalues and eigenfunctions for the SLPs:

    (a) $x^5 y'' + 5x^4 y' + \lambda x^3 y = 0$, $y(1) = y(e) = 0$, $1 \le x \le e$;

    (b) $[(3+x)^2 y']' + \lambda y = 0$, $y(-2) = y(1) = 0$, $-2 < x < 1$;

    (c) $[xy']' + [\lambda/x]y = 0$, $y(1) = y(2) = 0$, $1 \le x \le 2$.

4. If the linear operator L is defined by

$$Ly = c_0(x)y''(x) + c_1(x)y'(x) + c_2(x)y(x)$$

and the linear operator L^* defined so that

$$L^* y = [c_0(x)y(x)]'' - [c_1(x)y(x)]' + c_2(x)y(x)$$

then we say that L^* is the *adjoint* of L. If L and L^* are identical then operator L is *self-adjoint*.

    (a) Show that if

$$Ly = p(x)y''(x) + p'(x)y'(x) + [q(x) + \lambda r(x)]y(x)$$

    then L is self-adjoint.

    (b) For L defined in (a) show that

$$zLy - yLz = [p(zy' - z'y)]'$$

    is satisfied if all derivatives exits. This relation is credited to Lagrange. It is a form of (2.14).

    (c) If $Ly = (1 - x^2)y'' - 2xy' + n(n+1)y$ is L self-adjoint?

    (d) If $x^2 y'' + xy' + (x^2 - n^2)y = 0$, can the left member of the equation be transformed into an equivalent self-adjoint form?

2.6. UNIFORM CONVERGENCE OF SERIES

Assume that $\{u_n(x)\}$, $n \in \mathbf{N}$, is a set of functions on $[a, b]$. The nth partial sum is defined by

$$S_n(x) = \sum_{k=1}^{n} u_k(x)$$

The series

$$\sum_{n=1}^{\infty} u_n(x) \qquad (2.40)$$

is convergent in $[a, b]$ if

$$\lim_{n \to \infty} S_n(x) = S(x) \qquad (2.40a)$$

and $S(x)$ is called the sum of the series. We say that (2.40) converges to $S(x)$ if for $\varepsilon > 0$ and $x \in [a, b]$ we can find $M > 0$ so that

$$|S_n(x) - S(x)| < \varepsilon \quad \text{for all } n > M \qquad (2.41)$$

In this case $M(x, \varepsilon)$ is a function of x and ε and we refer to the condition as *pointwise convergence*. If $M(\varepsilon)$ is dependent on ε alone and not x, then the series is *uniformly convergent*.

The *Weierstrass M-test* for uniform convergence is stated as follows. Suppose that a sequence of positive constants $\{M_n\}$ can be found such that in some interval

$$(a) \; |u_n(x)| \leq M_n$$

and

$$(b) \; \sum_{n=1}^{\infty} M_n$$

converges. Then the series (2.40) is *uniformly and absolutely convergent* in the interval.

If all $u_n(x)$ are continuous functions on $[a, b]$ and the series (2.40) is uniformly convergent, then $S(x)$ in (2.40a) is a *continuous function*. Under these conditions the series can be *integrated termwise* over $[a, b]$ and the result is the integral of $S(x)$ over $[a, b]$. When u_n and u_n' are continuous and (2.40) converges and the series $\sum_{n=1}^{\infty} u_n'$ is uniformly convergent, then (2.40) is *termwise differentiable* and equal to $S'(x)$.

Example 2.7. Test the series $\sum_{n=1}^{\infty} (1/n^2) \cos nx$ for uniform convergence. Since

$$\left| \left(\frac{1}{n^2} \right) \cos nx \right| \leq \frac{1}{n^2} = M_n$$

and

$$\sum_{n=1}^{\infty} \left(\frac{1}{n^2} \right)$$

converges, then by the Weierstrass M-test the series is *uniformly convergent* for all real x.

Example 2.8. Investigate the uniform convergence of the series

$$\sum_{n=1}^{\infty} x^{n-1}(1-x), \quad x \in [-\tfrac{2}{3}, \tfrac{2}{3}]$$

The nth partial sum is

$$S_n(x) = 1 - x + x - x^2 + x^2 - + \cdots - x^{n-1} + x^{n-1} - x^n = 1 - x^n$$

From (2.40a)

$$S(x) = \lim_{n \to \infty} S_n(x) = 1$$

when $-\tfrac{2}{3} \leqslant x \leqslant \tfrac{2}{3}$.

$$|S_n(x) - S(x)| = |(1 - x^n) - 1| = |x^n| \leqslant (\tfrac{2}{3})^n < \varepsilon$$

Therefore,

$$n \ln (\tfrac{2}{3}) < \ln \varepsilon, \quad n > \frac{\ln \varepsilon}{\ln (\tfrac{2}{3})}$$

and

$$M(\varepsilon) = \frac{\ln \varepsilon}{\ln (\tfrac{2}{3})}$$

M does not depend on x in the interval. This result satisfies the *definition* for *uniform convergence*.

Example 2.9. Assume that

$$S(x) = \sum_{n=1}^{\infty} \left(\frac{1}{n^3} \right) \cos nx, \quad -\pi \leqslant x \leqslant \pi$$

(a) Is $S(x)$ continuous when $-\pi \leqslant x \leqslant \pi$?
(b) If $\int_0^{\pi/2} S(x)\, dx$ exists, compute it.
(c) Find $S'(\pi/2)$ if it exists.

Since $|(1/n^3)\cos nx| \leqslant 1/n^3$ we see from the M-test that the series is uniformly convergent for all real x.

(a) Since the series is uniformly convergent for all real x, it is uniformly convergent for $x \in [-\pi, \pi]$, and $\{(1/n^3)\cos nx\}$ contains only continuous functions for $n \in \mathbf{N}$ on the interval, then $S(x)$ is continuous on $[-\pi, \pi]$.

(b) The series satisfies the condition for *termwise integration*. Therefore we compute

$$\int_0^{\pi/2} S(x)\, dx = \int_0^{\pi/2} \sum_{n=1}^{\infty} \left(\frac{1}{n^3}\right) \cos nx\, dx = \sum_{n=1}^{\infty} \int_0^{\pi/2} \left(\frac{1}{n^3}\right) \cos nx\, dx$$

$$= \sum_{n=1}^{\infty} \left(\frac{1}{n^4}\right) \sin \left(\frac{n\pi}{2}\right) = \sum_{n=1}^{\infty} (-1)^{n-1} \left[\frac{1}{(2n-1)^4}\right]$$

(c) The series of derivatives

$$- \sum_{n=1}^{\infty} \left(\frac{1}{n^2}\right) \sin nx$$

is uniformly convergent as we can verify by the M-test. The original series is *convergent* since it is uniformly convergent. Therefore, the series is *termwise differentiable*. We compute

$$S'\left(\frac{\pi}{2}\right) = - \sum_{n=1}^{\infty} \left(\frac{1}{n^2}\right) \sin \left(\frac{n\pi}{2}\right)$$

$$= \sum_{n=1}^{\infty} \frac{(-1)^n}{(2n-1)^2}$$

2.7. SERIES OF ORTHOGONAL FUNCTIONS

Finding a Taylor series expansion for a specified function f is an experience encountered early in the study of elementary calculus. Here we examine an analogous venture of representing a given function with an infinite linear combination or series of orthogonal functions. Consider the set $\{g_n(x)\}$, $n \in \mathbf{N}$, $a \leqslant x \leqslant b$, orthogonal relative to a weight function $r(x)$. We assume that the given function f can be represented by a uniformly convergent series—a sufficient condition on the series to permit the procedure that follows. Let

$$f(x) = \sum_{n=1}^{\infty} C_n g_n(x) \tag{2.42}$$

To find the constants C_n, we multiply both sides of (2.42) by $r(x)g_m(x)$ and integrate termwise over the interval $[a, b]$. This results in

$$\int_a^b r(x)f(x)g_m(x)\,dx = \sum_{n=1}^{\infty} C_n \int_a^b r(x)g_n(x)g_m(x)\,dx \qquad (2.43)$$

Because of the orthogonality property, all terms in the series are zeros except when $n = m$. Then (2.43) becomes

$$\int_a^b r(x)f(x)g_n(x)\,dx = C_n \int_a^b r(x)g_n^2(x)\,dx$$

Solving for C_n one finds

$$C_n = \frac{1}{\|g_n(x)\|^2} \int_a^b r(x)f(x)g_n(x)\,dx \qquad (2.44)$$

where $\|g_n(x)\|^2$ is the square of the norm $\int_a^b r(x)g_n^2(x)\,dx$.

When the set $\{g_n(x)\}$ is orthogonal in the ordinary sense with $r(x) = 1$, then (2.44) becomes

$$C_n = \frac{1}{\|g_n(x)\|^2} \int_a^b f(x)g_n(x)\,dx \qquad (2.44a)$$

If the set $\{g_n(x)\}$ is orthonormal relative to the weight function $r(x)$, then (2.44) has the form

$$C_n = \int_a^b r(x)f(x)g_n(x)\,dx \qquad (2.44b)$$

The representation (2.42) for f is called an *orthonormal series* or a *generalized Fourier series*. The coefficients C_n are the *Fourier coefficients*. If $\|g_n(x)\| = 1$, then we say the expansion is an *orthonormal series*. When $\{g_n(x)\}$ is a set of eigenfunctions for a SLP, the terminology *Sturm–Liouville series* or *eigenfunction series* describes the expansion.

The series (2.42), with coefficients (2.44), (2.44a), or (2.44b), is assumed to be uniformly convergent. This limitation is more severe than we wish in certain situations. To avoid asserting that the series converges to f we may use a correspondence notation $\sim$. Then (2.42) is written

$$f(x) \sim \sum_{n=1}^{\infty} C_n g_n(x)$$

with coefficients C_n given by (2.44), (2.44a), or (2.44b).

Example 2.10. Determine the expansion for $f(x) = 1$, $0 < x < \pi$, in a series of eigenfunctions of the SLP

$$y'' + \lambda y = 0, \quad y(0) = y(\pi) = 0$$

First, we determine the solution of the SLP. If $\lambda = \alpha^2$, then

$$y'' + \alpha^2 y = 0$$

$$y = C_1 \cos \alpha x + C_2 \sin \alpha x$$

$$y(0) = C_1 = 0$$

$$y(\pi) = C_2 \sin \alpha \pi = 0$$

If $C_2 \neq 0$, $\sin \alpha \pi = 0$, and $\alpha \pi = n\pi$. Therefore $\alpha = n$ and the set of eigen-values includes $\lambda_n = \alpha_n^2 = n^2$. If $\lambda < 0$ or $\lambda = 0$ the solutions are trivial. Thus the eigenvalues are

$$\lambda_n = n^2, \quad n \in N$$

The matching set of eigenfunctions is

$$\{\sin nx\}, \quad n \in N, \quad 0 < x < \pi$$

Next, we write the series representing the function

$$f(x) \sim \sum_{n=1}^{\infty} C_n \sin nx$$

where

$$C_n = \frac{1}{\|\sin nx\|^2} \int_0^\pi 1 \cdot \sin nx \, dx$$

$$\|\sin nx\|^2 = \int_0^\pi \sin^2 nx \, dx = \frac{\pi}{2}$$

Therefore

$$C_n = \frac{2}{\pi} \int_0^\pi \sin nx \, dx = \left(\frac{2}{n\pi}\right)(1 - \cos n\pi)$$

$$= \frac{2}{n\pi} \begin{cases} 0 & \text{if } n \text{ is even} \\ 2 & \text{if } n \text{ is odd} \end{cases}$$

If we avoid writing the zero coefficients then

$$C_{2n-1} = \frac{4}{(2n-1)\pi}$$

and

$$1 \sim \frac{4}{\pi} \sum_{n=1}^{\infty} \left(\frac{1}{2n-1}\right) \sin(2n-1)x$$

2.8. APPROXIMATION BY LEAST SQUARES

We define the function f to be *square integrable* (SI) if f and f^2 are both integrable.

The set $\{g_n(x)\}$, $n \in \mathbf{N}$, $a \le x \le b$, is assumed orthogonal relative to a weight function $r(x) > 0$. The set and f are SI. The idea is to approximate f with a linear combination of these orthogonal functions

$$Q_n(x) = \sum_{k=1}^{n} \alpha_k g_k(x)$$

so that the error

$$E = \int_a^b r(x)[f(x) - Q_n(x)]^2 \, dx \qquad (2.45)$$

is as small as possible. This procedure is referred to as an *approximation by least squares*, *total square error*, or *best approximation in the mean*. We wish to determine the coefficients α_k of Q_n so that E is a minimum. If the *Fourier coefficients* (2.44) are written

$$C_n = \frac{(f, g_n)}{\|g_n\|^2}$$

where the inner products

$$(f, g_n) = \int_a^b r(x)f(x)g_n(x) \, dx$$

then

$$E = \int_a^b r(x)\left[f(x) - \sum_{k=1}^{n} \alpha_k g_k(x) \right]^2 dx$$

$$= \int_a^b r(x)f^2(x) \, dx + \sum_{k=1}^{n} \alpha_k^2 (g_k, g_k) - 2 \sum_{k=1}^{n} \alpha_k (f, g_k)$$

$$+ \text{(zero inner products)}$$

$$= \int_a^b r(x)f^2(x) \, dx + \sum_{k=1}^{n} \alpha_k^2 \|g_k\|^2 - 2 \sum_{k=1}^{n} \alpha_k C_k \|g_k\|^2$$

$$E = \int_a^b r(x)f^2(x) \, dx + \sum_{k=1}^{n} (\alpha_k - C_k)^2 \|g_k\|^2 - \sum_{k=1}^{N} C_k^2 \|g_k\|^2 \qquad (2.46)$$

Since $E \ge 0$ in (2.45) and E is smallest in (2.46) when $\alpha_k = C_k$, $k = 1, \ldots, n$ we see that the error is least when

$$Q_n(x) = \sum_{k=1}^{n} C_k g_k(x)$$

Our work can be stated as a theorem.

Theorem 2.7. Let the set $\{g_n(x)\}$, $n \in \mathbf{N}$, $a \leqslant x \leqslant b$, be orthogonal relative to a weight function $r(x) > 0$. Assume that the set and f are SI. If $Q_n(x)$ is a linear combination of the set, and if f is approximated by $Q_n(x)$ in the sense of least square errors, then the error

$$E = \int_a^b r(x)[f(x) - Q_n(x)]^2 \, dx$$

is least when the coefficients of $Q_n(x)$ are the Fourier coefficients C_n given by (2.44).

2.9. COMPLETENESS OF SETS

One may recall the necessity that all terms be included in a Taylor series representation of a function. As an example, if the constant term 1 is omitted in the expansion of e^x, then

$$e^x \neq \sum_{n=1}^{\infty} \frac{x^n}{n!} \tag{2.47}$$

but

$$e^x = 1 + \sum_{n=1}^{\infty} \frac{x^n}{n!} \tag{2.48}$$

The differentiable set upon which the Taylor series is based in (2.48) is *complete*, but not with the term 1 missing as in (2.47). In this section we pursue the idea of completeness for sets of orthogonal functions. Our main concern is related to *complete* sets upon which we build series.

We return to the material of the preceding section. From (2.46), the minimum value of E allows us to write

$$\int_a^b r(x)f^2(x) \, dx - \sum_{k=1}^{n} C_k^2 \|g_k\|^2 \geqslant 0$$

Therefore

$$\sum_{k=1}^{n} C_k^2 \|g_k\|^2 \leqslant \int_a^b r(x)f^2(x) \, dx \tag{2.49}$$

This result is known as *Bessel's inequality*. Since $\int_a^b r(x)f^2(x) \, dx$ is independent of n, in (2.49) $\sum_{k=1}^{n} C_k^2 \|g_k\|^2$ is bounded. The series

$$\sum_{n=1}^{\infty} C_n^2 \|g_n\|^2 \tag{2.50}$$

is bounded also. Therefore it is convergent. Since (2.50) converges

$$\lim_{n \to \infty} C_n \| g_n \| = 0$$

If the set $\{g_n(x)\}$ is orthonormal as assumed for (2.44b) then the series

$$\sum_{n=1}^{\infty} C_n^2$$

converges, and

$$\lim_{n \to \infty} C_n = 0 \tag{2.51}$$

When $Q_n(x)$ contains the Fourier coefficients, and

$$\lim_{n \to \infty} \int_a^b r(x)[f(x) - Q_n(x)]^2 \, dx = 0$$

we say that the sequence $Q_n(x)$ *converges in the mean* to $f(x)$ relative to the weight function $r(x)$. If the series $Q_n(x)$ converges in the mean to $f(x)$, then

$$\lim_{n \to \infty} \int_a^b r(x) \left[f(x) - \sum_{k=1}^{n} C_k g_k(x) \right]^2 \, dx = 0$$

Instead of obtaining Bessel's inequality, we have

$$\sum_{n=1}^{\infty} C_n^2 \| g_n \|^2 = \int_a^b r(x) f^2(x) \, dx \tag{2.52}$$

This is referred to as *Parseval's identity* or the *completeness relation*. The condition that the set $\{g_n(x)\}$ of orthogonal SI functions be *complete* is that the generalized Fourier series for any SI function f *converges to f in the mean*. This implies that if the set is complete then Parseval's identity (2.52) is satisfied. For a class of SI functions on $[a, b]$ it is known that the set of eigenfunctions of the SLP (2.9) and (2.10) is *complete*. It can be shown that a generalized Fourier series *converges in the mean* to a *single function only* (except possibly at a finite set of points). *Mean convergence* fails to assure *ordinary* or *pointwise convergence*. Parseval's identity is not equivalent to

$$f(x) = \sum_{n=1}^{\infty} C_n g_n(x)$$

Tolstov [47, pp. 54–60] defines completeness differently than we have here, but a number of additional properties are included in his discussion.

Exercises 2.4

1. Is the series $\sum_{n=1}^{\infty} (1/n^4) \sin nx$ convergent? What domain is appropriate?

2. Show that $(1-x)^{-1} = \sum_{n=1}^{\infty} x^{n-1}$ converges uniformly on the interval $x \in [-a, a]$ where $0 < a < 1$.

3. (a) If the interval in Example 2.8 is changed to $0 \leqslant x \leqslant 3$, what could be said about uniform convergence of the series?

   (b) For the interval $-1 < x < 1$, is the series in Example 2.8 uniformly convergent?

4. Test the series $\sum_{n=1}^{\infty} e^{-nx}$ for uniform convergence if $x \geqslant a > 0$.

5. Compute A_1, A_2, and A_3 so that the function

$$A_1 \sin \frac{\pi x}{2} + A_2 \sin \frac{2\pi x}{2} + A_3 \sin \frac{3\pi x}{2}$$

is the best approximation in the sense of least squares best fit to the function $f(x) = 1$ over the interval $(0, 2)$.

6. If $f(x) = |x|$ for $-\pi < x < \pi$, find the coefficients of the approximating function

$$Q(x) = \frac{\alpha_0}{2} + \alpha_1 \cos x + \beta_1 \sin x + \alpha_2 \cos 2x + \beta_2 \sin 2x$$

so that the square error is least.

7. Assume that

$$1 - x = \frac{8}{\pi^2} \sum_{n=1}^{\infty} \frac{1}{(2n-1)^2} \cos \frac{(2n-1)\pi x}{2}, \quad 0 \leqslant x \leqslant 2$$

   (a) Is termwise integration justified? If it is justified find the integral from 0 to x.

   (b) Using Parseval's identity show that

$$\frac{\pi^4}{96} = \sum_{n=1}^{\infty} \frac{1}{(2n-1)^4}$$

8. Find the expansion of $f(x) = x$, $0 < x < 1$, formally, in a series of eigenfunctions of the SLP

$$y'' + \lambda y = 0, \quad y(0) = y(1) = 0$$

9. Expand the function $f(x) = 1$, $0 < x < \pi$, formally, in a series of eigenfunctions of the SLP

$$y'' + \lambda y = 0, \quad y(0) = y'(\pi) = 0$$

10. Compute the limit if $n \in \mathbf{N}$

$$\lim_{n \to \infty} \int_a^b r(x)f(x)g_n(x) \, dx$$

[See (2.44b) and (2.51).]

3

FOURIER SERIES

In this chapter we are concerned with the series based on orthogonal sets of functions, primarily orthogonal sets of trigonometric functions. This series is named after the French mathematical physicist Joseph Fourier (1768–1830). Although Fourier's work generally failed to consider the validity of the series representation, it did much to create interest in the trigonometric series. Cajori [12, pp. 269–271] offers a short summary on the life of Fourier. Lanczos [31, p. 1] gives in a few brief paragraphs some hint of the early debates on the series. Langer [32, Chapter 5] indicates some of the controversy existing among d'Alembert, Euler and Bernoulli. The entire Langer paper is helpful for understanding the development of the theory. Our ultimate goal is to use the Fourier series effectively in the solution of BVPs. This chapter will introduce the mathematical background necessary to understand and utilize Fourier series in applied problems.

3.1. PIECEWISE CONTINUOUS FUNCTIONS

A function f is *sectionally continuous* or *piecewise continuous* (PWC) in $[a, b]$ if it is continuous at all points on the interval except at most at a finite number of points where finite jump discontinuities may exist. We assume f is a real valued function of a single variable x. The *left hand limit* of f at x_0, represented by $f(x_0 -)$, is the limit of f as $x \to x_0$ from the left of x_0. Symbolically if $h > 0$

$$f(x_0 -) = \lim_{\substack{h \to 0 \\ h > 0}} f(x_0 - h)$$

The *right hand limit* is defined similarly so that as $x \to x_0$ from the right of x_0, the limit of f is

$$f(x_0 +) = \lim_{\substack{h \to 0 \\ h > 0}} f(x_0 + h)$$

If f, a PWC function, has discontinuities at

$$a < x_1 < x_2 < \cdots < x_{n-1} < b$$

then

$$f(a+), \quad f(x_1-), \quad f(x_1+), \quad f(x_2-),$$

$$f(x_2+), \ldots, f(x_{n-1}-), \quad f(x_{n-1}+), \quad f(b-)$$

must all exist and be finite. The function f is continuous at x_0 if

$$f(x_0-) = f(x_0+) = f(x_0)$$

If $f(x_0-)$ and $f(x_0+)$ are unequal but both exist we say there is a *jump discontinuity* at x_0 and the *jump* is defined as $f(x_0+) - f(x_0-)$. The negative of $f(x_0+) - f(x_0-)$ is sometimes given as the jump in the function.

A PWC function on a closed interval is bounded and integrable on the interval. If f is PWC, then

$$\int_a^b f(x)\,dx = \int_a^{x_1} f(x)\,dx + \int_{x_1}^{x_2} f(x)\,dx + \cdots + \int_{x_{n-1}}^b f(x)\,dx$$

with at most finite discontinuities at

$$a, x_1, x_2, \ldots, x_{n-1}, b$$

If f_1 and f_2 are PWC functions on $[a, b]$, then there is a way of subdividing the interval so that the product $f_1 f_2$, the linear combination $c_1 f_1 + c_2 f_2$ and the square f_1^2 are all PWC. Therefore, the integrals of these combinations over the interval must exist.

If a PWC function has a jump discontinuity at x_0, then the derivative fails to exist at that point. We define one sided derivatives for these situations. Assume that f is a function whose limit from the left of x_0 exists. If $h > 0$ the left hand derivative at x_0 is defined by

$$f'_-(x_0) = \lim_{\substack{h \to 0 \\ h > 0}} \frac{f(x_0-) - f(x_0 - h)}{h}$$

In a similar way, if $f(x_0+)$ exists, the *right hand derivative* at x_0 is

$$f'_+(x_0) = \lim_{\substack{h \to 0 \\ h > 0}} \frac{f(x_0 + h) - f(x_0+)}{h}$$

Right and left hand derivatives are sometimes defined for continuous functions only so that $f(x_0 -)$ and $f(x_0 +)$ are replaced with $f(x_0)$. The definitions given here are more useful for some of our applications with certain PWC functions.

The function f is said to be *smooth* on the interval $[a, b]$ if it possesses a continuous derivative on the interval. Geometrically this means that the graph of a smooth function has a continuous curve that has a tangent which turns continuously as the tangent goes along the curve. There are no points where right and left hand derivatives differ. The function is *piecewise smooth* (PWS) on $[a, b]$ if it is PWC and has a PWC derivative on the interval. The graph of a PWS function is either a continuous curve or one that can have a finite number of points where the function or its derivative may have jump discontinuities.

Example 3.1. Given

$$f(x) = \begin{cases} x^2 - 1 & \text{if } x < 0 \\ x^2 + 1 & \text{if } x > 0 \end{cases}$$

discuss continuity and smoothness for the function. Compute jumps if they exist.

The function f is PWC with a jump discontinuity at $x = 0$. The jump $f(0 +) - f(0 -)$ is 2. The derivative of f exists for all x except $x = 0$. Since $f(0)$ is not defined, the definition of f' at $x = 0$ is not satisfied at this point, even though

$$f'_+(0) = f'_-(0) = 0$$

The jump at $x = 0$ for the derivative function is zero. The function f is PWS. See Figures 3.1 and 3.2.

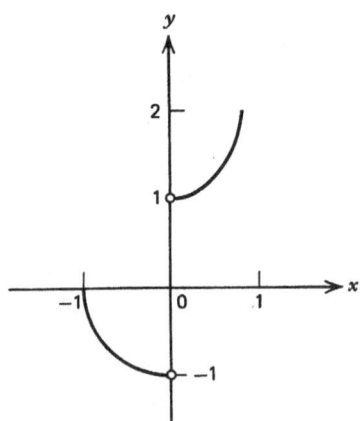

Figure 3.1. The PWC function f.

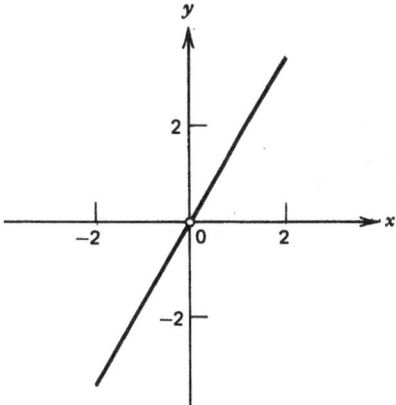

Figure 3.2. The derivative of f.

Example 3.2. The function is defined by

$$f(x) = \begin{cases} x^2 \sin\left(\dfrac{1}{x}\right) & \text{if } x \neq 0 \\ 0 & \text{if } x = 0 \end{cases}$$

(a) Find the right and left hand derivatives for f at $x = 0$.

(b) Determine $f'(0)$ if it exists.

For all values of $x \neq 0$, the function has the derivative formula

$$f'(x) = 2x \sin\left(\frac{1}{x}\right) - \cos\left(\frac{1}{x}\right)$$

(a) If $x = 0$, we investigate the definitions for the one sided derivatives

$$f'_-(0) = \lim_{\substack{h \to 0 \\ h > 0}} \frac{f(0-) + h^2 \sin(1/h)}{h}$$

$$= \lim_{\substack{h \to 0 \\ h > 0}} h \sin\left(\frac{1}{h}\right) = 0$$

$$f'_+(0) = \lim_{\substack{h \to 0 \\ h > 0}} \frac{h^2 \sin(1/h) - f(0+)}{h}$$

$$= \lim_{\substack{h \to 0 \\ h > 0}} h \sin\left(\frac{1}{h}\right) = 0$$

(b) Since $f(0)$ is defined,

$$f'(0) = \lim_{h \to 0} \frac{h^2 \sin(1/h) - f(0)}{h}$$

$$= \lim_{h \to 0} h \sin\left(\frac{1}{h}\right) = 0$$

Exercises 3.1

1. Graph the function

$$f(x) = \begin{cases} x & \text{if } x < 0 \\ 1 & \text{if } x = 0 \\ x^2 & \text{if } x > 0 \end{cases}$$

Determine whether the function is PWC, continuous, PWS, and smooth.

2. (a) If $f(x) = |x|$, is $f(x)$ continuous at $x = 0$?
   (b) Determine $f'_+(0)$ and $f'_-(0)$.
   (c) Is $f(x)$ differentiable at $x = 0$?

3. Suppose

$$f(x) = \begin{cases} \sqrt{-x} & \text{if } x < -1 \\ 0 & \text{if } x = -1 \\ x^3 & \text{if } x > -1 \end{cases}$$

Draw a graph of the function and compute the jump in the function at $x = -1$.

4. Consider the function

$$f(x) = \begin{cases} e^{-x} & \text{if } x \leq 0 \\ e^x & \text{if } x \geq 0 \end{cases}$$

   (a) Find $f'_+(0)$ and $f'(0+)$.
   (b) Determine $f'_-(0)$ and $f'(0-)$.
   (c) Compute $f'(0)$ if it exists.

5. Assume that

$$f(x) = \begin{cases} x^3 \sin(1/x) & \text{if } x \neq 0 \\ 0 & \text{if } x = 0 \end{cases}$$

   (a) Compute $f'(x)$ for all values of x where the derivative exists.
   (b) Is $f'(x)$ differentiable at $x = 0$?

3.2. A BASIC FOURIER SERIES

In Example 2.6, the set

$$\left\{ 1, \sin\left(\frac{n\pi x}{L}\right), \cos\left(\frac{m\pi x}{L}\right) \right\}, \quad n, m \in \mathbf{N}, \quad -L < x < L$$

is orthogonal. Using the procedure of Section 2.7, we construct the series

$$f(x) \sim \frac{a_0}{2} \cdot 1 + \sum_{n=1}^{\infty} \left[a_n \cos \left(\frac{n\pi x}{L} \right) + b_n \sin \left(\frac{n\pi x}{L} \right) \right]$$

Employing (2.44a) we write the coefficients

$$\frac{a_0}{2} = \frac{1}{\|1\|^2} \int_{-L}^{L} f \cdot 1 \, dx$$

$$a_n = \frac{1}{\|\cos (n\pi x/L)\|^2} \int_{-L}^{L} f \cos \left(\frac{n\pi x}{L} \right) dx$$

$$b_n = \frac{1}{\|\sin (n\pi x/L)\|^2} \int_{-L}^{L} f \sin \left(\frac{n\pi x}{L} \right) dx$$

For the squares of the norms,

$$\|1\|^2 = \int_{-L}^{L} 1^2 \, dx = 2L$$

$$\left\| \cos \left(\frac{n\pi x}{L} \right) \right\|^2 = \int_{-L}^{L} \cos^2 \left(\frac{n\pi x}{L} \right) dx = L$$

$$\left\| \sin \left(\frac{n\pi x}{L} \right) \right\|^2 = \int_{-L}^{L} \sin^2 \left(\frac{n\pi x}{L} \right) dx = L$$

As a result

$$\frac{a_0}{2} = \frac{1}{2L} \int_{-L}^{L} f \, dx$$

and

$$a_0 = \frac{1}{L} \int_{-L}^{L} f \, dx$$

$$a_n = \frac{1}{L} \int_{-L}^{L} f \cos \left(\frac{n\pi x}{L} \right) dx$$

$$b_n = \frac{1}{L} \int_{-L}^{L} f \sin \left(\frac{n\pi x}{L} \right) dx$$

Therefore, the series may be written

$$f(x) \sim \frac{a_0}{2} + \sum_{n=1}^{\infty} \left[a_n \cos \left(\frac{n\pi x}{L} \right) + b_n \sin \left(\frac{n\pi x}{L} \right) \right] \tag{3.1}$$

where

$$a_n = \frac{1}{L} \int_{-L}^{L} f \cos\left(\frac{n\pi x}{L}\right) dx, \quad n \in \mathbf{N}_0$$

$$b_n = \frac{1}{L} \int_{-L}^{L} f \sin\left(\frac{n\pi x}{L}\right) dx, \quad n \in \mathbf{N} \tag{3.2}$$

We refer to the series (3.1) with the coefficients (3.2) as a *basic Fourier series* or *Fourier trigonometric expansion* corresponding to the function f. By writing the constant $a_0/2$ instead of a_0, no separate formula is needed in (3.2).

Each term of the series is periodic with a period $2L$. As a result when the series converges to f on the fundamental interval $(-L, L)$, it converges to a periodic function with a period $2L$, a function that agrees with f on the fundamental interval. In this case we say that the series represents the *periodic extension* of f for all x.

A popular form of the Fourier series is obtained when $L = \pi$ in (3.1) and (3.2). With this substitution we have

$$f(x) \sim \frac{a_0}{2} + \sum_{n=1}^{\infty} (a_n \cos nx + b_n \sin nx), \quad -\pi < x < \pi \tag{3.1a}$$

where

$$a_n = \frac{1}{\pi} \int_{-\pi}^{\pi} f \cos nx\, dx, \quad n \in \mathbf{N}_0$$

$$b_n = \frac{1}{\pi} \int_{-\pi}^{\pi} f \sin nx\, dx, \quad n \in \mathbf{N} \tag{3.2a}$$

The formulas for a_n and b_n in (3.2) or (3.2a) are known as in the *Euler formulas* for the series.

Example 3.3

(a) Draw a graph of the function

$$f(x) = \begin{cases} 0 & \text{if } -2 < x < 0 \\ 1 & \text{if } 0 < x < 2 \end{cases}$$

where the period of the function is 4.

(b) Determine the Fourier coefficients and write the Fourier series corresponding to the function in (a).

The graph for this function is shown in Figure 3.3. For this function $L = 2$,

$$a_n = \frac{1}{2} \int_{-2}^{2} f \cos\left(\frac{n\pi x}{2}\right) dx$$

Figure 3.3. Graph of f.

$$= \frac{1}{2} \int_{-2}^{0} 0 \cdot \cos\left(\frac{n\pi x}{2}\right) dx + \frac{1}{2} \int_{0}^{2} 1 \cdot \cos\left(\frac{n\pi x}{2}\right) dx$$

$$= \left(\frac{1}{2}\right)\left(\frac{2}{n\pi}\right)\left[\sin\left(\frac{n\pi x}{2}\right)\right]_{0}^{2} = 0 \quad \text{if } n \neq 0$$

$$a_0 = \frac{1}{2} \int_{-2}^{2} f \, dx = \frac{1}{2} \int_{0}^{2} dx = 1$$

$$b_n = \frac{1}{2} \int_{-2}^{2} f \sin\left(\frac{n\pi x}{2}\right) dx = \frac{1}{2} \int_{0}^{2} 1 \cdot \sin\left(\frac{n\pi x}{2}\right) dx$$

$$= \frac{1}{n\pi} (1 - \cos n\pi) = \frac{1}{n\pi} \begin{cases} 2 & \text{if } n \text{ is odd} \\ 0 & \text{if } n \text{ is even} \end{cases}$$

Therefore,

$$b_{2n-1} = \frac{2}{(2n-1)\pi}, \quad b_{2n} = 0 \quad \text{if } n \in \mathbf{N}$$

The Fourier series may be expressed as

$$f(x) \sim \frac{1}{2} + \frac{1}{\pi} \sum_{n=1}^{\infty} \frac{1 - \cos n\pi}{n} \sin\left(\frac{n\pi x}{2}\right) \tag{3.3}$$

or, if the zero coefficients are omitted,

$$f(x) \sim \frac{1}{2} + \frac{2}{\pi} \sum_{n=1}^{\infty} \frac{1}{2n-1} \sin\left[\frac{(2n-1)\pi x}{2}\right] \tag{3.3a}$$

In Example 3.3, the function is undefined at points $x = 0, \pm 2, \pm 4, \ldots$. If the series converges at these jump discontinuities it cannot converge to f. This question is considered in a theorem stated without proof.

Theorem 3.1 (A Fourier convergence theorem). Assume that f is periodic with a period $2L$ and PWS on the interval $-L \leq x \leq L$. Then the corresponding Fourier series

$$\frac{a_0}{2} + \sum_{n=1}^{\infty} \left[a_n \cos\left(\frac{n\pi x}{L}\right) + b_n \sin\left(\frac{n\pi x}{L}\right) \right]$$

where

$$a_n = \frac{1}{L} \int_{-L}^{L} f \cos\left(\frac{n\pi x}{L}\right) dx, \quad n \in \mathbf{N}_0$$

$$b_n = \frac{1}{L} \int_{-L}^{L} f \sin\left(\frac{n\pi x}{L}\right) dx, \quad n \in \mathbf{N}$$

converges to the average

$$\frac{f(x+) + f(x-)}{2} \tag{3.4}$$

A proof of the theorem when $L = \pi$ is given by Young [54, pp. 179–186].

The function f in Example 3.3 satisfies the hypothesis of Theorem 3.1. At points $x = 0, \pm 2, \pm 4, \ldots$, the convergence given by (3.4) is 1/2. It should be emphasized that the Fourier series representing f, as defined in Theorem 3.1, may or may not converge to the function. At any point where

$$f(x+) = f(x-) = f(x)$$

and the hypothesis of the theorem is met, the Fourier series converges to f. At other points where the function is not defined or the function is defined differently than the arithmetic mean of $f(x+)$ and $f(x-)$, the series converges to

$$\frac{f(x+) + f(x-)}{2}$$

The conditions of the theorem are sufficient conditions only. Surprisingly, continuity of f alone is not adequate to establish convergence of the corresponding Fourier series. Illustrations of this situation are difficult to construct and are of a rather "abnormal" structure.

The convergence theorem provides us with a reason for the practical importance of the Fourier series. Functions with finite jump discontinuities can be expanded in Fourier series, but they would fail to meet the differentiability requirement of a Taylor series expansion. In physics and engineering a significant class of problems involves periodic finite jumps. Consequently, the Fourier series is a vital tool for studying these processes.

3.3. EVEN AND ODD FUNCTIONS

A function f having the property that

$$f(-x) = f(x)$$

is an *even function*, and a function satisfying the condition that

$$f(-x) = -f(x)$$

is an *odd function*.

Example 3.4. Test the functions for even or odd properties using the definitions x^2, $\cos x$, $\sin x$, x^3, e^x, $e^x + e^{-x}$, $e^x - e^{-x}$, $x^3 + x^2$, 1, and 0.

The functions x^2, $\cos x$, $e^x + e^{-x}$, and 1 are all even functions. Functions $\sin x$, x^3, $e^x - e^{-x}$ are all odd functions. The polynomial $x^3 + x^2$ and e^x are neither even nor odd, but 0 satisfies the definitions of both even and odd functions. Polynomials having only odd degree terms are odd, and those containing only even degree terms are even.

If f and g are both even functions, then the product fg is even. If f is even and g odd, the product is odd. For f and g both odd, the product is even. Conclusions involving other operations are easy to formulate from the definitions. We are especially interested in even and odd functions over *symmetric intervals* of the type $(-L, L)$. The graph of an even function is symmetric relative to the functional axis, and the graph of an odd function is symmetric with respect to the origin. Two properties associated with integrals of even and odd functions over symmetric intervals are essential for our current discussion. We observe that if f is an even function, then

$$\int_{-L}^{L} f \, dx = 2 \int_{0}^{L} f \, dx \tag{3.5}$$

and if f is odd,

$$\int_{-L}^{L} f \, dx = 0 \tag{3.6}$$

We assume in (3.5) and (3.6) that f is integrable.

3.4. FOURIER SINE AND COSINE SERIES

Fourier sine and cosine series frequently are referred to as *half range series* since only half of a symmetric interval is employed in the integrals defining the coefficients. To obtain these series one assumes that the function f is an even or an odd function.

We observe that if f is even, then $f \cos(n\pi x/L)$ is also even. The coefficient a_n in (3.2) has an even integrand on $(-L, L)$. Using the property (3.5) we write twice the integral over half the interval and obtain

$$a_n = \frac{2}{L} \int_{0}^{L} f \cos\left(\frac{n\pi x}{L}\right) dx$$

Since $f \sin (n\pi x/L)$ is odd and b_n has an odd integrand over a symmetric interval, by employing (3.6) in (3.2), we obtain

$$b_n = 0$$

With f even, we write (3.1) in its new form

$$f(x) \sim \frac{a_0}{2} + \sum_{n=1}^{\infty} a_n \cos \left(\frac{n\pi x}{L}\right) \tag{3.7}$$

where

$$a_n = \frac{2}{L} \int_0^L f \cos \left(\frac{n\pi x}{L}\right) dx, \quad n \in N_0 \tag{3.8}$$

The interval in this case is $(0, L)$, but the *even periodic extension* of f presumes a period of $2L$. This series is called the *Fourier cosine series* or the *half range Fourier cosine series*.

If f is an odd function, then $f \sin (n\pi x/L)$ is an even function. In this case

$$b_n = \frac{2}{L} \int_0^L f \sin \left(\frac{n\pi x}{L}\right) dx$$

The product $f \cos (n\pi x/L)$ is odd, and

$$a_n = 0$$

As a result we may write

$$f(x) \sim \sum_{n=1}^{\infty} b_n \sin \left(\frac{n\pi x}{L}\right) \tag{3.9}$$

where

$$b_n = \frac{2}{L} \int_0^L f \sin \left(\frac{n\pi x}{L}\right) dx \tag{3.10}$$

Again the interval is $(0, L)$ and a period of $2L$ is assumed when the *odd periodic extension* of f is considered. This is a *Fourier sine series*. Specialized theorems are stated for the convergence of series (3.7) and (3.9).

Theorem 3.2 (A Fourier convergence theorem for the cosine series). Assume that f is an even periodic function and PWS on the interval $(0, L)$. Then the corresponding Fourier series

$$\frac{a_0}{2} + \sum_{n=1}^{\infty} a_n \cos \left(\frac{n\pi x}{L}\right)$$

where

$$a_n = \frac{2}{L} \int_0^L f \cos\left(\frac{n\pi x}{L}\right) dx, \quad n \in N_0$$

converges to

$$\frac{f(x+) + f(x-)}{2}$$

Theorem 3.3 (A Fourier convergence theorem for the sine series). Assume that f is an odd periodic function and PWS on the interval $(0, L)$. Then the corresponding Fourier series

$$\sum_{n=1}^{\infty} b_n \sin\left(\frac{n\pi x}{L}\right)$$

where

$$b_n = \frac{2}{L} \int_0^L f \sin\left(\frac{n\pi x}{L}\right) dx, \quad n \in N$$

converges to

$$\frac{f(x+) + f(x-)}{2}$$

Example 3.5. (a) Find the Fourier series for the function

$$f(x) = \begin{cases} -\cos x & \text{if } -\pi < x < 0 \\ \cos x & \text{if } 0 < x < \pi \end{cases}$$

(b) Find the convergence at all jump discontinuities.

(a) Although a graph is not requested in this example, a picture of the function is helpful. See Figure 3.4.

Since f is odd we can construct a sine series representation for the function. In this case $L = \pi$.

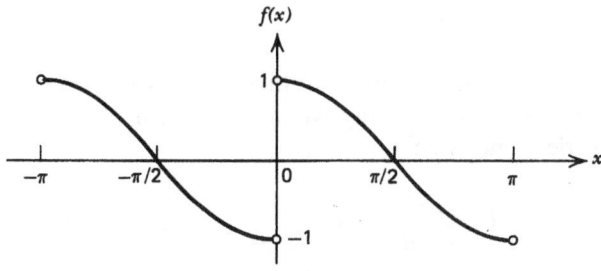

Figure 3.4. Graph of f.

$$b_n = \frac{2}{\pi} \int_0^\pi \cos x \sin nx \, dx$$

$$= \left(\frac{2}{\pi}\right)\left(\frac{1}{2}\right) \int_0^\pi [\sin(1+n)x - \sin(1-n)x] \, dx$$

$$= \frac{1}{\pi}\left[\frac{-1}{1+n}\cos(1+n)x + \frac{1}{1-n}\cos(1-n)x\right]_0^\pi$$

$$= \left[\frac{2n}{(n^2-1)\pi}\right][1+\cos n\pi] \quad \text{if } n \neq 1$$

$$= \left[\frac{2n}{(n^2-1)\pi}\right]\begin{cases} 0 & \text{if } n \text{ is odd} \\ 2 & \text{if } n \text{ is even} \end{cases}$$

$$b_{2n} = \frac{8n}{(4n^2-1)\pi}$$

$$b_1 = \frac{2}{\pi} \int_0^\pi \cos x \sin x \, dx = 0$$

Therefore, the series is

$$f(x) \sim \frac{8}{\pi} \sum_{n=1}^\infty \frac{n \sin 2nx}{4n^2-1}$$

(b) A jump discontinuity in the function exists at $x = 0$. If the odd periodic extension is considered for f, then jump discontinuities exist at $x = n\pi$, $n \in \mathbf{Z}$. At each discontinuity the convergence is zero.

3.5. COMPLEX FOURIER SERIES

In Example 2.3 we determined that the set of exponential functions $\{\exp[2n\pi ix/(b-a)]\}$, $n \in \mathbf{Z}$, $a < x < b$, is orthogonal in the Hermitian sense. If adequate convergence conditions are assumed and we write a series for f based on this set, then

$$f(x) = \sum_{n=-\infty}^\infty c_n \exp\left(\frac{2n\pi ix}{b-a}\right)$$

where c_n may be determined by multiplying by $\exp[-2m\pi ix/(b-a)]$ and then integrating over (a, b). Only when $n = m$ will we obtain a nonzwero term in the series, and then

$$\int_a^b f \exp\left(\frac{-2n\pi ix}{b-a}\right) dx = c_n \int_a^b \exp 0 \, dx = c_n(b-a)$$

Therefore,

$$c_n = \frac{1}{b-a} \int_a^b f \exp\left(\frac{-2n\pi ix}{b-a}\right) dx$$

Proceeding as we did earlier with relaxed conditions, we write the correspondence

$$f(x) \sim \sum_{n=-\infty}^{\infty} c_n \exp\left(\frac{2n\pi ix}{b-a}\right) \tag{3.11}$$

where

$$c_n = \frac{1}{b-a} \int_a^b f \exp\left(\frac{-2n\pi ix}{b-a}\right) dx \tag{3.12}$$

This is a *complex form* of the Fourier series. It also may be called the *exponential form* of the series. Convergence follows the pattern of previous series.

If $a = -\pi$ and $b = \pi$, the series (3.11) has the form

$$f(x) \sim \sum_{n=-\infty}^{\infty} c_n e^{inx} \tag{3.13}$$

where

$$c_n = \frac{1}{2\pi} \int_{-\pi}^{\pi} f e^{-inx} dx \tag{3.14}$$

Example 3.6. Determine the Fourier complex series for the function $f(x) = e^{2x}$, $-\pi < x < \pi$.

Series (3.13) with coefficients (3.14) fit this problem exactly.

$$e^{2x} \sim \sum_{n=-\infty}^{\infty} c_n e^{inx}$$

where

$$c_n = \frac{1}{2\pi} \int_{-\pi}^{\pi} e^{2x} e^{-inx} dx = \frac{1}{2\pi} \int_{-\pi}^{\pi} e^{(2-in)x} dx$$

$$= \frac{1}{2\pi(2-in)} [e^{(2-in)x}]_{-\pi}^{\pi}$$

$$= \frac{1}{2\pi(2-in)} [e^{(2-in)\pi} - e^{-(2-in)\pi}]$$

$$= \frac{(-1)^n(2+in)\sinh 2\pi}{\pi(4+n^2)}$$

Therefore the series is

$$\frac{1}{\pi} \sum_{n=-\infty}^{\infty} \frac{(-1)^n (2 + in) \sinh 2\pi \, e^{inx}}{4 + n^2}$$

By inserting (3.12) in the series (3.11), employing Euler's identity, displaying the series as an isolated term plus two series, changing an index, and finally combining the two series again, one obtains

$$f(x) \sim \frac{1}{(b-a)} \int_a^b f(t) \, dt + \frac{2}{(b-a)} \sum_{n=1}^{\infty} \int_a^b f(t) \cos\left[\frac{2n\pi(x-t)}{b-a}\right] dt \tag{3.15}$$

3.6. HARMONIC ANALYSIS

Included in the topic of *Harmonic Analysis* is the problem of approximating a function, assumed to be periodic, by a linear combination of orthogonal functions. If a periodic function is to be approximated over a complete period or more than one period, it is appropriate to use an approximating function which has the same period as $f(x)$. Although there are other ways to determine an approximation, the procedure we propose here follows Theorem 2.7.

We assume that the linear combination of orthogonal functions $Q_n(x)$ of the theorem is composed of sines and cosines. The approximation for $f(x)$ has the form

$$f(x) \sim \frac{a_0}{2} + \sum_{k=1}^{n} (a_k \cos kx + b_k \sin kx) \tag{3.16}$$

where the coefficients a_k and b_k are to be found so that the *integrated square error* is least. We restrict our investigation to the interval $(0, 2\pi)$ and assume that the function $f(x)$ has a period of 2π, so that $f(x + 2\pi) = f(x)$. The requirement of the theorem specifies that

$$E = \int_0^{2\pi} \left[f(x) - \frac{a_0}{2} - \sum_{k=1}^{n} (a_k \cos kx + b_k \sin kx) \right]^2 dx$$

is smallest when the coefficients of (3.16) are the Fourier coefficients. Therefore,

$$a_k = \frac{1}{\pi} \int_0^{2\pi} f(x) \cos kx \, dx \, , \quad b_k = \frac{1}{\pi} \int_0^{2\pi} f(x) \sin kx \, dx \, , \quad k \in \mathbf{N}_0 \tag{3.17}$$

Typical of the *least squares* approximation by orthogonal functions is that each coefficient is computed independently of the others. Each coefficient has a value not depending on the number of *harmonics* to be included in the approximation.

Example 3.7. Let

$$f(x) = \begin{cases} x & \text{if } 0 \leqslant x \leqslant \pi, \\ 0 & \text{if } \pi < x \leqslant 2\pi, \end{cases} \quad f(x + 2\pi) = f(x)$$

Use (3.16) and (3.17) to determine the least squares approximation.
    Let $f(x)$ be represented by the linear combination

$$f(x) \sim \frac{a_0}{2} + \sum_{k=1}^{n} (a_k \cos kx + b_k \sin kx)$$

For the "best fit" by least squares

$$a_0 = \frac{1}{\pi} \int_0^{2\pi} f(x) \, dx = \frac{1}{\pi} \int_0^{\pi} x \, dx + \frac{1}{\pi} \int_{\pi}^{2\pi} 0 \, dx = \frac{\pi}{2}$$

$$a_k = \frac{1}{\pi} \int_0^{2\pi} f(x) \cos kx \, dx = \frac{1}{\pi} \int_0^{\pi} x \cos kx \, dx + 0 = \frac{1}{\pi} \left(\frac{\cos k\pi - 1}{k^2} \right)$$

$$b_k = \frac{1}{\pi} \int_0^{2\pi} f(x) \sin kx \, dx = \frac{1}{\pi} \int_0^{\pi} x \sin kx \, dx + 0 = - \frac{\cos k\pi}{k}$$

If we choose $n = 10$ in (3.16), $f(x)$ is represented by

$$\hat{f}(x) = \frac{\pi}{4} - \frac{2}{\pi} \cos x + \sin x - \frac{1}{2} \sin 2x - \frac{2}{9\pi} \cos 3x + \frac{1}{3} \sin 3x - \frac{1}{4} \sin 4x$$

$$- \frac{2}{25\pi} \cos 5x + \frac{1}{5} \sin 5x - \frac{1}{6} \sin 6x - \frac{2}{49\pi} \cos 7x + \frac{1}{7} \sin 7x$$

$$- \frac{1}{8} \sin 8x - \frac{2}{81\pi} \cos 9x + \frac{1}{9} \sin 9x - \frac{1}{10} \sin 10x$$

A comparison of the given function f and its truncated series approximation $\hat{f}$ at selected points is shown in Figure 3.5. The asterisks represent the values of $\hat{f}$ for $x = 0, \pi/10, 2\pi/10, \ldots, 2\pi$. Notice the jump discontinuity at $x = \pi$. From Theorem 3.1 the Fourier series converges to

$$\frac{f(x+) + f(x-)}{2} = \frac{\pi}{2}$$

at this discontinuity. Note also the slight overshoot of the approximation to the left of the discontinuity; this is evidence of *Gibbs's phenomenon* which will be discussed presently.
    There are other important problems associated with *harmonic analysis*. The determination of an approximating formula from a discrete set of

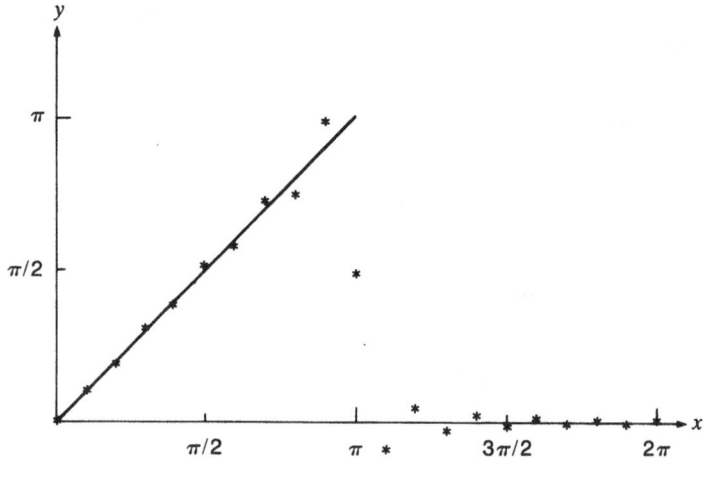

Figure 3.5. Graph of f and $\hat{f}$.

periodic data has many applications. Here we assume that the values of the function are known for certain equally spaced points. The problem is then to approximate the data with a truncated trigonometric series, either in the least squares sense or by interpolation. It will be shown in Chapter 7 that this problem can be solved by means of the discrete Fourier transform.

Exercises 3.2

For Exercises 1–5 (a) sketch the graph of f, (b) determine the Fourier series corresponding to f, and (c) indicate the convergence at the given points. It is assumed that the functions are periodic and one period is given.

1. $f(x) = \begin{cases} -2 & \text{if } -2 < x < 0 \\ 2 & \text{if } \ \ 0 < x < 2 \end{cases}$

   Find the convergence at $x = 0$.

2. $f(x) = \begin{cases} 0 & \text{if } -2 < x < -1 \\ 2 & \text{if } -1 < x < \ \ 1 \\ 0 & \text{if } \ \ 1 < x < \ \ 2 \end{cases}$

   Find the convergence at $x = -1$.

3. $f(x) = x$, $-1 < x < 1$. Find the convergence at $x = 1$.

4. $f(x) = \begin{cases} \ \ 0 & \text{if } -2 < x < -1 \\ 2 + x & \text{if } -1 < x < \ \ 0 \\ 2 - x & \text{if } \ \ 0 < x < \ \ 1 \\ \ \ 0 & \text{if } \ \ 1 < x < \ \ 2 \end{cases}$

   Find the convergence at $x = 1$ and $x = 2$.

5. $f(x) = e^{-x}$, $-1 < x < 1$. Find the convergence at $x = -1$ and $x = 1$.

6. Prove that the sum of two odd functions is odd.

7. Show that if f is odd, then $|f|$ and f^2 are even functions.

8. Show that if f is defined for all x, then
   (a) $g(x) = [f(x) + f(-x)]/2$ is even,
   (b) $h(x) = [f(x) - f(-x)]/2$ is odd.

9. If
$$f(x) = \begin{cases} 1 & \text{when } -\pi/2 < x < \pi/2 \\ 0 & \text{when } \pi/2 < x < 3\pi/2 \end{cases}$$
with a period 2π, (a) find the Fourier series for f, and (b) show that
$$\frac{\pi}{4} = \sum_{n=1}^{\infty} \frac{(-1)^{n-1}}{2n-1}$$

10. (a) Find the Fourier series for $f(x) = |\sin x|$, $-\pi < x < \pi$
    (b) Show that
$$\frac{1}{2} = \sum_{n=1}^{\infty} \frac{1}{4n^2 - 1}$$
    (c) Show that
$$\frac{1}{2} - \frac{\pi}{4} = \sum_{n=1}^{\infty} \frac{(-1)^n}{4n^2 - 1}$$

11. Write (a) the Fourier cosine series and (b) the Fourier sine series for $f(x) = \cos x$, $0 < x < \pi$.

12. Write the Fourier cosine series for the function
$$f(x) = \begin{cases} \cos x & \text{when } 0 < x \leqslant \pi/2 \\ 0 & \text{when } \pi/2 \leqslant x < \pi \end{cases}$$

13. If $f(x) = x$ for $-\pi < x < \pi$, find the Fourier series for the function. If the series represents $f(x)$ on the given interval show graphically the function represented by the series for all x.

14. Write the Fourier cosine series for $f(x) = x$, $0 \leqslant x \leqslant \pi$. If the series represents $f(x)$ on the given interval, show graphically the function represented by the series for all x. What differences do you notice in the extensions of the functions of Exercises 13 and 14?

15. If $f(x) = x^2$, $0 < x < \pi$, find the Fourier sine series and draw a graph of the function with its periodic extension.

16. If $f(x) = x^2$, $-L < x < L$, write the Fourier series corresponding to f.

17. Find the complex form of the Fourier series for the function
$$f(x) = \begin{cases} -1 & \text{when } -\pi < x < 0 \\ 1 & \text{when } 0 < x < \pi \end{cases}$$
and $f(x + 2\pi) = f(x)$.

18. From (3.2a) and (3.14),
    (a) Express c_n and $\bar{c}_n$ in terms of a_n and b_n.
    (b) Find a_n and b_n in terms of c_n and $\bar{c}_n$.
    (c) Determine c_{-n} and $\bar{c}_{-n}$ in terms of a_n and b_n.
    (d) How are c_{-n} and c_n related?

19. Determine the Fourier complex series for the function $f(x) = e^{3x}$, $-\pi < x < \pi$, and $f(x + 2\pi) = f(x)$.

20. Derive the form of the Fourier series indicated in (3.15).

21. Compute Exercise 9(a) using the formula derived in Exercise 20.

22. Using (3.15), find the Fourier series for $f(x) = e^{-x}$, $0 < x < 1$.

23. Determine the complex form of the Fourier series for $f(x) = \cosh x$, $-1 < x < 1$.

24. In Exercise 17, the function is odd. The sine expansion for the function is

$$f(x) = \frac{4}{\pi} \sum_{k=1}^{\infty} \frac{\sin(2k-1)x}{2k-1}$$

The graph of the function (Figure 3.6) is compared for one period with graphs of the partial sums $S_{2n-1}(x)$ for $n = 1, 2, 3$ and a larger number. When f has a finite jump discontinuity, as this one does, the Fourier series for the function cannot converge uniformly to f on an interval containing the discontinuity. While successive graphs of $S_{2n-1}(x)$ apparently fit the function better and better, well inside the subintervals between jumps, the approximation is poor for values of x near discontinuities. For example, the distance from the minimum nearest $x = 0$ to the maximum nearest $x = 0$ is considerably greater than the actual jump in the function. This excess prevails for all $S_{2n-1}(x)$. The condition is characteristic for partial sums of Fourier series near points of discontinuity. The behavior of the approximating Fourier curves for the function with discontinuities resembles that of a diver as he goes through his preliminary motions before jumping from the spring board. The greatest oscillation takes place immediately before the "jump." At jump discontinuities the "overshooting" or "undershooting" is known as *Gibb's phenomenon*. The demonstration function used here is referred to as the *square wave function*.

   To compute the extent of this condition we follow the procedure frequently employed by many investigators. The sum of the first n terms of the Fourier sine series of the function of Exercise 17 is

$$S_n(x) = \frac{4}{\pi} \sum_{k=1}^{n} \frac{\sin(2k-1)x}{2k-1} \tag{3.18}$$

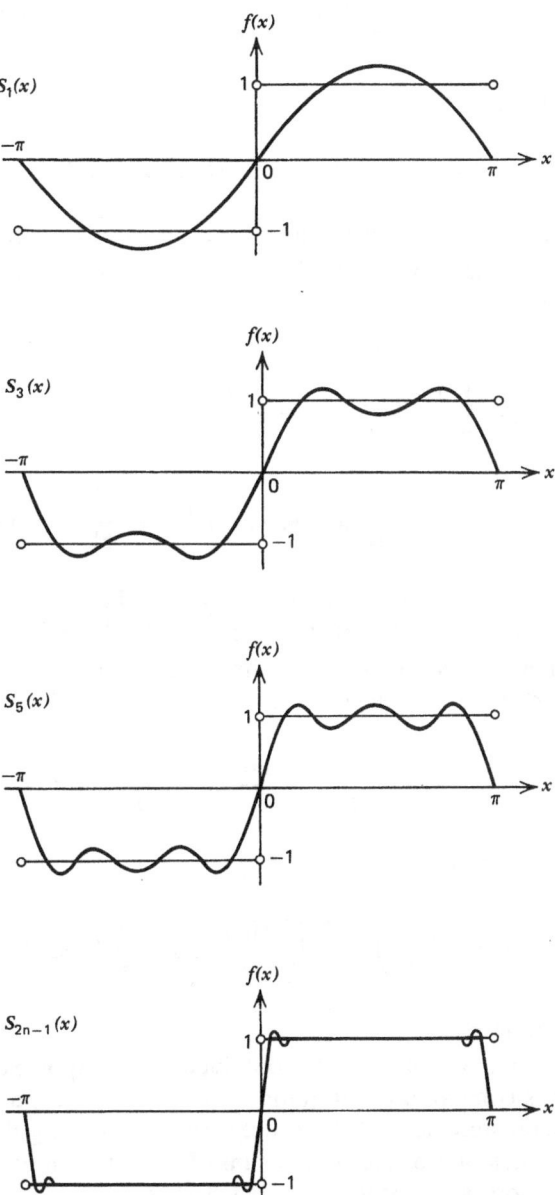

Figure 3.6. Graphical comparison of the square-wave function and $S_{2n-1}(x)$. (Reproduced in modified form from Carslaw [13, p. 300], by permission of Dover Publications, Inc.)

(a) Show that

$$S'_n(x) = \frac{4}{\pi} \sum_{k=1}^{n} \cos(2k-1)x = \frac{2}{\pi} \frac{\sin 2nx}{\sin x} \qquad (3.19)$$

The identity

$$2 \sin x \cos(2k-1)x = \sin 2kx - \sin(2k-2)x \qquad (3.20)$$

will be useful to form a telescoping series in (3.19).

(b) Find extrema points from $S'_n(x) = 0$, and show that the extremum at $x = \pi/2n$ is a maximum.

(c) From (3.19) show that

$$S_n(x) - \frac{2}{\pi} \int_0^x \frac{\sin 2nt}{t} \, dt = \frac{2}{\pi} \int_0^x \sin 2nt(t \csc t)\left(\frac{1}{t} - \frac{\sin t}{t^2}\right) dt \qquad (3.21)$$

$$= \frac{2}{\pi} \int_0^x \sin 2nt(t \csc t)\left(\frac{t}{3!} - \frac{t^3}{5!} + \cdots\right) dt \qquad (3.22)$$

(d) Let $\phi(t) = t \csc t$ for $t \neq 0$ and $\phi(0) = 1$. Show that ϕ is continuous and increasing on $[0, \pi/2]$. Thus $\max |\phi(t)| = \pi/2$ for $0 \leq t \leq \pi/2$.

(e) By changing variables in the first integral of (3.21) and observing that $0 < t/3! - t^3/5! + \cdots < t/3!$ when $0 < t \leq \pi/2$, verify that

$$\left| S_n(x) - \frac{2}{\pi} \int_0^{2nx} \frac{\sin \alpha}{\alpha} \, d\alpha \right| < \frac{1}{3!} \int_0^x t \, dt = \frac{1}{12} x^2$$

(f) Finally show that

$$\lim_{n \to \infty} S_n\left(\frac{\pi}{2n}\right) = \frac{2}{\pi} \int_0^{\pi} \frac{\sin \alpha}{\alpha} \, d\alpha = \frac{2}{\pi} (1.852) = 1.179$$

approximately.

Instead of a maximum 1, we obtained 1.179 approximately. If the overshoot is compared to the jump in the function it is about 9%. For a more general statement of the process for determining the approximate maximum deviation of the partial sum of the function near a finite jump in the function, see Bôcher [4, pp. 131–132].

25. The function $f(x) = |x|$, $-\pi \leq x \leq \pi$, $f(x + 2\pi) = f(x)$ is called the *sawtooth function*. It is an even function and has a cosine series

$$\frac{\pi}{2} - \frac{4}{\pi} \sum_{k=1}^{\infty} \frac{\cos(2k-1)x}{(2k-1)^2}$$

(a) Are there any discontinuities in f?

(b) Write $S_0(x)$ and $S_{2n-1}(x)$, $n = 1, 2, 3$.

(c) Graph and compare $S_0(x)$, $S_1(x)$, and $S_3(x)$ with $|x|$ over one period. We see by Theorem 3.4 that this series is uniformly convergent.

(d) If a computer is available, write a program to compute $S_{2n-1}(x)$, $x = 0, \pm 0.1\pi, \pm 0.2\pi, \ldots, \pm \pi$ for $n = 1, 3, 8$. Graph these partial sums (by hand or computer) and compare with $|x|$.

3.7. UNIFORM CONVERGENCE OF FOURIER SERIES

As a matter of simplifying notation we investigate the special series (3.1a) when $L = \pi$ in this section. We assume that f is a PWS function on $(-\pi, \pi)$ and $f(-\pi) = f(\pi)$. The derivative function f' is a PWC function, since f is PWS. The coefficients of the derivative series are

$$a'_n = \frac{1}{\pi} \int_{-\pi}^{\pi} f' \cos nx \, dx , \quad n \in N_0 \tag{3.23}$$

$$b'_n = \frac{1}{\pi} \int_{-\pi}^{\pi} f' \sin nx \, dx , \quad n \in N \tag{3.24}$$

Since $f(-\pi) = f(\pi)$,

$$a'_0 = \frac{1}{\pi} \int_{-\pi}^{\pi} f' \, dx = 0 \tag{3.25}$$

Integrating by parts (3.23) and (3.24)

$$a'_n = \frac{1}{\pi} [f \cos nx]_{-\pi}^{\pi} + \frac{n}{\pi} \int_{-\pi}^{\pi} f \sin nx \, dx$$

$$= nb_n \tag{3.26}$$

$$b'_n = \frac{1}{\pi} [f \sin nx]_{-\pi}^{\pi} - \frac{n}{\pi} \int_{-\pi}^{\pi} f \cos nx \, dx$$

$$= -na_n \tag{3.27}$$

From (3.26) and (3.27)

$$|a_n| = \frac{|b'_n|}{n} \quad \text{and} \quad |b_n| = \frac{|a'_n|}{n}$$

We observe that

$$\left[|a_n'| - \frac{1}{n}\right]^2 \geq 0$$

or

$$(a_n')^2 - \frac{2|a_n'|}{n} + \frac{1}{n^2} \geq 0$$

Thus

$$(a_n')^2 + \frac{1}{n^2} \geq \frac{2|a_n'|}{n}$$

Similarly, we find that

$$(b_n')^2 + \frac{1}{n^2} \geq \frac{2|b_n'|}{n}$$

Therefore,

$$(a_n')^2 + (b_n')^2 + \frac{2}{n^2} \geq \frac{2}{n}\left[|a_n'| + |b_n'|\right]$$

$$\geq 2[|a_n| + |b_n|]$$

and

$$|a_n| + |b_n| \leq \frac{1}{2}[(a_n')^2 + (b_n')^2] + \frac{1}{n^2}$$

Using Bessel's inequality for f' and (3.25), we have

$$\sum_{k=1}^{n} [(a_n')^2 + (b_n')^2] \leq \frac{1}{\pi} \int_{-\pi}^{\pi} [f']^2 \, dx \qquad (3.28)$$

The corresponding series in (3.28)

$$\sum_{n=1}^{\infty} [(a_n')^2 + (b_n')^2]$$

is convergent. The series

$$\sum_{n=1}^{\infty} \frac{1}{n^2}$$

also converges. Then

$$\sum_{n=1}^{\infty} [|a_n| + |b_n|] \tag{3.29}$$

converges. We see that

$$|a_n \cos nx + b_n \sin nx| \le |a_n \cos nx| + |b_n \sin nx|$$
$$\le |a_n| + |b_n|$$

Using the Weierstrass M-test, knowing that (3.29) converges, we have

$$\frac{a_0}{2} + \sum_{n=1}^{\infty} (a_n \cos nx + b_n \sin nx)$$

converges absolutely and uniformly for $-\pi \le x \le \pi$. Young [54, pp. 190–192] gives the proof. We state the results as a theorem.

Theorem 3.4.* Assume that f is a continuous PWS function of period 2π on $-\pi \le x \le \pi$ with $f(-\pi) = f(\pi)$. Then the Fourier series corresponding to f

$$\frac{a_0}{2} + \sum_{n=1}^{\infty} (a_n \cos nx + b_n \sin nx) \tag{3.30}$$

where

$$a_n = \frac{1}{\pi} \int_{-\pi}^{\pi} f \cos nx \, dx \,, \quad n \in \mathbf{N}_0$$

$$b_n = \frac{1}{\pi} \int_{-\pi}^{\pi} f \sin nx \, dx \,, \quad n \in \mathbf{N}$$

is convergent absolutely and uniformly to f for $-\pi \le x \le \pi$.

The periodic extension of f is continuous and PWS. Conditions of the theorem assure absolute and uniform convergence on any interval to the periodic extension of the function.

If f is continuous PWS on $0 \le x \le \pi$, $f(0) = f(\pi)$, then

$$\frac{a_0}{2} + \sum_{n=1}^{\infty} a_n \cos nx$$

where

$$a_n = \frac{2}{\pi} \int_0^{\pi} f \cos nx \, dx \,, \quad n \in \mathbf{N}_0$$

*From Young [54], by permission of the author.

is convergent absolutely and uniformly to f for $0 \leqslant x \leqslant \pi$ and to an even periodic extension of f for other x. Also

$$\sum_{n=1}^{\infty} b_n \sin nx$$

where

$$b_n = \frac{2}{\pi} \int_0^{\pi} f \sin nx \, dx \,, \quad n \in \mathbf{N}$$

is convergent absolutely and uniformly to f for $0 \leqslant x \leqslant \pi$ and to an odd periodic extension of f for other x. These are specializations for the cosine and sine series.

Let f satisfy the conditions of Theorem 3.4. If we multiply the Fourier series (3.30) for f by the function f, the result is a uniformly convergent series. We integrate this result.

$$\int_{-\pi}^{\pi} f^2 \, dx = \frac{a_0}{2} \int_{-\pi}^{\pi} f \, dx + \sum_{n=1}^{\infty} \left[a_n \int_{-\pi}^{\pi} f \cos nx \, dx + b_n \int_{-\pi}^{\pi} f \sin nx \, dx \right] \tag{3.31}$$

after identification and rearrangement, we write (3.31) in the form

$$\frac{a_0^2}{2} + \sum_{n=1}^{\infty} (a_n^2 + b_n^2) = \frac{1}{\pi} \int_{-\pi}^{\pi} f^2 \, dx \tag{3.32}$$

Equation (3.32) is *Parseval's identity*. Thus the Fourier series converges to f in the mean and this implies that the set

$$\{1, \cos nx, \sin mx\} \,, \quad m, n \in \mathbf{N} \,, \quad -\pi \leqslant x \leqslant \pi$$

is *complete*.

3.8. DIFFERENTIATION OF FOURIER SERIES

In some cases term-by-term differentiation of a Fourier series fails to converge to the derivative of the convergence of the original series. Consider the following.

Example 3.8. Differentiate the series

$$\cos x = \frac{8}{\pi} \sum_{n=1}^{\infty} \frac{n \sin 2nx}{(4n^2 - 1)}$$

and investigate the possibility of the newly formed series converging to the function $-\sin x$.

By termwise differentiation we have the series

$$\frac{16}{\pi} \sum_{n=1}^{\infty} \frac{n^2 \cos 2nx}{4n^2 - 1}$$

presumably the representation for the function $-\sin x$. Upon investigating the limit of $(16/\pi)(n^2/(4n^2 - 1)) \cos nx$ as $n \to \infty$, we find it is not zero. Therefore the new series is divergent and cannot be the convergence of $-\sin x$.

If the function f is replaced by f' in Theorem 3.1 with $L = \pi$, then we are assured that the series corresponding to f' converges. If f' is periodic with a period 2π and PWS on $-\pi \leq x \leq \pi$, then the corresponding Fourier series

$$\frac{a_0'}{2} + \sum_{n=1}^{\infty} (a_n' \cos nx + b_n' \sin nx)$$

where

$$a_n' = \frac{1}{\pi} \int_{-\pi}^{\pi} f' \cos nx \, dx, \quad n \in \mathbf{N}_0$$

$$b_n' = \frac{1}{\pi} \int_{-\pi}^{\pi} f' \sin nx \, dx, \quad n \in \mathbf{N}$$

converges to

$$\frac{f'(x+) + f'(x-)}{2}$$

If we add that $f(-\pi) = f(\pi)$ and make f a continuous function with f' PWS, then both f' and f are PWC. Coefficients

$$a_0' = 0, \quad a_n' = nb_a, \quad b_n' = -na_n$$

have been determined. The derivative f' is continuous where f'' exists. For the values of x where f'' exists.

$$f'(x) = f'(x+) = f'(x-)$$

and

$$\frac{f'(x+) + f'(x-)}{2} = f'(x)$$

or

$$f'(x) = \sum_{n=1}^{\infty} (nb_n \cos nx - na_n \sin nx)$$

The following theorem contains the results.

Theorem 3.5. Assume that f is a continuous function of period 2π on the interval $-\pi \leqslant x \leqslant \pi$ with $f(-\pi) = f(\pi)$. Let f', also a periodic function of period 2π, be PWC on the interval. Then at every point where f'' exists, f is termwise differentiable and the series converges to f'. The series

$$f(x) = \frac{a_0}{2} + \sum_{n=1}^{\infty} (a_n \cos nx + b_n \sin nx)$$

has the derivative

$$f'(x) = \sum_{n=1}^{\infty} (nb_n \cos nx - na_n \sin nx)$$

where

$$a_n = \frac{1}{\pi} \int_{-\pi}^{\pi} f \cos nx \, dx$$

$$b_n = \frac{1}{\pi} \int_{-\pi}^{\pi} f \sin nx \, dx$$

When f'' fails to exist but $f''_+(x)$ and $f''_-(x)$ exists, differentiation is valid in the sense that the series for f' converges to

$$\frac{f'(x+) + f'(x-)}{2}$$

For other types of Fourier series, Theorem 3.5 applies if the natural modifications are made in the theorem. If f is continuous and f' is PWC on $-L \leqslant x \leqslant L$, then where f'' exists the Fourier series for f is differentiable.

3.9. INTEGRATION OF FOURIER SERIES

We let f be PWC on $[-\pi, \pi]$ so that

$$h(x) = \int_a^x \left[f - \frac{a_0}{2} \right] dt$$

The derivative

$$h'(x) = f(x) - \frac{a_0}{2}$$

Therefore h is continuous everywhere that f exists. Even at discontinuous points it can be shown that

$$|h(x) - h(x_0)| < \varepsilon \quad \text{for all } 0 < |x - x_0| < \delta$$

if x is to the right or left of the discontinuous point x_0. Thus h is a continuous function. We compute

$$h(x + 2\pi) = \int_a^x \left[f - \frac{a_0}{2} \right] dt + \int_x^{x+2\pi} \left[f - \frac{a_0}{2} \right] dt$$

$$= h(x) + \int_{-\pi}^{\pi} \left[f - \frac{a_0}{2} \right] dt = h(x)$$

Since we have shown that h is periodic with period 2π, then

$$h(-\pi) = h(\pi)$$

Summarizing properties for h, we find that h is a continuous PWS function of period 2π, $-\pi \leqslant x \leqslant \pi$, with $h(-\pi) = h(\pi)$. Therefore, according to Theorem 3.4, h can be represented by a uniformly and absolutely convergent Fourier series

$$h(x) = \frac{A_0}{2} + \sum_{n=1}^{\infty} (A_n \cos nx + B_n \sin nx) \tag{3.33}$$

with coefficients

$$A_n = \frac{1}{\pi} \int_{-\pi}^{\pi} h \cos nx \, dx, \quad n \in \mathbf{N}_0$$

$$B_n = \frac{1}{\pi} \int_{-\pi}^{\pi} h \sin nx \, dx, \quad n \in \mathbf{N}$$

Integrating by parts, we obtain

$$A_n = \frac{1}{\pi} \left\{ \left[\frac{h \sin nx}{n} \right]_{-\pi}^{\pi} - \frac{1}{n} \int_{-\pi}^{\pi} \left[f - \frac{a_0}{2} \right] \sin nx \, dx \right\}$$

$$= -\frac{1}{n} \left[\frac{1}{\pi} \int_{-\pi}^{\pi} f \sin nx \, dx - \frac{a_0}{2\pi} \int_{-\pi}^{\pi} \sin nx \, dx \right]$$

$$A_n = -\frac{b_n}{n}, \quad n \in \mathbf{N} \tag{3.34}$$

Using a similar procedure, we find that

$$B_n = \frac{a_n}{n}, \quad n \in \mathbf{N} \tag{3.35}$$

After substituting coefficients (3.34) and (3.35) in (3.33) and observing that

$$\int_a^x f \, dt = \int_0^x f \, dt - \int_0^a f \, dt$$

we have

$$\int_a^x f\, dt = \left(\frac{a_0}{2}\right)(x-a) + \sum_{n=1}^{\infty} \left[\frac{a_n(\sin nx - \sin na)}{n} - \frac{b_n(\cos nx - \cos na)}{n}\right]$$
(3.36)

Formula (3.36) is exactly what one finds by termwise integration of the Fourier series. We are able at this time to state the result as a theorem.

Theorem 3.6. Assume that f is PWC and periodic with period 2π on $-\pi \leqslant x \leqslant \pi$. Then whether or not the Fourier series for f,

$$\frac{a_0}{2} + \sum_{n=1}^{\infty} (a_n \cos nx + b_n \sin nx)$$

is convergent, the series can be integrated termwise over any interval with the result (3.36).

Example 3.9. Show that the Fourier series for

$$x^2 = \frac{\pi^2}{3} + 4 \sum_{n=1}^{\infty} \frac{(-1)^n \cos nx}{n^2}, \quad -\pi < x < \pi$$

can be integrated from 0 to x when $-\pi \leqslant x \leqslant \pi$ and obtain a converging series

$$x^3 - \pi^2 x = 12 \sum_{n=1}^{\infty} \frac{(-1)^n \sin nx}{n^3}$$

The function x^2 satisfies the hypothesis of Theorem 3.6 so that termwise integration is permitted.

$$\int_0^x t^2\, dt = \frac{\pi^2}{3} \int_0^x dt + 4 \sum_{n=1}^{\infty} \frac{(-1)^n}{n^2} \int_0^x \cos nt\, dt$$

$$\left[\frac{t^3}{3}\right]_0^x = \frac{\pi^2}{3} [t]_0^x + 4 \sum_{n=1}^{\infty} \frac{(-1)^n}{n^3} [\sin nt]_0^x$$

$$\frac{x^3}{3} = \pi^2 \frac{x}{3} + 4 \sum_{n=1}^{\infty} \frac{(-1)^n \sin nx}{n^3}$$

or

$$x^3 - \pi^2 x = 12 \sum_{n=1}^{\infty} \frac{(-1)^n \sin nx}{n^3}$$
(3.37)

The series for the polynomial converges to the polynomial. If $F(x) = x^3 - \pi^2 x$, then $F(-\pi) = F(\pi)$ and all other conditions of the hypothesis of Theorem 3.4 are satisfied. Therefore the series of (3.37) is uniformly and absolutely convergent to $F(x)$ for $-\pi \leqslant x \leqslant \pi$.

3.10. DOUBLE FOURIER SERIES

Fourier series for functions of two variables are similar to expansions that we considered in Section 2.7. As in our earlier work we assume that adequate convergence conditions exist and $f(x, y)$ is defined so that all suggested operations are permissible. Myint-U and Debnath [36, pp. 125–126] gives adequate conditions for a problem similar to the one we consider here. Assume that $f(x, y)$ is *smooth* in the domain $-K < x < K$, $-L < y < L$, and let the series be uniformly convergent. Then if y is held constant

$$f(x, y) = \frac{a_0(y)}{2} + \sum_{m=1}^{\infty} \left[a_m(y) \cos\left(\frac{m\pi x}{K}\right) + b_m(y) \sin\left(\frac{m\pi x}{K}\right) \right]$$
(3.38)

The coefficients, functions of y, are

$$a_m(y) = \frac{1}{K} \int_{-K}^{K} f(x, y) \cos\left(\frac{m\pi x}{K}\right) dx, \quad m \in N_0$$

$$b_m(y) = \frac{1}{K} \int_{-K}^{K} f(x, y) \sin\left(\frac{m\pi x}{K}\right) dx, \quad m \in N$$

Coefficients $a_m(y)$ and $b_m(y)$ are smooth, so we expand them in uniformly convergent series

$$a_m(y) = \frac{a_{m0}}{2} + \sum_{n=1}^{\infty} \left[a_{mn} \cos\left(\frac{n\pi y}{L}\right) + b_{mn} \sin\left(\frac{n\pi y}{L}\right) \right]$$

$$b_m(y) = \frac{c_{m0}}{2} + \sum_{n=1}^{\infty} \left[c_{mn} \cos\left(\frac{n\pi y}{L}\right) + d_{mn} \sin\left(\frac{n\pi y}{L}\right) \right]$$

The coefficients of the last two series are

$$a_{mn} = \frac{1}{L} \int_{-L}^{L} a_m(y) \cos\left(\frac{n\pi y}{L}\right) dy$$

$$= \frac{1}{L} \int_{-L}^{L} \left[\frac{1}{K} \int_{-K}^{K} f(x, y) \cos\left(\frac{m\pi x}{K}\right) dx \right] \cos\left(\frac{n\pi y}{L}\right) dy$$

$$= \frac{1}{KL} \int_{-L}^{L} \int_{-K}^{K} f(x, y) \cos\left(\frac{m\pi x}{K}\right) \cos\left(\frac{n\pi y}{L}\right) dx \, dy$$

$$b_{mn} = \frac{1}{KL} \int_{-L}^{L} \int_{-K}^{K} f(x, y) \cos\left(\frac{m\pi x}{K}\right) \sin\left(\frac{n\pi y}{L}\right) dx \, dy$$

$$c_{mn} = \frac{1}{KL} \int_{-L}^{L} \int_{-K}^{K} f(x, y) \sin\left(\frac{m\pi x}{K}\right) \cos\left(\frac{n\pi y}{L}\right) dx \, dy$$

$$d_{mn} = \frac{1}{KL} \int_{-L}^{L} \int_{-K}^{K} f(x, y) \sin\left(\frac{m\pi x}{K}\right) \sin\left(\frac{n\pi y}{L}\right) dx \, dy$$

By substituting a_m and b_m into (3.38) we obtain

$$
\begin{aligned}
f(x, y) = {} & \frac{a_{00}}{4} + \frac{1}{2} \sum_{n=1}^{\infty} \left[a_{0n} \cos\left(\frac{n\pi y}{L}\right) + b_{0n} \sin\left(\frac{n\pi y}{L}\right) \right] \\
& + \frac{1}{2} \sum_{m=1}^{\infty} \left[a_{m0} \cos\left(\frac{m\pi x}{K}\right) + c_{m0} \sin\left(\frac{m\pi x}{K}\right) \right] \\
& + \sum_{m=1}^{\infty} \sum_{n=1}^{\infty} \left[a_{mn} \cos\left(\frac{m\pi x}{K}\right) \cos\left(\frac{n\pi y}{L}\right) + b_{mn} \cos\left(\frac{m\pi x}{K}\right) \sin\left(\frac{n\pi y}{L}\right) \right. \\
& \left. + c_{mn} \sin\left(\frac{m\pi x}{K}\right) \cos\left(\frac{n\pi y}{L}\right) + d_{mn} \sin\left(\frac{m\pi x}{K}\right) \sin\left(\frac{n\pi y}{L}\right) \right]
\end{aligned}
$$

$$(3.39)$$

Result (3.39) is referred to as a *double Fourier series*.

Conditions of symmetry in $z = f(x, y)$ relative to coordinate planes allow simplifications in the double series (3.39). If $f(-x, y) = f(x, y)$ and $f(x, -y) = f(x, y)$, a_{mn} is the only nonzero set of coefficients. The double series becomes a cosine series

$$
\begin{aligned}
f(x, y) = {} & \frac{a_{00}}{4} + \frac{1}{2} \sum_{n=1}^{\infty} a_{0n} \cos \frac{n\pi y}{L} + \frac{1}{2} \sum_{m=1}^{\infty} a_{m0} \cos \frac{m\pi x}{K} \\
& + \sum_{m=1}^{\infty} \sum_{n=1}^{\infty} a_{mn} \cos \frac{m\pi x}{K} \cos \frac{n\pi y}{L}
\end{aligned}
$$

$$
a_{mn} = \frac{4}{KL} \int_0^L \int_0^K f(x, y) \cos \frac{m\pi x}{K} \cos \frac{n\pi y}{L} \, dx \, dy
$$

If $f(-x, y) = f(x, y)$ and $f(x, -y) = -f(x, y)$, the only nonzero coefficients are b_{mn}.

$$
f(x, y) = \frac{1}{2} \sum_{n=1}^{\infty} b_{0n} \sin\left(\frac{n\pi y}{L}\right) + \sum_{m=1}^{\infty} \sum_{n=1}^{\infty} b_{mn} \cos\left(\frac{m\pi x}{K}\right) \sin\left(\frac{n\pi y}{L}\right)
$$

$$
b_{mn} = \frac{4}{KL} \int_0^L \int_0^K f(x, y) \cos\left(\frac{m\pi x}{K}\right) \sin\left(\frac{n\pi y}{L}\right) dx \, dy
$$

If $f(-x, y) = -f(x, -y)$ and $f(x, -y) = f(x, y)$, c_{mn} is the only nonzero set of coefficients.

$$
f(x, y) = \frac{1}{2} \sum_{m=1}^{\infty} c_{m0} \sin\left(\frac{m\pi x}{K}\right) + \sum_{m=1}^{\infty} \sum_{n=1}^{\infty} c_{mn} \sin\left(\frac{m\pi x}{K}\right) \cos\left(\frac{n\pi y}{L}\right)
$$

$$
c_{mn} = \frac{4}{KL} \int_0^L \int_0^K f(x, y) \sin\left(\frac{m\pi x}{K}\right) \cos\left(\frac{n\pi y}{L}\right) dx \, dy
$$

If $f(-x, y) = -f(x, y)$ and $f(x, -y) = -f(x, y)$, d_{mn} is the only nonzero set of coefficients. The double series is a sine series in this case.

$$f(x, y) = \sum_{m=1}^{\infty} \sum_{n=1}^{\infty} d_{mn} \sin\left(\frac{m\pi x}{K}\right) \sin\left(\frac{n\pi y}{L}\right)$$

$$d_{mn} = \frac{4}{KL} \int_0^L \int_0^K f(x, y) \sin\left(\frac{m\pi x}{K}\right) \sin\left(\frac{n\pi y}{L}\right) dx\, dy$$

Example 3.10. If $f(x, y) = xy$, $0 < x < 1$, $0 < y < 2$, determine the double series representation.

The function $f(x, y)$ satisfies the condition

$$f(-x, y) = -xy = -f(x, y)$$
$$f(x, -y) = -xy = -f(x, y)$$

Therefore, we adopt the sine series representation

$$f(x, y) = \sum_{m=1}^{\infty} \sum_{n=1}^{\infty} d_{mn} \sin\left(\frac{m\pi x}{1}\right) \sin\left(\frac{n\pi x}{2}\right)$$

where

$$d_{mn} = \frac{4}{1(2)} \int_0^2 \int_0^1 xy \sin(m\pi x) \sin\left(\frac{n\pi y}{2}\right) dx\, dy$$

$$= 2 \int_0^2 \left[\frac{\sin m\pi x}{m^2 \pi^2} - \frac{x \cos m\pi x}{m\pi}\right]_0^1 y \sin\left(\frac{n\pi y}{2}\right) dy$$

$$= \frac{-2(-1)^m}{m\pi} \int_0^2 y \sin\left(\frac{n\pi y}{2}\right) dy$$

$$= \frac{-2(-1)^m}{m\pi} \left(\frac{-4(-1)^n}{n\pi}\right)$$

$$d_{mn} = \frac{8(-1)^{m+n}}{mn\pi^2}$$

Exercises 3.3

1. In No. 12 of Exercises 3.2

$$f(x) = \begin{cases} \cos x & \text{when } 0 \le x \le \pi/2 \\ 0 & \text{when } \pi/2 \le x \le \pi \end{cases}$$

and

$$f(x) \sim \frac{1}{\pi} + \frac{1}{2}\cos x + \frac{2}{\pi} \sum_{n=1}^{\infty} \frac{(-1)^{n+1}}{4n^2 - 1} \cos 2nx$$

(a) Show directly from the Fourier series for f that the series converges uniformly for all x.

(b) Investigate the conditions imposed upon f in Theorem 3.4. Is the hypothesis satisfied?

(c) Is the series termwise differentiable except for isolated points? What are these points?

2. If $f(x) = x$, $0 < x < \pi$, No. 14 of Exercises 3.2, differentiate the Fourier series for f to obtain the expansion for $f'(x) = 1$ on the interval. Does the derived series converge to 1?

3. If $f(x) = x$, $-\pi < x < \pi$, No. 13 of Exercises 3.2, the Fourier series for x can be expressed

$$2 \sum_{n=1}^{\infty} \frac{(-1)^{n-1} \sin nx}{n}$$

Differentiate the series and x. Does the derived series here converge to 1? Do you see basic differences in Exercises 2 and 3?

4. If $f(x) = x$, $-\pi < x < \pi$, Exercise 3, integrate the series. Is termwise integration valid 0 to x?

5. (a) Write the Fourier sine series for the function

$$f(x) = 1, \quad 0 < x < \pi$$

(b) Integrate the series obtained in (a). Does this new series converge to x?

(c) Differentiate the series determined in (a). Does this derived series converge to 0?

6. Assume that f and g are PWC functions, both with $-\pi \leqslant x \leqslant \pi$, and both periodic with a period 2π. The Fourier coefficients for f are a_n, b_n, and for g they are A_n, B_n. Determine *Parseval's identity for the inner product*:

$$\frac{1}{\pi} \int_{-\pi}^{\pi} fg \, dx = \frac{a_0 A_0}{2} + \sum_{n=1}^{\infty} (a_n A_n + b_n B_n)$$

7. Assume that the Fourier coefficients a_n and b_n are in the series for f in Theorem 3.5. Show that
   (a) $na_n \to 0$ as $n \to \infty$.
   (b) $nb_n \to 0$ as $n \to \infty$.

8. Write a theorem similar to Theorem 3.4 on uniform convergence if $-L \leqslant x \leqslant L$.

9. Write a theorem similar to Theorem 3.5 on differentiation if the function has a period $2L$.

10. Write a theorem on integration similar to Theorem 3.6 if f has a period $2L$.

11. Determine the double Fourier sine series for

$$f(x, y) = 1, \quad 0 < x < a, \quad 0 < y < b$$

12. Find the double Fourier series if

$$f(x, y) = xy^2, \quad -\pi < x < \pi, \quad -\pi < y < \pi$$

13. Write the double Fourier series if

$$f(x, y) = x^2 y^2, \quad -\pi < x < \pi, \quad -\pi < y < \pi$$

14. Expand as a double Fourier series,

$$f(x, y) = x \cos y, \quad -1 < x < 1, \quad -2 < y < 2$$

15. The Fourier series for x^2, $-\pi \leqslant x \leqslant \pi$, is given in Example 3.9. The integral of the series from 0 to x is computed.
(a) show that

$$\frac{\pi^3}{32} = \sum_{n=1}^{\infty} \frac{(-1)^{n+1}}{(2n-1)^3}$$

(b) Using Parseval's identity, show that

$$\frac{\pi^6}{945} = \sum_{n=1}^{\infty} \frac{1}{n^6}$$

4

FOURIER INTEGRALS

In the field of function representations, Fourier series are extremely useful for periodic functions. When the functions that we wish to represent are not periodic, Fourier integrals serve a similar special need. We seek a representation analogous to a Fourier series on $(-L, L)$ with an infinite L. Before investigating these Fourier integrals, we include some of the basic mathematics to study improper integrals.

4.1. UNIFORM CONVERGENCE OF INTEGRALS

The improper integral $\int_a^\infty u(x, t)\, dt$ has a strong analogy with the infinite series $\sum_{n=1}^\infty u_n(x)$. Essentially the variable of integration t replaces the index n. When the series and integral are each convergent, we say they define a function $S(x)$. Definitions and discussions parallel closely the ideas concerning series.

We define the improper integral

$$\int_a^\infty u(x, t)\, dt$$

to be *uniformly convergent to* $S(x)$ for a domain of x, if for $\varepsilon > 0$, a number P can be found so that

$$\left| \int_a^q u(x, t)\, dt - S(x) \right| < \varepsilon \quad \text{for all } q > P \tag{4.1}$$

where $P(\varepsilon)$ is dependent on ε alone. If the inequalities of (4.1) require $P(x, \varepsilon)$ dependent on both x and ε, then the integral is simply *convergent*.

An M-test for improper integrals similar to the one for series is stated. We let $M(t)$ be continuous for $a \leq t < \infty$, and $u(x, t)$ be continuous as a function of t for $a \leq t < \infty$ for each x in a domain D. If

$$|u(x, t)| \leq M(t)$$

for x in D and

$$\int_a^\infty M(t)\, dt$$

converges, then

$$\int_a^\infty u(x, t)\, dt$$

converges uniformly and absolutely for x in D.
Suppose $u(x, t)$ is continuous for $a \leq t < \infty$, $b \leq x \leq c$, and

$$\int_a^\infty u(x, t)\, dt$$

is uniformly convergent to $S(x)$ on $b \leq x \leq c$; then $S(x)$ *is continuous on* $[b, c]$, and

$$\lim_{x \to x_0} \int_a^\infty u(x, t)\, dt = \int_a^\infty \lim_{x \to x_0} u(x, t)\, dt$$

Under the assumptions stated at the beginning of this paragraph, the improper integral and its convergence $S(x)$ may be integrated over any two points, $b \leq x_b \leq x \leq x_c \leq c$, and the order of integration interchanged.

$$\int_{x_b}^{x_c} S(x)\, dx = \int_{x_b}^{x_c} \int_a^\infty u(x, t)\, dt\, dx = \int_a^\infty \int_{x_b}^{x_c} u(x, t)\, dx\, dt$$

Now suppose that $u(x, t)$ and its partial derivative u_x are continuous in t and x for $a \leq t < \infty$, $b \leq x \leq c$ and the integral

$$\int_a^\infty u(x, t)\, dt$$

converges, while

$$\int_a^\infty u_x(x, t)\, dt$$

converges uniformly for $b \leq x \leq c$. Then for any x on the interval

$$S'(x) = \frac{d}{dx} \int_a^\infty u(x, t)\, dt = \int_a^\infty u_x(x, t)\, dt$$

Example 4.1. Show that the integral

$$\int_1^\infty \frac{\sin xt}{t^2} \, dt$$

converges uniformly for all real x.

For any real x,

$$\left| \frac{\sin xt}{t^2} \right| \le \frac{1}{t^2}$$

The integral

$$\int_1^\infty \frac{1}{t^2} \, dt = \left[-\frac{1}{t} \right]_1^\infty = \lim_{q \to \infty} \left[-\frac{1}{t} \right]_1^q = 1$$

From the result of the M-test, the given integral converges uniformly.

Example 4.2. Investigate the uniform convergence of

$$S(x) = \int_0^\infty x e^{-tx} \, dt$$

for $0 < a \le x$, and show that $S(x) = 1$.

The convergence

$$S(x) = \lim_{q \to \infty} \int_0^q x e^{-tx} \, dt = \lim_{q \to \infty} [-e^{-tx}]_{t=0}^q = 1$$

if $x > 0$.

Using the idea expressed in the definition for uniform convergence, if $\varepsilon > 0$ we can find a P which is dependent on ε but not x, so that

$$\left| 1 - \int_0^q x e^{-tx} \, dt \right| = |1 - (1 - e^{-qx})| = e^{-qx} < \varepsilon$$

for all $q > (1/a) \ln(1/\varepsilon)$. Therefore uniform convergence follows. We should point out that as $a \to 0$, P increases without limit and the integral fails to converge uniformly for $x > 0$.

Example 4.3. Evaluate the integral

$$\int_0^\infty \frac{\sin t}{t} \, dt$$

First, we show that if

$$S(x) = \int_0^\infty \frac{e^{-xt} \sin t}{t} \, dt$$

with $x > 0$, then

$$\left| \frac{e^{-xt} \sin t}{t} \right| \le e^{-xt}$$

The integral

$$\int_0^\infty e^{-xt} \, dt$$

converges on any interval $0 < a \le x < \infty$. Therefore, by the M-test

$$S(x) = \int_0^\infty \frac{e^{-xt} \sin t}{t} \, dt \le \int_0^\infty e^{-at} \, dt \tag{4.2}$$

converges uniformly for $x \ge a$. Thus the derivative may be computed

$$S'(x) = - \int_0^\infty e^{-xt} \sin t \, dt \tag{4.3}$$

If $0 < a \le x < \infty$, then

$$\left| e^{-xt} \sin t \right| \le e^{-at}$$

and

$$\int_0^\infty e^{-at} \, dt$$

converges. Therefore (4.3) converges uniformly on $a \le x < \infty$. Integrating the improper integral (4.3) one obtains

$$S'(x) = - \left[\frac{e^{-xt}(-x \sin t - \cos t)}{x^2 + 1} \right]_0^\infty = - \frac{1}{x^2 + 1}$$

Therefore,

$$S(x) = -\arctan x + C \tag{4.4}$$

Now from (4.2)

$$|S(x)| \le \int_0^\infty \left| \frac{e^{-xt} \sin t}{t} \right| \, dt \le \int_0^\infty e^{-xt} \, dt = \frac{1}{x}$$

for any $x > 0$. As $x \to \infty$, $S(x) \to 0$. In (4.4) as $x \to \infty$, $S(x) \to -\pi/2 + C$, and $C = \pi/2$. Formula (4.4) may be written

$$S(x) = -\arctan x + \frac{\pi}{2} \tag{4.5}$$

It can be shown (see Fulks [21, p. 598]) that the integral of (4.2) is uniformly convergent for $x \ge 0$, hence

$$\lim_{x \to 0^+} S(x) = \int_0^\infty \lim_{x \to 0^+} \left(\frac{e^{-xt} \sin t}{t} \right) dt = \int_0^\infty \frac{\sin t}{t} dt \qquad (4.6)$$

Finally, as $x \to 0^+$ in (4.5), $S(x) \to \pi/2$. Therefore,

$$\int_0^\infty \frac{\sin t}{t} dt = \frac{\pi}{2} \qquad (4.7)$$

Exercises 4.1

1. Show that

$$\int_0^\infty \frac{\cos xt}{1 + t^2} dt$$

converges uniformly for all x.

2. Verify that

$$\int_0^\infty e^{-t} \cos xt \, dt$$

is uniformly convergent for all x.

3. (a) If $0 < a < b$, show that $\int_0^\infty e^{-xt} dt$ is uniformly convergent on $[a, b]$, and then evaluate the given integral.
   (b) Integrate the result found in (a) relative to x over $[a, b]$ and show that

$$\ln \frac{b}{a} = \int_0^\infty \frac{e^{-at} - e^{-bt}}{t} dt$$

4. (a) The evaluation of the integral in Exercise 3(a) is

$$\frac{1}{x} = \int_0^\infty e^{-xt} dt$$

for $x > 0$. Differentiation of the integral is permissible. Why?
   (b) Show that

$$\frac{n!}{x^{n+1}} = \int_0^\infty t^n e^{-xt} dt$$

5. (a) Show that if the integral in Exercise 2 converges to $S(x)$, then

$$S(x) = \frac{1}{1 + x^2}$$

   (b) Establish the result

$$\int_0^\infty \frac{e^{-t} \sin xt}{t} dt = \arctan x$$

4.2. A GENERALIZATION OF THE FOURIER SERIES

If the series (3.1) converges to f and the coefficients (3.2) replace a_n and b_n in the series, then

$$f(x) = \frac{1}{2L} \int_{-L}^{L} f(t) \, dt$$

$$+ \frac{1}{L} \sum_{n=1}^{\infty} \left[\int_{-L}^{L} f(t) \left(\cos \frac{n\pi t}{L} \cos \frac{n\pi x}{L} + \sin \frac{n\pi t}{L} \sin \frac{n\pi x}{L} \right) dt \right]$$

$$= \frac{1}{2L} \int_{-L}^{L} f(t) \, dt + \frac{1}{L} \sum_{n=1}^{\infty} \int_{-L}^{L} f(t) \cos \frac{n\pi(t-x)}{L} \, dt \qquad (4.8)$$

If $\lim_{L\to\infty} \int_{-L}^{L} |f(t)| \, dt$ exists, then $\lim_{L\to\infty} 1/2L \int_{-L}^{L} |f(t)| \, dt$ becomes zero. We say that f is *absolutely integrable* (AI) on $-\infty < x < \infty$ when $\int_{-\infty}^{\infty} |f(t)| \, dt$ converges. The remainder of (4.8) is

$$f(x) = \lim_{L\to\infty} \sum_{n=1}^{\infty} \frac{1}{L} \int_{-L}^{L} f(t) \cos \frac{n\pi(t-x)}{L} \, dt \qquad (4.9)$$

If we allow $\alpha_n = n\pi/L$, then $\Delta\alpha_n = \alpha_{n+1} - \alpha_n = \pi/L$. Now the series (4.9) is

$$f(x) = \lim_{L\to\infty} \sum_{n=1}^{\infty} \frac{1}{\pi} \left[\int_{-L}^{L} f(t) \cos \alpha_n(t-x) \, dt \right] \Delta\alpha_n \qquad (4.10)$$

or

$$f(x) = \lim_{L\to\infty} \sum_{n=1}^{\infty} F_L(\alpha_n, x) \, \Delta\alpha_n \qquad (4.11)$$

where

$$F_L(\alpha_n, x) = \frac{1}{\pi} \int_{-L}^{L} f(t) \cos \alpha_n(t-x) \, dt \qquad (4.12)$$

The sum in (4.11) is analogous to a definite integral. The limit as $L \to \infty$ may suggest an improper integral in (4.12). It is at least suggestive that the series (4.10) as $\Delta\alpha_n \to 0$ and $L \to \infty$ resembles an improper integral of the form

$$f(x) = \frac{1}{\pi} \int_0^{\infty} \left[\int_{-\infty}^{\infty} f(t) \cos \alpha(t-x) \, dt \right] d\alpha \qquad (4.13)$$

The result given in (4.13) is a *Fourier integral formula* or *Fourier integral representation* for f. Our analogy developed from a few manipulations preceding (4.10) does not provide a mathematical justification that the integral converges to f.

If we write (4.13) in the form

$$f(x) = \frac{1}{\pi} \int_0^\infty \left\{ \left[\int_{-\infty}^\infty f(t) \cos \alpha t \, dt \right] \cos \alpha x + \left[\int_{-\infty}^\infty f(t) \sin \alpha t \, dt \right] \sin \alpha x \right\} d\alpha$$

then we can express the Fourier integral

$$f(x) = \int_0^\infty [A(\alpha) \cos \alpha x + B(\alpha) \sin \alpha x] \, d\alpha , \quad -\infty < x < \infty \quad (4.14)$$

where

$$A(\alpha) = \frac{1}{\pi} \int_{-\infty}^\infty f(t) \cos \alpha t \, dt$$

$$B(\alpha) = \frac{1}{\pi} \int_{-\infty}^\infty f(t) \sin \alpha t \, dt \quad (4.15)$$

Convergence of the integral to f has been assumed up to this point in this section. Even though the integral may converge, conditions may exist so that the integral fails to converge to the function. In some situations the* correspondence notation may be appropriate in place of the equal sign. We state a theorem, without proof, that supplies conditions for convergence.

Theorem 4.1 (A Fourier integral convergence theorem). Assume that f is PWS on every finite interval on the x axis and let f be AI for all real x. Then for every x on the entire axis

$$\frac{1}{\pi} \int_0^\infty \int_{-\infty}^\infty f(t) \cos \alpha(t - x) \, dt \, d\alpha = \frac{f(x+) + f(x-)}{2} \quad (4.16)$$

As with the Fourier series, if f is continuous and all of the other conditions of the hypothesis of the theorem are satisfied, the Fourier integral converges to f. If f is defined so that it matches $[f(x+) + f(x-)]/2$ and all other conditions are satisfied, the Fourier integral converges to f.

Example 4.4. (a) Draw a graph for the function

$$f(x) = \begin{cases} 0 & \text{when } x < 0 \\ x & \text{when } 0 < x < 1 \\ 0 & \text{when } x > 1 \end{cases}$$

(b) Find the Fourier integral representing f of part (a).
(c) Determine the convergence of the integral at $x = 1$.

(a) See Figure 4.1.
(b) The integral representation of f is

$$f(x) \sim \int_0^\infty [A(\alpha) \cos \alpha x + B(\alpha) \sin \alpha x] \, d\alpha \quad (4.17)$$

where

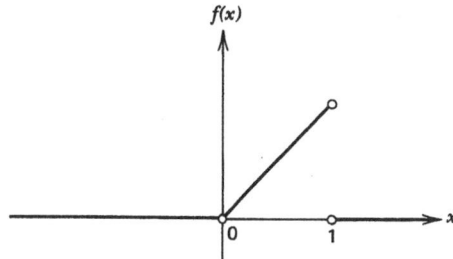

Figure 4.1. Graph of f.

$$A(\alpha) = \frac{1}{\pi} \int_{-\infty}^{\infty} f(t) \cos \alpha t \, dt$$

$$= \frac{1}{\pi} \int_{-\infty}^{0} 0 \cdot \cos \alpha t \, dt + \frac{1}{\pi} \int_{0}^{1} t \cos \alpha t \, dt + \int_{1}^{\infty} 0 \cdot \cos \alpha t \, dt$$

$$= \frac{1}{\pi} \left[\frac{1}{\alpha^2} \cos \alpha t + \frac{1}{\alpha} \sin \alpha t \right]_{0}^{1} = \frac{1}{\pi} \left[\frac{\cos \alpha + \alpha \sin \alpha - 1}{\alpha^2} \right]$$

$$B(\alpha) = \frac{1}{\pi} \int_{-\infty}^{\infty} f(t) \sin \alpha t \, dt = \frac{1}{\pi} \int_{0}^{1} t \sin \alpha t \, dt$$

$$= \frac{1}{\pi} \left[\frac{1}{\alpha^2} \sin \alpha t - \frac{t}{\alpha} \cos \alpha t \right]_{0}^{1} = \frac{1}{\pi} \left[\frac{\sin \alpha - \alpha \cos \alpha}{\alpha^2} \right]$$

Replacing $A(\alpha)$ and $B(\alpha)$ in (4.17) by their computed values, we have

$$f(x) \sim \frac{1}{\pi} \int_{0}^{\infty} \left[\frac{\cos \alpha + \alpha \sin \alpha - 1}{\alpha^2} \cos \alpha x + \frac{\sin \alpha - \alpha \cos \alpha}{\alpha^2} \sin \alpha x \right] d\alpha$$

$$= \frac{1}{\pi} \int_{0}^{\infty} \frac{\cos \alpha (1 - x) + \alpha \sin \alpha (1 - x) - \cos \alpha x}{\alpha^2} d\alpha$$

(c) At $x = 1$, the integral fails to converge to the function. In fact, the function is not defined at this point. The convergence at $x = 1$ is

$$\frac{f(1+) + f(1-)}{2} = \frac{0 + 1}{2} = \frac{1}{2}$$

4.3. FOURIER SINE AND COSINE INTEGRALS

Even and odd functions play a simplifying role for Fourier integral representations. In (4.14) if f is even, the integral in (4.15) for $A(\alpha)$ has an integrand which is even. Therefore,

$$A(\alpha) = \frac{2}{\pi} \int_{0}^{\infty} f(t) \cos \alpha t \, dt$$

Since the integrand of $B(\alpha)$ in (4.15) is odd

$$B(\alpha) = 0$$

Rewriting (4.14) and (4.15), we have

$$f(x) \sim \int_0^\infty A(\alpha) \cos \alpha x \, d\alpha , \quad 0 < x < \infty \qquad (4.18)$$

where

$$A(\alpha) = \frac{2}{\pi} \int_0^\infty f(t) \cos \alpha t \, dt \qquad (4.19)$$

If f is odd in (4.14), then $f(t) \cos \alpha t$ is odd and

$$A(\alpha) = 0$$

and

$$B(\alpha) = \frac{2}{\pi} \int_0^\infty f(t) \sin \alpha t \, dt$$

It follows that if f is odd

$$f(x) \sim \int_0^\infty B(\alpha) \sin \alpha x \, d\alpha , \quad 0 < x < \infty \qquad (4.20)$$

where

$$B(\alpha) = \frac{2}{\pi} \int_0^\infty f(t) \sin \alpha t \, dt \qquad (4.21)$$

A convergence theorem can be altered to fit the even and odd functions and their integrals.

Theorem 4.2 (A Fourier sine and cosine integral convergence theorem). Assume that f is PWS on every finite interval on the positive x axis and let f be AI for all real $x > 0$. Then f may be represented by either:

(a) Fourier sine integral

$$\int_0^\infty B(\alpha) \sin \alpha x \, d\alpha , \quad 0 < x < \infty \qquad (4.20\text{a})$$

where

$$B(\alpha) = \frac{2}{\pi} \int_0^\infty f(t) \sin \alpha t \, dt \qquad (4.21\text{a})$$

(b) Fourier cosine integral

$$\int_0^\infty A(\alpha) \cos \alpha x \, d\alpha, \quad 0 < x < \infty \tag{4.18a}$$

where

$$A(\alpha) = \frac{2}{\pi} \int_0^\infty f(t) \cos \alpha t \, dt \tag{4.19a}$$

Each integral (4.18) and (4.20) converges to

$$\frac{f(x+) + f(x-)}{2} \tag{4.22}$$

The odd extension of f in (a) when represented by a sine integral will be an odd function over the entire real axis. Similarly, the even extension of f represented in (b) by a cosine integral will be even over the entire real axis. Extensions here are not periodic as we discussed with Fourier series.

Example 4.5. (a) Draw a graph of the function

$$f(x) = \begin{cases} 0 & \text{when } -\infty < x < -\pi \\ -1 & \text{when } -\pi < x < 0 \\ 1 & \text{when } 0 < x < \pi \\ 0 & \text{when } \pi < x < \infty \end{cases}$$

(b) Determine the Fourier integral for the function described in (a).
(c) To what number does the integral found in (b) converge at $x = -\pi$?

(a) See Figure 4.2.
(b) Since f is an odd function and AI and PWS, we use (4.20) with the coefficients (4.21) for its representation.

$$f(x) \sim \int_0^\infty B(\alpha) \sin \alpha x \, dx$$

where

Figure 4.2. Graph of f.

$$B(\alpha) = \frac{2}{\pi} \int_0^\infty f(t) \sin \alpha t \, dt = \frac{2}{\pi} \int_0^\pi 1 \cdot \sin \alpha t \, dt + \frac{2}{\pi} \int_\pi^\infty 0 \cdot \sin \alpha t \, dt$$

$$= \frac{2}{\pi} \left[-\frac{\cos \alpha t}{\alpha} \right]_0^\pi = \frac{2}{\pi \alpha} (1 - \cos \alpha \pi)$$

Therefore,

$$f(x) \sim \frac{2}{\pi} \int_0^\infty \left(\frac{1 - \cos \alpha \pi}{\alpha} \right) \sin \alpha x \, d\alpha$$

(c) According to the convergence theorem and (4.22) we conclude that the integral converges to $-\frac{1}{2}$ at $x = -\pi$.

4.4. THE EXPONENTIAL FOURIER INTEGRAL

The Fourier integral for f can be expressed

$$f(x) = \frac{1}{\pi} \int_0^\infty \int_{-\infty}^\infty f(t) \cos \alpha(x - t) \, dt \, d\alpha \tag{4.23}$$

Since $\cos \alpha(x - t) = \cos \alpha(t - x)$, this representation agrees with (4.13). From Euler's relation

$$\cos \alpha(x - t) = \frac{e^{i\alpha(x-t)} + e^{-i\alpha(x-t)}}{2} \tag{4.24}$$

Inserting (4.24) in (4.23), we have

$$f(x) = \frac{1}{\pi} \int_0^\infty \int_{-\infty}^\infty f(t) \frac{e^{i\alpha(x-t)} + e^{-i\alpha(x-t)}}{2} \, dt \, d\alpha \tag{4.25}$$

The representation (4.25) is equivalent to

$$f(x) = \frac{1}{2\pi} \int_0^\infty \int_{-\infty}^\infty f(t) e^{i\alpha(x-t)} \, dt \, d\alpha + \frac{1}{2\pi} \int_0^\infty \int_{-\infty}^\infty f(t) e^{-i\alpha(x-t)} \, dt \, d\alpha \tag{4.26}$$

If α is replaced with $-\alpha$ in the second integral of (4.26)

$$f(x) = \frac{1}{2\pi} \int_0^\infty \int_{-\infty}^\infty f(t) e^{i\alpha(x-t)} \, dt \, d\alpha + \frac{1}{2\pi} \int_0^{-\infty} \int_{-\infty}^\infty f(t) e^{i\alpha(x-t)} \, dt \, d\alpha \tag{4.27}$$

We express (4.27) in the form

$$f(x) = \frac{1}{2\pi} \int_0^\infty \int_{-\infty}^\infty f(t)e^{i\alpha(x-t)} \, dt \, d\alpha + \frac{1}{2\pi} \int_{-\infty}^0 \int_{-\infty}^\infty f(t)e^{i\alpha(x-t)} \, dt \, d\alpha \tag{4.28}$$

Combining the integrals of (4.28) we have

$$f(x) = \frac{1}{2\pi} \int_{-\infty}^\infty \int_{-\infty}^\infty f(t)e^{i\alpha(x-t)} \, dt \, d\alpha, \quad -\infty < x < \infty \tag{4.29}$$

But (4.29) is the same as

$$f(x) = \frac{1}{2\pi} \int_{-\infty}^\infty \int_{-\infty}^\infty f(t)e^{-i\alpha t} e^{i\alpha x} \, dt \, d\alpha \tag{4.30}$$

Therefore, we write (4.30) in the form

$$f(x) \sim \int_{-\infty}^\infty C(\alpha)e^{i\alpha x} \, d\alpha, \quad -\infty < x < \infty \tag{4.31}$$

where

$$C(\alpha) = \frac{1}{2\pi} \int_{-\infty}^\infty f(t)e^{-i\alpha t} \, dt \tag{4.32}$$

Either (4.29) or (4.31) with coefficients (4.32) are *exponential forms of the Fourier integral*. The reader should be aware that some references are made to a slightly different basic integral than (4.29). Some require the *Cauchy principal value of an integral*. The *Cauchy principal value* of $\int_{-\infty}^\infty f(x) \, dx$ is defined by

$$\int_{-\infty}^\infty f(x) \, dx = \lim_{s \to \infty} \int_{-s}^s f(x) \, dx$$

This differs from the usual definition of the improper integral

$$\int_{-\infty}^\infty f(x) \, dx = \lim_{\substack{b \to \infty \\ a \to -\infty}} \int_a^b f(x) \, dx$$

The Cauchy principal value of the integral may exist even though the regular improper integral fails to converge. As a simple example, note that

$$\lim_{s \to \infty} \int_{-s}^s x \, dx$$

exists and is zero, while

$$\int_a^b x \, dx$$

fails to converge as $a \to -\infty$ and $b \to \infty$. If the regular improper integral exists, then so does the Cauchy principal value and the two integrals have the same value. In this context we agree to use the Cauchy principal value of the integral.

Exercises 4.2

1. Find the Fourier integral representation for the function

$$f(x) = \begin{cases} 0 & \text{if } x < -2 \text{ and } x > 2 \\ 1 & \text{if } -2 < x < 2 \end{cases}$$

2. Determine the Fourier integral representing

$$f(x) = \begin{cases} \sin x & \text{when } -\pi < x < \pi \\ 0 & \text{when } x < -\pi \text{ and } x > \pi \end{cases}$$

3. (a) Assume that

$$f(x) = \begin{cases} 1 & \text{when } 0 < x < 2 \\ 0 & \text{when } x < 0 \text{ and } x > 2 \end{cases}$$

Show that

$$f(x) \sim \frac{1}{\pi} \int_0^\infty \frac{\sin[\alpha(2-x)] + \sin \alpha x}{\alpha} \, d\alpha$$

(b) When $x = 1$, show that

$$\int_0^\infty \frac{\sin \alpha}{\alpha} \, d\alpha = \frac{\pi}{2}$$

4. (a) The function

$$f(x) = \begin{cases} 0 & \text{if } x < 0 \text{ and } x > \pi \\ \cos x & \text{if } 0 < x < \pi \end{cases}$$

Show that

$$f(x) \sim \frac{1}{\pi} \int_0^\infty \frac{\alpha}{1-\alpha^2} [\sin \alpha(\pi - x) - \sin \alpha x] \, d\alpha$$

(b) If the result of part (a) is valid, show that

$$\int_0^\infty \frac{\alpha \sin \alpha \pi}{1-\alpha^2} \, d\alpha = \frac{\pi}{2}$$

(c) How can one define $f(0)$ and $f(\pi)$ so that the integral converges to $f(x)$ for all real x?

5. (a) Show that for $f(x) = e^{-x}$, $0 < x < \infty$, the Fourier cosine integral is

$$e^{-x} = \frac{2}{\pi} \int_0^\infty \frac{\cos \alpha x}{1 + \alpha^2} \, d\alpha$$

(b) For the function of (a), show that the Fourier sine integral is

$$e^{-x} = \frac{2}{\pi} \int_0^\infty \frac{\alpha \sin \alpha x}{1 + \alpha^2} \, d\alpha$$

(c) Determine the Fourier integral for

$$f(x) = \begin{cases} e^{-x} & \text{when } 0 < x \\ 0 & \text{when } x < 0 \end{cases}$$

and show that

$$f(x) \sim \frac{1}{\pi} \int_0^\infty \frac{\cos \alpha x + \alpha \sin \alpha x}{1 + \alpha^2} \, d\alpha$$

(d) If the results of (a) and (b) are added what result does one obtain for e^{-x}? Is this compatible with the result in (c)? Show graphically the actual functions and extensions of functions involved if we consider the entire real axis.

(e) If f is defined as in part (c), show that the exponential form of the Fourier integral is

$$f(x) \sim \frac{1}{2\pi} \int_{-\infty}^\infty \frac{1 - i\alpha}{1 + \alpha^2} e^{i\alpha x} \, d\alpha$$

To what number do you believe the integral converges if $x = 0$?

6. Let

$$f(x) = \begin{cases} |x| & \text{when } -2 < x < 2 \\ 0 & \text{when } x < -2 \text{ and } x > 2 \end{cases}$$

(a) Show that

$$f(x) \sim \frac{2}{\pi} \int_0^\infty \frac{\cos 2\alpha + 2\alpha \sin 2\alpha - 1}{\alpha^2} \cos \alpha x \, d\alpha$$

(b) Find the convergence of the integral in (a) when $x = 2$.

7. (a) Assume that

$$f(x) = \begin{cases} 1 - x & \text{if } 0 < x < 1 \\ 0 & \text{if } 1 < x < \infty \end{cases}$$

and determine the Fourier sine integral

$$f(x) \sim \frac{2}{\pi} \int_0^\infty \frac{\alpha - \sin \alpha}{\alpha^2} \sin \alpha x \, d\alpha$$

(b) Show that

$$\frac{\pi}{4} = \int_0^\infty \frac{(\alpha - \sin \alpha) \sin (\alpha/2)}{\alpha^2} \, d\alpha$$

8. Let $f(x) = e^{-x} \cos x$, $0 < x < \infty$.
   (a) Show that the Fourier sine integral is

   $$f(x) \sim \frac{2}{\pi} \int_0^\infty \frac{\alpha^3 \sin \alpha x}{\alpha^4 + 4} \, d\alpha$$

   (b) Determine that the Fourier cosine integral is

   $$f(x) \sim \frac{2}{\pi} \int_0^\infty \frac{(\alpha^2 + 2) \cos \alpha x}{\alpha^4 + 4} \, d\alpha$$

9. Show that if $f(x) = 1$ on the positive real axis then the Fourier cosine integral fails to exist.

10. Show that an exponential form of the Fourier integral may be written

$$f(x) = \int_{-\infty}^\infty C(\alpha) e^{-i\alpha x} \, d\alpha, \quad -\infty < x < \infty \qquad (4.33)$$

where

$$C(\alpha) = \frac{1}{2\pi} \int_{-\infty}^\infty f(t) e^{i\alpha t} \, dt \qquad (4.34)$$

if (4.23) is written in an equivalent form

$$f(x) = \frac{1}{\pi} \int_0^\infty \left[\int_{-\infty}^\infty f(t) \cos \alpha(t - x) \, dt \right] d\alpha$$

11. Write an appropriate convergence theorem for the exponential Fourier integral.

5

BESSEL FUNCTIONS

Most of our geometrical discussions up to this point have been concerned with rectangular coordinates. Certain models or physical systems have geometrical properties that fit cylindrical coordinates better than rectangular coordinates. It is for these problems that we are especially concerned with the *Bessel differential equation* and its solution set the *Bessel functions*. In Chapter 1 we considered the *Frobenius series solution* of the Bessel equation

$$x^2 y'' + xy' + (x^2 - n^2)y = 0 , \quad x > 0 \tag{5.1}$$

In addition to differential forms of cylindrical and polar coordinates, we discuss orthogonality of Bessel functions and the resulting *Fourier–Bessel series*. We begin this chapter with a brief sketch of the gamma function.

5.1. THE GAMMA FUNCTION AND THE BESSEL FUNCTION

The *gamma function* is defined by

$$\Gamma(\alpha) = \int_0^\infty e^{-s} s^{\alpha-1} \, ds , \quad \alpha > 0 \tag{5.2}$$

It can be demonstrated using integration by parts that the recurrence formula

$$\Gamma(\alpha + 1) = \alpha \Gamma(\alpha) \tag{5.3}$$

is valid. If (5.3) is written in the form

$$\Gamma(\alpha) = \frac{\Gamma(\alpha + 1)}{\alpha} , \quad \alpha \neq 0, -1, -2, \ldots$$

then $\Gamma(\alpha)$ is a real valued function defined for all nonintegral negative α and $\alpha \neq 0$.

A sketch of the gamma function appears in Figure 5.1. This graph is a modification of Fulks [21]. When $\alpha = 1$,

$$\Gamma(1) = \int_0^\infty e^{-s}\, ds = [-e^{-s}]_0^\infty = 1$$

Using (5.3), one obtains sequentially

$$\Gamma(2) = \Gamma(1) = 1!$$
$$\Gamma(3) = 2\Gamma(2) = 2!$$
$$\Gamma(4) = 3\Gamma(3) = 3!$$
$$\Gamma(5) = 4\Gamma(4) = 4!$$
$$- - - - - - - - - -$$
$$\Gamma(n+1) = n!$$

For $n \in \mathbf{N}$, the integral (5.2) and the factorial have common values.
For $\alpha = \frac{1}{2}$, (5.2) becomes

$$\Gamma(\tfrac{1}{2}) = \int_0^\infty e^{-s} s^{-1/2}\, ds$$

If we let $s = t^2$, then

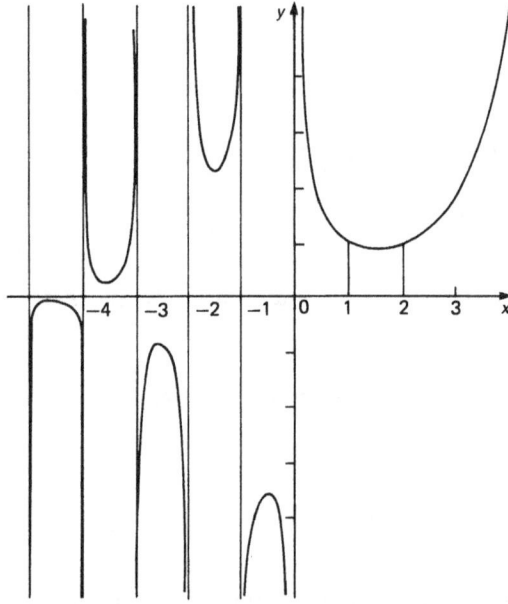

Figure 5.1. The gamma function. (From Fulks [21] by permission of John Wiley & Sons, Inc.)

$$\Gamma(\tfrac{1}{2}) = 2 \int_0^\infty e^{-t^2}\, dt \qquad (5.4)$$

When $b = 1$ in (9.67), the integral for $w(0)$ agrees with the integral of (5.4). Therefore,

$$\Gamma(\tfrac{1}{2}) = \sqrt{\pi}$$

By replacing $(n + k)!$ by $\Gamma(n + k + 1)$ in (1.45), we now define the *Bessel function of the first kind of order n* without the restriction that $n \in N_0$

$$J_n(x) = \sum_{k=0}^\infty \frac{(-1)^k}{k!\,\Gamma(n + k + 1)} \left(\frac{x}{2}\right)^{n+2k} \qquad (5.5)$$

In (5.5) we need only restrict $n + k \neq -1, -2, \dots$. Since $k \in N_0$ in (5.5), the definition is adequate for all $n \neq -1, -2, \dots$.

If x is replaced by $-x$ in (5.5) it is easy to see that

$$J_n(-x) = (-1)^n J_n(x) \qquad (5.6)$$

In (5.6) one finds that $J_n(x)$ is an odd function if n is an odd number, and $J_n(x)$ is an even function of n is an even number.

We observe that if n is replaced by $-n$ in (5.5), then

$$J_{-n}(x) = \sum_{k=0}^\infty \frac{(-1)^k}{k!\,\Gamma(-n + k + 1)} \left(\frac{x}{2}\right)^{-n+2k}$$

$$= \sum_{k=0}^{n-1} \frac{(-1)^k}{k!\,\Gamma(-n + k + 1)} \left(\frac{x}{2}\right)^{-n+2k}$$

$$+ \sum_{k=n}^\infty \frac{(-1)^k}{k!\,\Gamma(-n + k + 1)} \left(\frac{x}{2}\right)^{-n+2k}$$

As $k \to 0, 1, \dots, n - 1$, $\Gamma(-n + k + 1)$ either increases or decreases without bound when $n \in N$. Thus the first of the two sums for $J_{-n}(x)$ goes to zero. If we let $k = n + q$ in the second sum, then

$$J_{-n}(x) = \sum_{n+q=n}^\infty \frac{(-1)^{n+q}}{(n + q)!\,\Gamma(q + 1)} \left(\frac{x}{2}\right)^{-n+2(n+q)}$$

$$= \sum_{q=0}^\infty \frac{(-1)^n(-1)^q}{\Gamma(n + q + 1)q!} \left(\frac{x}{2}\right)^{n+2q}$$

$$= (-1)^n J_n(x) \qquad (5.7)$$

when $n \in N$. It is apparent in (5.7) that $J_{-n}(x)$ and $J_n(x)$ are linearly dependent. When $n > 0$ and $n \notin N$, $J_{-n}(x)$ has no bound and $J_n(x) \to 0$ as

$x \to 0$. In this case the two functions $J_{-n}(x)$ and $J_n(x)$ are linearly indepen-dent. Therefore, for $n > 0$ and $n \notin \mathbf{N}$,

$$y = C_1 J_n(x) + C_2 J_{-n}(x) \tag{5.8}$$

is a general solution of (5.1).

5.2. ADDITIONAL BESSEL FUNCTIONS

If the conditions of (5.8) are not met, we need to find a second independent solution for Bessel's differential equation. By introducing

$$y_2 = v(x) J_n(x)$$

in (5.1) we find a second solution

$$y_2 = J_n(x) \int \frac{dx}{x J_n^2(x)} \tag{5.9}$$

for (5.1) which is linearly independent.

As another approach to the problem, we define

$$Y_n(x) = \frac{J_n(x) \cos n\pi - J_{-n}(x)}{\sin n\pi} \tag{5.10}$$

If n is not an integer, then $Y_n(x)$ is a solution of (5.1), since it is a linear combination of $J_n(x)$ and $J_{-n}(x)$. If n is an integer, (5.10) is undefined and we consider

$$Y_n(x) = \lim_{r \to n} \frac{J_r(x) \cos r\pi - J_{-r}(x)}{\sin r\pi} \tag{5.11}$$

Details are omitted, but the limit of (5.11) exists and can be written

$$Y_n(x) = \frac{2}{\pi} J_n(x) \left(\ln \frac{x}{2} + \gamma \right) - \frac{1}{\pi} \sum_{k=0}^{n-1} \frac{(n-k-1)!}{k!} \left(\frac{2}{x} \right)^{n-2k}$$

$$- \frac{1}{\pi} \sum_{k=0}^{\infty} \frac{(-1)^k}{k!(n+k)!} [h_k + h_{k+n}] \left(\frac{x}{2} \right)^{n+2k}$$

where

$$h_k = 1 + \frac{1}{2} + \cdots + \frac{1}{k}$$

and

$$\gamma = \lim_{k \to \infty} \left(1 + \frac{1}{2} + \cdots + \frac{1}{k} - \ln k \right) = 0.577215 \ldots$$

is the Euler or Mascheroni constant. The function $Y_n(x)$ is the *Bessel function of the second kind of order n*. The reader may consult Watson [50, pp. 57–73] for additional information on a second solution of Bessel's equation.

We mention two other Bessel functions. The *modified Bessel function of the first kind of order n* is defined by

$$I_n(x) = i^{-n}J_n(ix) \tag{5.12}$$

The *modified Bessel function of the second kind of order n* is given by

$$K_n(x) = \frac{\pi}{2} \frac{I_{-n}(x) - I_n(x)}{\sin n\pi}$$

if $n \in N_0$, and

$$K_n(x) = \frac{\pi}{2} \lim_{r \to n} \frac{I_{-r}(x) - I_r(x)}{\sin r\pi} \tag{5.13}$$

if $n \in N_0$. The limit indicated in (5.13) exists. In (5.12) the definition for $I_n(x)$ is a real function. $I_n(x)$ and $K_n(x)$ are solutions of the modified Bessel differential equation

$$x^2 y'' + xy' - (x^2 + n^2)y = 0 \tag{5.14}$$

If $n \in Z$, then

$$I_{-n}(x) = I_n(x)$$

and these Bessel functions are dependent. $I_n(x)$ and $I_{-n}(x)$ are linearly independent functions, and

$$y = C_1 I_n(x) + C_2 I_{-n}(x)$$

is a general solution of (5.14) if $n \notin Z$. If n is an integer, then $I_n(x)$ and $K_n(x)$ are linearly independent and

$$y = C_1 I_n(x) + C_2 K_n(x) \tag{5.15}$$

is a general solution. If n is not an integer, $K_n(x)$ is not dependent on $I_n(x)$ and (5.15) is still a solution of (5.14). Figure 5.2 is an adaptation from Abramowitz and Stegun [1, pp. 359 and 374]. Resembling a bit the cosine and sine functions, $J_n(x)$ and $Y_n(x)$ oscillate about zero with a decreasing amplitude. $I_n(x)$ and $K_n(x)$ fail to be oscillatory functions. Their behavior is somewhat like exponential functions. The series for $J_n(x)$ converges for all values of x if $n \geq 0$. $J_n(x) = 0$ has infinitely many real roots. It can be shown that the difference between successive roots for $J_n(x) = 0$ approaches π as the roots increase. $Y_n(x)$ is not defined for $x = 0$. Roots for $J_n(x) = 0$ are between those for $J_{n-1}(x) = 0$ and $J_{n+1}(x) = 0$. In our applications we use $J_n(x)$ and $Y_n(x)$ primarily.

Figure 5.2. Bessel functions.(*a*) First kind; (*b*) second kind; (*c*) modified first kind; (*d*) modified second kind. (Adapted from Abramowitz and Stegun [1], by permission of the author.)

5.3. DIFFERENTIAL EQUATIONS SOLVABLE WITH BESSEL FUNCTIONS

There are numerous ODEs which can be transformed into Bessel equations. After writing the solution of the Bessel equation, we employ the transformation again and obtain the solution of the original equation.

Example 5.1. Given the equation

$$x^2y'' + xy' + 4(x^4 - n^2)y = 0 \tag{5.16}$$

with the transformation $x^2 = t$. Find $y(x)$.

Using the transformation we change the given equation with $y(x)$ to a new equation with $y(t)$. Derivative transformations follow:

$$\frac{dy}{dx} = 2x\frac{dy}{dt} = 2t^{1/2}\frac{dy}{dt}$$

$$\frac{d^2y}{dx^2} = 4t\frac{d^2y}{dt^2} + 2\frac{dy}{dt}$$

The differential equation (5.16) becomes after substitution and simplification

$$t^2\frac{d^2y}{dt^2} + t\frac{dy}{dt} + (t^2 - n^2)y = 0 \tag{5.17}$$

The solution of the new equation (5.17) is

$$y(t) = AJ_n(t) + BY_n(t)$$

Since $x^2 = t$,

$$y(x) = AJ_n(x^2) + BY_n(x^2)$$

is the solution of (5.16).

Example 5.2. Show that the differential equation

$$x^2y'' + xy' + (\lambda^2x^2 - n^2)y = 0 \tag{5.18}$$

can be transformed into a Bessel differential equation if $\lambda x = t$.

Using the transformation to change independent variables, we write

$$\frac{dy}{dx} = \lambda\frac{dy}{dt}, \quad \frac{d^2y}{dx^2} = \lambda^2\frac{d^2y}{dt^2}$$

Equation (5.18) becomes

$$t^2\frac{d^2y}{dt^2} + t\frac{dy}{dt} + (t^2 - n^2)y = 0 \tag{5.19}$$

Since (5.19) is a Bessel differential equation

$$y(t) = AJ_n(t) + BY_n(t)$$

is a solution. The original equation (5.18) has the solution

$$y(x) = AJ_n(\lambda x) + BY_n(\lambda x) \tag{5.20}$$

Equation (5.18) is referred to as a *Bessel differential equation of order n with a parameter* λ. This equation and its solution (5.20) have special significance when we discuss orthogonality properties of Bessel functions. For more details of the reduction to Bessel's equation see Brand [8, pp. 495–496].

5.4. SPECIAL BESSEL FUNCTIONS AND IDENTITIES

We have shown the form of a few Bessel functions and examined some identities while discussing dependency. By a set of examples and problems we wish to expand our capability for using Bessel functions.

Example 5.3. Establish that

$$\frac{d}{dx}[x^n J_n(x)] = x^n J_{n-1}(x) \tag{5.21}$$

Using termwise differentiation of the series for $x^n J_n(x)$, we obtain

$$\frac{d}{dx}[x^n J_n(x)] = \frac{d}{dx} \sum_{k=0}^{\infty} \frac{(-1)^k x^{2n+2k}}{2^{n+2k}k!\Gamma(n+k+1)}$$

$$= \sum_{k=0}^{\infty} \frac{(-1)^k 2(n+k)x^{2n+2k-1}}{2^{n+2k}k!(n+k)\Gamma(n+k)}$$

$$= x^n \sum_{k=0}^{\infty} \frac{(-1)^k x^{(n-1)+2k}}{2^{(n-1)+2k}k!\Gamma((n-1)+k+1)}$$

$$= x^n J_{n-1}(x)$$

Example 5.4. Show that

$$\frac{d}{dx}[x^{-n} J_n(x)] = -x^{-n} J_{n+1}(x) \tag{5.22}$$

The procedure is similar to the process used in Example 5.3.

$$\frac{d}{dx}\left[x^{-n}J_n(x)\right] = \frac{d}{dx}\sum_{k=0}^{\infty}\frac{(-1)^k x^{2k}}{2^{n+2k}k!\Gamma(n+k+1)}$$

$$= \sum_{k=1}^{\infty}\frac{(-1)^k(2k)x^{2k-1}}{2^{n+2k}k!\Gamma(n+k+1)}$$

$$= -x^{-n}\sum_{m=0}^{\infty}\frac{(-1)^m x^{2m+(n+1)}}{2^{2m+(n+1)}m!\Gamma((n+1)+m+1)}, \quad k = m+1$$

$$= -x^{-n}J_{n+1}(x)$$

Example 5.5. Find $J_n'(x)$ in terms of $J_{n-1}(x)$ and $J_{n+1}(x)$.
From (5.21) and (5.22),

$$\frac{d}{dx}\left[x^n J_n(x)\right] = x^n J_n'(x) + nx^{n-1}J_n(x) = x^n J_{n-1}(x) \tag{5.23}$$

and

$$\frac{d}{dx}\left[x^{-n}J_n(x)\right] = x^{-n}J_n'(x) - nx^{-(n+1)}J_n(x) = -x^{-n}J_{n+1}(x) \tag{5.24}$$

In (5.23) we find that

$$J_n'(x) = J_{n-1}(x) - \frac{n}{x}J_n(x) \tag{5.25}$$

and from (5.24)

$$J_n'(x) = -J_{n+1}(x) + \frac{n}{x}J_n(x) \tag{5.26}$$

Adding (5.25) and (5.26), we find that

$$2J_n'(x) = J_{n-1}(x) - J_{n+1}(x)$$

or

$$J_n'(x) = \frac{J_{n-1}(x) - J_{n+1}(x)}{2} \tag{5.27}$$

Example 5.6. Find an identity involving $J_{n-1}(x)$, $J_n(x)$, and $J_{n+1}(x)$.
If one subtracts (5.26) from (5.25), derivative terms vanish and

$$J_{n+1}(x) + J_{n-1}(x) = \frac{2n}{x}J_n(x) \tag{5.28}$$

Example 5.7. Show that

$$J_{1/2}(x) = \sqrt{\frac{2}{\pi x}} \sin x$$

Substituting $n = 1/2$ in (5.5), one finds that

$$J_{1/2}(x) = \sum_{k=0}^{\infty} \frac{(-1)^k}{k!\Gamma(k + \frac{3}{2})} \left(\frac{x}{2}\right)^{2k+1/2}$$

$$= \frac{1}{\Gamma(\frac{3}{2})} \left(\frac{x}{2}\right)^{1/2} - \frac{1}{1!\Gamma(\frac{5}{2})} \left(\frac{x}{2}\right)^{5/2} + \frac{1}{2!\Gamma(\frac{7}{2})} \left(\frac{x}{2}\right)^{9/2} - + \cdots$$

$$= \frac{1}{(\frac{1}{2})\sqrt{\pi}} \left(\frac{a}{2}\right)^{1/2} - \frac{1}{1!(\frac{3}{2})(\frac{1}{2})\sqrt{\pi}} \left(\frac{x}{2}\right)^{5/2}$$

$$+ \frac{1}{2!(\frac{5}{2})(\frac{3}{2})(\frac{1}{2})\sqrt{\pi}} \left(\frac{x}{2}\right)^{9/2} - + \cdots$$

$$= \frac{(x/2)^{1/2}}{(\frac{1}{2})\sqrt{\pi}} \left(\frac{1}{x}\right) \left(x - \frac{x^3}{3!} + \frac{x^5}{5!} - + \cdots\right)$$

$$= \left(\frac{4}{2\pi x}\right)^{1/2} \sin x$$

$$= \sqrt{\frac{2}{\pi x}} \sin x$$

Is it safe to conclude that the differential equation

$$x^2 y'' + xy' + \left(x^2 - \frac{1}{4}\right) y = 0 \tag{5.29}$$

has a solution

$$J_{1/2}(x) = \sqrt{\frac{2}{\pi x}} \sin x?$$

Bessel functions of the order $\pm(n + \frac{1}{2})$, $n \in \mathbf{N}_0$, are often referred to as *half order Bessel functions* or *spherical Bessel functions*. From the *recurrence relation* (5.28), we are able to write sequentially $J_{(n+1/2)}(x)$ and $J_{-(n+1/2)}(x)$. Substituting $n = \frac{1}{2}$ and then $n = -\frac{1}{2}$ in (5.28), we obtain

$$J_{3/2}(x) + J_{-1/2}(x) = \frac{1}{x} J_{1/2}(x)$$

and

$$J_{1/2}(x) + J_{-3/2}(x) = -\frac{1}{x} J_{-1/2}(x)$$

Once we have $J_{1/2}(x)$ and $J_{-1/2}(x)$, we can compute $J_{3/2}(x)$ and $J_{-3/2}(x)$. Substituting $n = \frac{3}{2}$ and then $n = -\frac{3}{2}$ in (5.28), we obtain

$$J_{5/2}(x) + J_{1/2}(x) = \frac{3}{x} J_{3/2}(x)$$

and

$$J_{-1/2}(x) + J_{-5/2}(x) = -\frac{3}{x} J_{-3/2}(x)$$

These spherical Bessel functions are elementary functions.

Example 5.8. Determine $\int x^n J_{n-1}(x)\, dx$.
From (5.21)

$$\int x^n J_{n-1}(x)\, dx = x^n J_n(x) + C \tag{5.30}$$

Example 5.9. Find $\int_0^2 x^4 J_3(x)\, dx$.
Employing (5.30),

$$\int_0^2 x^4 J_3(x)\, dx = [x^4 J_4(x)]_0^2 = 2^4 J_4(2)$$

Example 5.10. Show that if n is not an integer

$$\frac{d}{dx}[x^n Y_n(x)] = x^n Y_{n-1}(x) \tag{5.31}$$

According to the definition of $Y_n(x)$ and (5.21) and (5.22)

$$\frac{d}{dx}[x^n Y_n(x)] = \frac{d}{dx}\left[x^n \frac{J_n(x)\cos n\pi - J_{-n}(x)}{\sin n\pi}\right]$$

$$= x^n\left[\frac{J_{n-1}(x)\cos n\pi + J_{-n+1}(x)}{\sin n\pi}\right]$$

$$= x^n\left[\frac{J_{n-1}(x)(-\cos(n-1)\pi) + J_{-(n-1)}(x)}{-\sin(n-1)\pi}\right]$$

$$= x^n\left[\frac{J_{n-1}(x)\cos(n-1)\pi - J_{-(n-1)}(x)}{-\sin(n-1)\pi}\right]$$

$$= x^n Y_{n-1}(x)$$

The result (5.31) may be established also when n is an integer by using the limits in the definition.

Exercises 5.1

1. Show that $J_0'(x) = -J_1(x)$.
2. Show that

   (a) $J_1'(x) = \dfrac{xJ_0(x) - J_1(x)}{x}$,

   (b) $2J_2'(x) = J_1(x) - J_3(x)$.

3. Show that

$$\frac{d}{dx}\left[x^{-n}Y_n(x)\right] = -x^{-n}Y_{n+1}(x)$$

4. Demonstrate that

$$\int_0^1 J_1(x)\, dx = 1 - J_0(1)$$

5. Show that

$$\int_0^c xJ_0(x)\, dx = cJ_1(c)$$

6. Establish that

$$\int x^{-n}J_{n+1}(x)\, dx = -x^{-n}J_n(x) + C$$

7. Determine that

$$\int x^n Y_{n-1}(x)\, dx = x^n Y_n(x) + C$$

8. Demonstrate that

$$\int_0^1 xJ_1(x)\, dx = -J_0(1) + \int_0^1 J_0(x)\, dx$$

9. Observing that $x^{-1} = x^{-2}x$ and integrating by parts, show that

$$\int x^{-1}J_1(x)\, dx = -J_1(x) + \int J_0(x)\, dx$$

10. Establish that

$$\int J_2(x)\, dx = J_3(x) + 3\int \frac{J_3(x)}{x}\, dx$$

11. Show that

$$\int J_{n+1}(x)\, dx = \int J_{n-1}(x)\, dx - 2J_n(x)$$

12. (a) Establish that

$$J_{-1/2}(x) = \sqrt{\frac{2}{\pi x}} \cos x$$

(b) Write a general solution for Bessel's equation (5.29).

13. Demonstrate that

(a) $J_{3/2}(x) = \sqrt{\dfrac{2}{\pi x}} \left(\dfrac{\sin x}{x} - \cos x \right),$

(b) $J_{-3/2}(x) = -\sqrt{\dfrac{2}{\pi x}} \left(\dfrac{\cos x}{x} + \sin x \right),$

(c) $J_{5/2}(x) = \sqrt{\dfrac{2}{\pi x}} \left(\dfrac{3 - x^2}{x^2} \sin x - \dfrac{3 \cos x}{x} \right),$

(d) $J_{-5/2}(x) = \sqrt{\dfrac{2}{\pi x}} \left(\dfrac{3 \sin x}{x} + \dfrac{3 - x^2}{x^2} \cos x \right),$

(e) Write a general solution for the differential equation

$$x^2 y'' + xy' + \left(x^2 - \frac{9}{4} \right) y = 0$$

14. Show that the differential equation

$$xy'' - y' + xy = 0$$

has a solution

$$y = C_1 x J_1(x) + C_2 x Y_1(x)$$

if $x \neq 0$. Transform the equation by $y = vx$ so that v is the new dependent variable.

15. Show that $I_n(x)$ is a solution of the differential equation

$$x^2 y'' + xy' - (x^2 + n^2) y = 0$$

16. Show that

$$\int_0^{2\pi} e^{i\alpha \cos \beta} \, d\beta = 2\pi J_0(\alpha) \qquad (5.32)$$

Suggestions:

(a) Let $A(\alpha) = \int_0^{2\pi} \cos (\alpha \cos \beta) \, d\beta$.
(b) By employing the power series for $\cos (\alpha \cos \beta)$, show that

$$A(\alpha) = \sum_{n=0}^{\infty} \frac{(-1)^n \alpha^{2n}}{(2n)!} \int_0^{2\pi} \cos^{2n} \beta \, d\beta = 2\pi \sum_{n=0}^{\infty} \left(\frac{-\alpha^2}{4} \right)^n \frac{1}{(n!)^2}$$

$$(5.33)$$

(c) Compare the series (5.33) with the series for $J_0(\alpha)$ and show that $A(\alpha) = 2\pi J_0(\alpha)$.
(d) Show that $B(\alpha) = \int_0^{2\pi} \sin (\alpha \cos \beta) \, d\beta = 0$.
(e) Notice that $A(\alpha) + iB(\alpha) = \int_0^{2\pi} e^{i\alpha \cos \beta} \, d\beta$.
(f) Then $\int_0^{2\pi} e^{i\alpha \cos \beta} \, d\beta = 2\pi J_0(\alpha)$.

5.5. AN INTEGRAL FORM FOR $J_n(x)$

The exponential function $\exp [x(t - 1/t)/2]$ is called a *generating function* and is employed to obtain an integral form for $J_n(x)$. We observe that

$$\exp \left[\frac{x(t - 1/t)}{2} \right] = \exp \left(\frac{xt}{2} \right) \exp \left(-\frac{x}{2t} \right)$$

$$= \left[\sum_{r=0}^{\infty} \frac{x^r}{2^r r!} t^r \right] \left[\sum_{s=0}^{\infty} \frac{(-1)^s x^s}{2^s s!} \left(\frac{1}{t} \right)^s \right]$$

$$= \sum_{r=0}^{\infty} \sum_{s=0}^{\infty} \frac{(-1)^s x^{r+s}}{2^{r+s} r! s!} t^{r-s} \qquad (5.34)$$

by multiplying two absolutely convergent series. If we let $r = n + s$ or $n = r - s$ and $n \in \mathbf{Z}$, then (5.34) becomes

$$\exp\left[\frac{x(t-1/t)}{2}\right] = \sum_{n=-\infty}^{\infty}\left[\sum_{s=0}^{\infty}\frac{(-1)^s x^{n+2s}}{2^{n+2s}s!(n+s)!}\right]t^n$$

$$= \sum_{n=-\infty}^{\infty} J_n(x)t^n$$

$$= \sum_{n=1}^{\infty} J_{-n}(x)t^{-n} + J_0(x) + \sum_{n=1}^{\infty} J_n(x)t^n$$

$$= J_0(x) + \sum_{n=1}^{\infty} J_n(x)\left[t^n + \frac{(-1)^n}{t^n}\right] \tag{5.35}$$

If we let $t = e^{i\theta}$, then

$$\frac{1}{2}\left(t - \frac{1}{t}\right) = \frac{e^{i\theta} - e^{-i\theta}}{2} = i\sin\theta$$

and

$$\exp\left[\frac{x(t-1/t)}{2}\right] = e^{ix\sin\theta} = \cos(x\sin\theta) + i\sin(x\sin\theta) \tag{5.36}$$

If n is even, say $n = 2k$, then

$$t^n + \frac{(-1)^n}{t^n} = t^{2k} + \frac{1}{t^{2k}} = e^{i2k\theta} + e^{-i2k\theta} = 2\cos 2k\theta \tag{5.37}$$

If n is odd, say $n = 2k - 1$, then

$$t^n + \frac{(-1)^n}{t^n} = t^{2k-1} - \frac{1}{t^{2k-1}} = e^{i(2k-1)\theta} - e^{-i(2k-1)\theta} = 2i\sin(2k-1)\theta \tag{5.38}$$

From (5.35) and (5.36) along with (5.37) and (5.38) we have

$$\cos(x\sin\theta) = J_0(x) + 2\sum_{k=1}^{\infty} J_{2k}(x)\cos 2k\theta \tag{5.39}$$

and

$$\sin(x\sin\theta) = 2\sum_{k=1}^{\infty} J_{2k-1}(x)\sin(2k-1)\theta \tag{5.40}$$

by equating real and imaginary parts. Equation (5.39) indicates that $2J_{2k}(x)$ are the Fourier cosine coefficients of $\cos(x\sin\theta)$ considered as a function of θ. Therefore for $0 \le \theta \le \pi$,

$$2J_{2k}(x) = \frac{2}{\pi} \int_0^{\pi} \cos (x \sin \theta) \cos 2k\theta \ d\theta$$

or

$$J_{2k}(x) = \frac{1}{\pi} \int_0^{\pi} \cos (x \sin \theta) \cos 2k\theta \ d\theta$$

Similarly with (5.40)

$$J_{2k-1}(x) = \frac{1}{\pi} \int_0^{\pi} \sin (x \sin \theta) \sin (2k - 1)\theta \ d\theta$$

We see that

$$\int_0^{\pi} \cos n\theta \cos (x \sin \theta) \ d\theta = \begin{cases} \pi J_n(x) & \text{if } n \text{ is even} \\ 0 & \text{if } n \text{ is odd} \end{cases}$$

and

$$\int_0^{\pi} \sin n\theta \sin (x \sin \theta) \ d\theta = \begin{cases} \pi J_n(x) & \text{if } n \text{ is odd} \\ 0 & \text{if } n \text{ is even} \end{cases}$$

Therefore,

$$J_n(x) = \frac{1}{\pi} \int_0^{\pi} [\cos n\theta \cos (x \sin \theta) + \sin n\theta \sin (x \sin \theta)] \ d\theta$$

or

$$J_n(x) = \frac{1}{\pi} \int_0^{\pi} \cos (n\theta - x \sin \theta) \ d\theta \tag{5.41}$$

If we differentiate (5.41) m times we obtain

$$J_n^{(m)}(x) = \frac{1}{\pi} \int_0^{\pi} \sin^m\theta \cos \left(n\theta - x \sin \theta + \frac{m\pi}{2} \right) d\theta \tag{5.42}$$

Exercises 5.2

1. Show that $J_n'(x) = \dfrac{1}{\pi} \displaystyle\int_0^{\pi} \sin (n\theta - x \sin \theta) \sin \theta \ d\theta$.

2. Verify (5.42).
3. Demonstrate that $J_n(x)$ is a bounded function, so that $|J_n(x)| \leqslant 1$.
4. Show that $|J_n^{(m)}(x)| \leqslant 1$.
5. Using (5.39) and (5.40) show that
   (a) $\cos x = J_0(x) - 2J_2(x) + 2J_4(x) - \cdots$,
   (b) $\sin x = 2J_1(x) - 2J_3(x) + 2J_5(x) - \cdots$,
   (c) $1 = J_0(x) + 2J_2(x) + 2J_4(x) + \cdots$.

6. If $n \in \mathbb{N}$, show that

$$\lim_{n \to \infty} J_n(x) = 0$$

using Bessel's inequality and recognizing Bessel's functions as Fourier coefficients.

5.6. SINGULAR SLPs

In Section 2.5 we discussed the regular SLP. The equation (2.9) with end point conditions (2.10)

$$[p(x)y']' + [q(x) + \lambda r(x)]y = 0 \tag{5.43}$$
$$a_1 y(a) + a_2 y'(a) = 0$$
$$b_1 y(b) + b_2 y'(b) = 0 \tag{5.44}$$

comprise the SLP. The equation of our present concern fails to meet the specifications of Section 2.5. When the interval is infinite or semi-infinite, or when $p(x)$ or $r(x)$ vanish, or when one of the coefficients becomes infinite at one or both ends of a finite interval the SLDE is *singular*. A *singular SLP* is composed of a singular SLDE along with appropriate homogeneous linear end conditions of the type (5.44). We do not demonstrate all cases, but investigate briefly the two situations for $p(x)$ continuous where either $p(a)$ or both $p(a)$ and $p(b)$ vanish.

If $p(a) = 0$, instead of integrating (2.15) over $[a, b]$ as we did in Theorem 2.1, we consider the integral, $\varepsilon > 0$,

$$\int_{a+\varepsilon}^{b} \frac{d}{dx} \{p[y'_m y_n - y'_n y_m]\} \, dx$$
$$= p(b)[y'_m(b)y_n(b) - y'_n(b)y_n(b)]$$
$$\quad - p(a + \varepsilon)[y'_m(a + \varepsilon)y_n(a + \varepsilon) - y'_n(a + \varepsilon)y_m(a + \varepsilon)] \tag{5.45}$$

If we assume that $y(x)$ and $y'(x)$ are finite for all x and the second condition of (5.44), then as $\varepsilon \to 0$, the integral (5.45) vanishes. The result

$$\int_{a}^{b} r y_n y_m \, dx = 0 \quad \text{if } m \neq n$$

is immediate. If

$$p(a) = p(b) = 0$$

and $y(x)$ and $y'(x)$ are finite, the integral

$$\int_{a+\varepsilon}^{b-\varepsilon} \frac{d}{dx} \{p[y_m' y_n - y_n' y_m]\}\, dx$$

$$= p(b-\varepsilon)[y_m'(b-\varepsilon)y_n(b-\varepsilon) - y_n'(b-\varepsilon)y_m(b-\varepsilon)]$$

$$\qquad - p(a+\varepsilon)[y_m'(a+\varepsilon)y_n(a+\varepsilon) - y_n'(a+\varepsilon)y_m(a+\varepsilon)] \qquad (5.46)$$

As $\varepsilon \to 0$, the integral (5.46) vanishes and

$$\int_a^b r y_n y_m\, dx = 0 \quad \text{if } n \neq m \qquad (5.47)$$

For the singular case discussed first we used one end condition, the second of (5.44), to establish orthogonality. In the second singular case, we employed no boundary conditions to accompany the differential equation to show orthogonality. Eigenfunctions matching distinct eigenvalues of a singular SLP are orthogonal relative to $r(x)$ if the eigenfunctions are SI. The reader may wish to see Birkhoff and Rota [2, pp. 263–265] for added information on the singular SLP.

5.7. ORTHOGONALITY OF BESSEL FUNCTIONS

The differential equation (5.18) may be written in the form

$$[xy']' + \left[\lambda^2 x - \frac{n^2}{x}\right] y = 0 \qquad (5.48)$$

The equation is a SLDE given in (5.43) where $p(x) = r(x) = x$, $q(x) = -n^2/x$, and λ^2 replaces λ. As we observe from Example 5.2

$$y = AJ_n(\lambda x) + BY_n(\lambda x)$$

is a solution of (5.18) and therefore a solution of (5.48). On the interval $[0, b]$ the SLP composed of (5.48) and an end condition

$$b_1 y(b) + b_2 y'(b) = 0 \qquad (5.49)$$

is singular, since $p(0) = 0$ and $q(x)$ increases without bound as $x \to 0$. At $x = 0$, $Y(0)$ is undefined and we take $B = 0$. Therefore, the solution we consider is $J_n(\lambda x)$. There are three principal cases for discussion considering (5.49).

If $b_2 = 0$ in (5.49), then

$$y(b) = 0$$

is given and the eigenvalues λ are obtained from the roots of $J_n(\lambda_k b) = 0$.

The zeros occur at points where $\lambda_k b = \alpha_k$ or $\lambda_k = \alpha_k/b$ and $J_n(\alpha_k) = 0$. Therefore, the eigenvalues are the zeros of the Bessel function divided by the length of the interval b.

If b_1 is zero in (5.49), then

$$y'(b) = 0$$

In this case $\lambda_k = \alpha_k/b$ with $J'_n(\alpha_k) = 0$.

For the general case we multiply (5.49) by b/b_2, $b_2 \neq 0$, replace $b_1 b/b_2$ by h, and recognize that $y'(b) = \lambda J'_n(\lambda b)$. Therefore, the boundary condition is

$$hJ_n(\lambda b) + \lambda b J'_n(\lambda b) = 0 , \quad h \geq 0$$

and the eigenvalues are $\lambda_k = \alpha_k/b$ where

$$hJ_n(\alpha_k) + \alpha_k J'_n(\alpha_k) = 0$$

In all three end condition cases (5.49), the eigenfunctions matching eigenvalues λ_k are

$$J_n(\lambda_k x) = J_n\left(\alpha_k \frac{x}{b}\right) \tag{5.50}$$

According to Section 5.6 with the conclusion (5.47), $\{J_n(\lambda_k x)\}$ is orthogonal relative to x, and

$$\int_0^b x J_n(\lambda_k x) J_n(\lambda_m x) \, dx = 0 , \quad k \neq m$$

We are obliged to include all eigenfunctions of the problem. Zero functions are not eigenfunctions. From the identity

$$J_n(-x) = (-1)^n J_n(x)$$

it is apparent that negative values of λ_k may be excluded. If $\lambda_0 = 0$, then the Bessel function as given by (5.50) is zero except when the order n of the Bessel function is zero. If $n = 0$, only the second case permits the solution $\lambda_0 = 0$. It is required that we consider only the set $\{J_n(\lambda_k x)\}$ corresponding to $k \in \mathbf{N}$ in all cases except when $n = 0$ so that $J_0(\lambda_0 x) = 1$. This must be included in the set of eigenfunctions.

To use the set $\{J_n(\lambda_k x)\}$ in an orthogonal series it is necessary to discuss norms of the set. First, we multiply (5.48) by $2xy'$ and obtain

$$2xy'[xy']' + 2[\lambda^2 x^2 - n^2] yy' = 0$$

or

$$2x^2 y'y'' + 2x(y')^2 + 2[\lambda^2 x^2 - n^2]yy' = 0 \qquad (5.51)$$

We observe that

$$[(xy')^2]' = 2(xy')(xy'' + y') = 2x^2 y'y'' + 2x(y')^2$$

Hence (5.51) may be written

$$[(xy')^2]' + [\lambda^2 x^2 - n^2][y^2]' = 0 \qquad (5.51a)$$

Integrating (5.51a) over $(0, b)$ we have

$$\int_0^b [(xy')^2]' \, dx + \lambda^2 \int_0^b x^2 [y^2]' \, dx - n^2 \int_0^b [y^2]' \, dx = 0 \qquad (5.52)$$

In the second integral of (5.52) we integrate by parts and find that

$$[(xy')^2]_0^b + \lambda^2 [x^2 y^2]_0^b - 2\lambda^2 \int_0^b xy^2 \, dx - n^2 [y^2]_0^b = 0$$

Therefore,

$$2\lambda^2 \int_0^b xy^2 \, dx = [(xy')^2 + (\lambda^2 x^2 - n^2)y^2]_0^b$$

If $y(x) = J_n(\lambda x)$ is the solution of (5.48), then $y'(x) = \lambda J_n'(\lambda x)$ and $y'(b) = \lambda J_n(\lambda b)$. Therefore,

$$2\lambda^2 \int_0^b xJ_n^2(\lambda x) \, dx = [\{(x\lambda)J_n'(\lambda x)\}^2 + (\lambda^2 x^2 - n^2)J_n^2(\lambda x)]_0^b$$

$$= \lambda^2 b^2 [J_n'(\lambda b)]^2 + [\lambda^2 b^2 - n^2]J_n^2(\lambda b) \qquad (5.53)$$

If the boundary condition is

$$J_n(\lambda b) = 0$$

with $\lambda_k = \alpha_k/b$ and α_k the positive roots of $J_n(\alpha_k) = 0$, then

$$\int_0^b xJ_n^2(\lambda_k x) \, dx = \frac{b^2}{2} [J_n'(\lambda_k b)]^2 \qquad (5.54)$$

According to (5.26), we may write

$$\lambda_k bJ_n'(\lambda_k b) = nJ_n(\lambda_k b) - \lambda_k bJ_{n+1}(\lambda_k b)$$

With our boundary condition,

$$J_n'(\lambda_k b) = -J_{n+1}(\lambda_k b)$$

Therefore, (5.54) becomes

$$\int_0^b x J_n^2(\lambda_k x)\, dx = \frac{b^2}{2} J_{n+1}^2(\lambda_k b)$$

or

$$\|J_n(\lambda_k x)\|^2 = \frac{b^2}{2} J_{n+1}^2(\lambda_k b), \quad k \in \mathbf{N} \tag{5.55}$$

is the square of the norm.

If the boundary condition is

$$h J_n(\lambda b) + b J_n'(\lambda b) = 0$$

and $\lambda_k = \alpha_k / b$ with α_k the positive roots of

$$h J_n(\alpha_k) + \alpha_k J'(\alpha_k) = 0$$

then

$$h^2 J_n^2(\lambda_k b) = (\lambda_k b)^2 [J_n'(\lambda_k b)]^2$$

From (5.53) and the boundary condition,

$$\|J_n(\lambda_k x)\|^2 = \left[\frac{h^2 + (\lambda_k b)^2 - n^2}{2\lambda_k^2} \right] J_n^2(\lambda_k b), \quad k \in \mathbf{N} \tag{5.56}$$

The remaining condition is $J_0'(\lambda b) = 0$. Suppose $\lambda_0 = 0$, then $J_0(\lambda_0 x) = 1$ and

$$\|J_0(\lambda_0 x)\|^2 = \int_0^b x\, dx = \frac{b^2}{2}$$

From (5.56) with $n = h = 0$,

$$\|J_0(\lambda_k x)\|^2 = \frac{b^2}{2} J_0^2(\lambda_k b), \quad k \in \mathbf{N}$$

5.8. ORTHOGONAL SERIES OF BESSEL FUNCTIONS

We use the procedure of Section 2.7 to construct series based on the orthogonal set $\{J_n(\lambda_k x)\}$, $0 < x < b$, relative to a weight function x. In each case the set is accompanied by an appropriate boundary condition. The representation for $f(x)$ follows:

$$f(x) \sim \sum_{k=1}^{\infty} A_k J_n(\lambda_k x), \quad 0 < x < b \tag{5.57}$$

If the construction of Section 2.7 is used, the coefficients (2.44) become

$$A_k = \frac{1}{\|J_n(\lambda_k x)\|^2} \int_0^b xf(x) J_n(\lambda_k x)\, dx \tag{5.58}$$

if the end condition $J_n(\lambda b) = 0$ is given. Since the norm of (5.55) accompanies this end condition,

$$A_k = \frac{2}{b^2 J_{n+1}^2(\lambda_k b)} \int_0^b xf(x) J_n(\lambda_k x)\, dx, \quad k \in \mathbf{N} \tag{5.59}$$

If λ_k are the eigenvalues from the end condition

$$hJ_n(\lambda b) + \lambda b J_n'(\lambda b) = 0, \quad h \geq 0$$

then the norm of (5.56) accompanies the end condition and

$$A_k = \frac{2\lambda_k^2}{[h^2 + (\lambda_k b)^2 - n^2] J_n^2(\lambda_k b)} \int_0^b xf(x) J_n(\lambda_k x)\, dx, \quad k \in \mathbf{N} \tag{5.60}$$

If the boundary condition is $J_0'(\lambda b) = 0$, then we write the series

$$f(x) \sim A_0 + \sum_{k=1}^{\infty} A_k J_0(\lambda_k x), \quad 0 < x < b \tag{5.61}$$

where

$$A_0 = \frac{2}{b^2} \int_0^b xf(x)\, dx \tag{5.62}$$

and

$$A_k = \frac{2}{b^2 J_0^2(\lambda_k b)} \int_0^b xf(x) J_0(\lambda_k x)\, dx, \quad k \in \mathbf{N} \tag{5.63}$$

No special case needs to be discussed for the coefficients if $J_n'(\lambda b) = 0$ and $n \neq 0$. In this situation coefficients come from (5.60) when $h = 0$.

For the series (5.57) the coefficients (5.59) and (5.60) are certain determinations of (5.58). The special case (5.61) when $J_0'(\lambda b) = 0$ has a constant term A_0 given by (5.62) and the remaining coefficients (5.63) are special cases of (5.60). These series are referred to as *Fourier–Bessel series*.

The convergence theorem for the /Fourier–Bessel series representing a function f is established by Watson [50, pp. 591–592]. We include a similar theorem without proof.

Theorem 5.1. If f is sectionally smooth on the interval $(0, b)$, then the Fourier–Bessel series (5.57) or its special case (5.61) with appropriate coefficients converges to

$$\frac{f(x+) + f(x-)}{2} \tag{5.64}$$

Example 5.11. Expand $f(x) = 2$ over the interval $(0, 2)$ if $J_n(2\lambda) = 0$.

$$2 \sim \sum_{k=1}^{\infty} A_k J_n(\lambda_k x), \quad 0 < x < 2$$

where

$$A_k = \frac{2}{2^2 J_{n+1}^2(2\lambda_k)} \int_0^2 2x J_n(\lambda_k x) \, dx$$

or

$$A_k = \frac{1}{J_{n+1}^2(2\lambda_k)} \int_0^2 x J_n(\lambda_k x) \, dx$$

Example 5.12. Find the representation for $f(x) = 2$, $0 < x < 3$ if the end condition is

$$h J_3(3\lambda) + 3\lambda J_3'(3\lambda) = 0, \quad h > 0$$

$$2 \sim \sum_{k=1}^{\infty} A_k J_3(\lambda_k x), \quad 0 < x < 3$$

where

$$A_k = \frac{2\lambda_k^2}{[h^2 + (2\lambda_k)^2 - 3^2] J_3^2(3\lambda_k)} \int_0^3 2x J_3(\lambda_k x) \, dx$$

$$= \frac{4\lambda_k^2}{[h^2 + 9\lambda_k^2 - 9] J_3^2(3\lambda_k)} \int_0^3 x J_3(\lambda_k x) \, dx$$

Exercises 5.3

1. Expand $f(x) = 1$ over the interval $(0, 2)$ in terms of Bessel functions of the first kind order 0 which satisfy the end condition $J_0(2\lambda) = 0$.

2. If

$$f(x) = \begin{cases} 2 & \text{when } 0 < x < 2 \\ 0 & \text{when } 2 < x < 4 \\ 1 & \text{when } x = 2 \end{cases}$$

find the Fourier–Bessel series representation for $f(x)$ on $0 < x < 4$ given the condition $J_0(4\lambda) = 0$.

3. Show that if $h J_0(2\lambda) + (2\lambda) J_0'(2\lambda) = 0$, $h > 0$, and $f(x) = 1$, $0 < x < 2$, then

$$1 \sim 4 \sum_{k=1}^{\infty} \frac{\lambda_k J_1(2\lambda_k) J_0(\lambda_k x)}{[4\lambda_k^2 + h^2] J_0^2(2\lambda_k)}$$

4. If $f(x) = 1$, $0 < x < 3$, and $J_0'(3\lambda) = 0$, find the Fourier–Bessel series representation.

5. Show that the Fourier–Bessel series in $J_1(\lambda_k x)$ for $f(x) = x$, $0 < x < 2$, where $J_1(2\lambda) = 0$, is

$$x = 2 \sum_{k=1}^{\infty} \frac{J_1(\lambda_k x)}{\lambda_k J_2(2\lambda_k)}$$

5.9. BESSEL FUNCTIONS AND CYLINDRICAL GEOMETRY

The formulas relating cylindrical and rectangular coordinates are given by (see Figure 5.3)

$$x = r \cos \theta, \quad y = r \sin \theta, \quad z = z \tag{5.65}$$

In the xy plane the formulas are those of the rectangular-polar coordinate relations. The z coordinate remains the same in both rectangular and cylindrical coordinates.

In rectangular coordinates the Laplacian is

$$\nabla^2 u = u_{xx} + u_{yy} + u_{zz}$$

To find $\nabla^2 u$ in cylindrical coordinates we must determine u_{xx} and u_{yy} in polar coordinates. The term u_{zz} will remain unchanged in cylindrical coordinate representation. From (5.65)

$$r = (x^2 + y^2)^{1/2}, \quad \theta = \arctan \frac{y}{x}, \quad z = z \tag{5.66}$$

Using the chain rule,

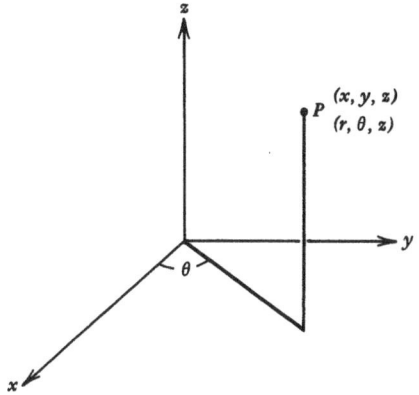

Figure 5.3. Rectangular and cylindrical coordinates for a point.

$$u_x = u_{rx} + u_\theta \theta_x = u_r x(x^2 + y^2)^{-1/2} - u_\theta y(x^2 + y^2)^{-1}$$

$$= u_r \cos\theta - u_\theta \frac{\sin\theta}{r}$$

$$u_{xx} = \left[u_r \cos\theta - u_\theta \frac{\sin\theta}{r} \right]_r r_x + \left[u_r \cos\theta - u_\theta \frac{\sin\theta}{r} \right]_\theta \theta_x$$

$$= u_{rr} \cos^2\theta - 2u_{r\theta} \frac{\sin\theta \cos\theta}{r} + u_r \frac{\sin^2\theta}{r} + 2u_\theta \frac{\sin\theta \cos\theta}{r^2} + u_{\theta\theta} \frac{\sin^2\theta}{r^2}.$$

Employing (5.66) and the chain rule again,

$$u_y = u_r \sin\theta + u_\theta \frac{\cos\theta}{r}$$

$$u_{yy} = u_{rr} \sin^2\theta + 2u_{r\theta} \frac{\sin\theta \cos\theta}{r} + u_r \frac{\cos^2\theta}{r} - 2u_\theta \frac{\sin\theta \cos\theta}{r^2} + u_{\theta\theta} \frac{\cos^2\theta}{r^2}$$

Therefore,

$$\nabla^2 u = u_{rr}(\sin^2\theta + \cos^2\theta) + u_r \left(\frac{\sin^2\theta + \cos^2\theta}{r} \right) + u_{\theta\theta} \left(\frac{\sin^2\theta + \cos^2\theta}{r^2} \right) + u_{zz}$$

or

$$\nabla^2 u = u_{rr} + \frac{1}{r} u_r + \frac{1}{r^2} u_{\theta\theta} + u_{zz} \tag{5.67}$$

In the xy or polar plane (5.67) becomes

$$\nabla^2 u(r, \theta) = u_{rr} + \frac{1}{r} u_r + \frac{1}{r^2} u_{\theta\theta}$$

Other specializations are considered in BVPs.

6

LEGENDRE POLYNOMIALS

Just as in Chapter 5, nonrectangular coordinates will motivate our discussion in this chapter. Some models of physical systems have geometrical properties that are spherical in nature. *Legendre polynomials* are solutions for *Legendre differential equations*, and *associated Legendre polynomials* are solutions of *associated Legendre differential equations*. All lend themselves to spherical geometry. We consider spherical coordinates and differential forms related to spherical coordinates, as well as properties of the polynomials and differential equations.

6.1. SOLUTIONS TO THE LEGENDRE EQUATION

In Chapter 1, we used a power series to express the solution of the *Legendre differential equation of degree n*

$$(1 - x^2)y'' - 2xy' + n(n + 1)y = 0 \tag{6.1}$$

The solution of the equation is repeated for our immediate use

$$y(x) = C_0 y_0(x) + C_1 y_1(x) \tag{6.2}$$

where

$$y_0(x) = 1 - \frac{n(n + 1)}{2!} x^2 + \frac{n(n + 1)(n - 2)(n + 3)}{4!} x^4$$
$$- \frac{n(n + 1)(n - 2)(n + 3)(n - 4)(n + 5)}{6!} x^6 + \cdots \tag{6.3}$$

and

$$y_1(x) = x - \frac{(n-1)(n+2)}{3!} x^3 + \frac{(n-1)(n+2)(n-3)(n+4)}{5!} x^5$$

$$- \frac{(n-1)(n+2)(n-3)(n+4)(n-5)(n+6)}{7!} x^7 + \cdots \quad (6.4)$$

If $n = 0$, $C_0 = 1$, and $C_1 = 0$, then 1 is a solution of Legendre's equation. If $n = 1$, $C_0 = 0$, and $C_1 = 1$, then x is a solution of the equation. If $n = 2$, $1 - 3x^2$ is a solution, and if $n = 3$, $x - 5x^3/3$ is a solution. While these polynomials are solutions of (6.1), they are not all Legendre polynomials. C_0 and C_1 are assigned so that the coefficients of the highest power of x have the value

$$\frac{(2n)!}{2^n(n!)^2} \quad (6.5)$$

The recurrence relation (1.29) may be written

$$C_k = - \frac{(k+2)(k+1)}{(n-k)(n+k+1)} C_{k+2} \quad (6.6)$$

If k is replaced by $n - 2$, then

$$C_{n-2} = - \frac{n(n-1)}{2(2n-1)} C_n$$

Continuing the use of (6.6), we may write

$$C_{n-2k} = \frac{(-1)^k n(n-1)(n-2)\ldots(n-2k+2)(n-2k+1)}{2^k k!(2n-1)(2n-3)\ldots(2n-2k+1)} C_n$$

Replacing C_n with (6.5), employing some factorial arithmetic and a few simplifications, we obtain

$$C_{n-2k} = \frac{(-1)^k(2n-2k)!}{2^n k!(n-k)!(n-2k)!} \quad (6.7)$$

Using the coefficients (6.7), we define the solution as the *Legendre polynomial of degree n* as follows:

$$P_n(x) = \frac{1}{2^n} \sum_{k=0}^{M} \frac{(-1)^k(2n-2k)!}{k!(n-k)!(n-2k)!} x^{n-2k}, \quad n \in N_0 \quad (6.8)$$

where $M = n/2$ or $M = (n-1)/2$, whichever makes M an integer. We observe that $P_n(x)$ is even or odd as n is an even or odd number. The solution of (6.1) involves no new form for the equation if n is replaced by $-(n+1)$. Therefore, it is adequate to consider only nonnegative integers in the Legendre equation. A few of the polynomials represented by (6.8) are listed as follows:

$$P_0(x) = 1, \qquad\qquad P_1(x) = x$$

$$P_2(x) = \frac{3x^2 - 1}{2}, \qquad P_3(x) = \frac{5x^3 - 3x}{2}$$

$$P_4(x) = \frac{35x^4 - 30x^2 + 3}{8}, \quad P_5(x) = \frac{63x^5 - 70x^3 + 15x}{8}$$

We observe that a general solution (6.2) of (6.1) is composed of a linear combination of a series (6.3) with even powers of x and a series (6.4) with odd powers of x. When n is an even number the series $y_0(x)$ terminates with x^n and $y_0(x)$ is a polynomial, while $y_1(x)$ is an infinite series. If n is an odd number then the series $y_1(x)$ terminates with x^n but $y_0(x)$ is an infinite series.

A general solution (6.2) contains a polynomial $P_n(x)$ and an infinite series that we shall denote as $Q_n(x)$. The definition

$$Q_n(x) = \begin{cases} y_0(1)y_1(x) & \text{when } n \text{ is even} \\ -y_1(1)y_0(x) & \text{when } n \text{ is odd} \end{cases} \qquad (6.8a)$$

for $|x| < 1$, is the *Legendre function of the second kind*. It can be shown that $Q_n(x)$ for $|x| < 1$ converges, and $Q_n(x)$ and $P_n(x)$ are linearly independent. Therefore,

$$y(x) = A_1 P_n(x) + A_2 Q_n(x)$$

is a general solution for (6.1).

If $|x| > 1$, then (6.8a) fails to converge. An alternate definition for $Q_n(x)$ in this case is given by Pipes and Harvill [39, p. 802], but it is neglected in our discussion. In fact, the polynomial $P_n(x)$ is more important for our immediate work than is $Q_n(x)$. However, we shall include the following Q_n functions:

$$Q_0(x) = x + \frac{x^3}{3} + \frac{x^5}{5} + \cdots = \frac{1}{2} \ln\left(\frac{1+x}{1-x}\right)$$

$$Q_1(x) = x\left(x + \frac{x^3}{3} + \frac{x^5}{5} + \cdots\right) - 1 = \frac{x}{2} \ln\left(\frac{1+x}{1-x}\right) - 1$$

$$Q_2(x) = \frac{3x^2 - 1}{4}\left(x + \frac{x^3}{3} + \frac{x^5}{5} + \cdots\right) - \frac{3x}{2} = \left(\frac{3x^2 - 1}{4}\right) \ln\left(\frac{1+x}{1-x}\right) - \frac{3x}{2}$$

Figure 6.1 is a graphical display of a few of the Legendre polynomials and functions adapted from Brand [8, p. 464] and Jahnke and Emde [28, p. 110].

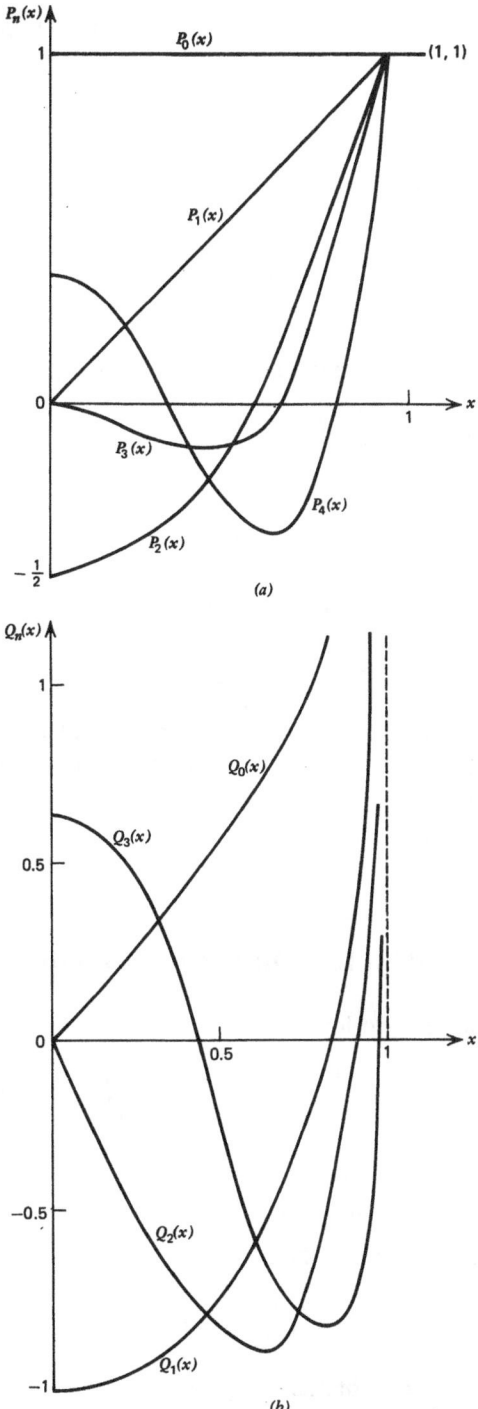

Figure 6.1. Legendre polynomials and functions. (*a*) Legendre polynomials $P_n(x)$; (*b*) Legendre functions $Q_n(x)$. (Adapted from Brand [8] and Jahnke and Emde [28], by permission of John Wiley & Sons, Inc., and Dover Publications, Inc.)

Exercises 6.1

1. Show that
   (a) $P_3(-x) = -P_3(x)$,
   (b) $P_4(-x) = P_4(x)$.
2. Determine from (6.8)
   (a) $P_0(x)$,
   (b) $P_6(x)$.
3. From the definition of a Legendre polynomial show that

   (a) $P_{2n}(0) = (-1)^n \dfrac{(2n)!}{2^{2n}(n!)^2}$,

   (b) $P_{2n+1}(0) = 0$.
4. By first differentiating the Legendre polynomial $P_{2n}(x)$, show that $P'_{2n}(0) = 0$.
5. (a) Determine the polynomial solution $CP_n(x)$ of the differential equation

   $$w = (x^2 - 1)^n$$

   (b) If the positive value is chosen for n in (a) and $y(0) = 2$, what is the solution of the resulting IVP assuming that the solution is a valid one when $x = 1$?
6. Extend the graph for $P_0(x)$, $P_1(x)$, $P_2(x)$, $P_3(x)$, and $P_4(x)$ in Figure 6.1(a) for $-1 \leqslant x < 0$.

6.2. RODRIGUES' FORMULA FOR LEGENDRE POLYNOMIALS

To develop this formula we let

$$(1 - x^2)y'' - 2xy' + 6y = 0$$

Differentiating, we have

$$\frac{dw}{dx} = 2nx(x^2 - 1)^{n-1}$$

This we multiply by $x^2 - 1$, so that

$$(1 - x^2)\frac{dw}{dx} + 2nxw = 0 \tag{6.9}$$

Differentiating (6.9), one obtains

$$(1 - x^2)\frac{d^2w}{dx^2} + 2(n - 1)x\frac{dw}{dx} + 2nw = 0$$

Continuing the differentiation, we see that

$$(1-x^2)\frac{d^3w}{dx^3} + 2(n-2)x\frac{d^2w}{dx^2} + 2(2n-1)\frac{dw}{dx} = 0$$

and

$$(1-x^2)\frac{d^4w}{dx^4} + 2(n-3)x\frac{d^3w}{dx^3} + 3(2n-2)\frac{d^2w}{dx^2} = 0$$

Differentiating $k+1$ times, we find that

$$(1-x^2)\frac{d^2w^{(k)}}{dx^2} + 2(n-k-1)x\frac{dw^{(k)}}{dx} + (k+1)(2n-k)w^{(k)} = 0 \tag{6.10}$$

where

$$w^{(k)} = \frac{d^kw}{dx^k}$$

If we let $k = n$ in (6.10), then

$$(1-x^2)\frac{d^2w^{(n)}}{dx^2} - 2x\frac{dw^{(n)}}{dx} + n(n+1)w^{(n)} = 0 \tag{6.11}$$

From (6.11) we see that $Kw^{(n)}$ is a solution for the Legendre differential equation. Assuming that the Legendre polynomials form a unique polynomial solution set for (6.1) except for multiplicative constants,

$$P_n(x) = K\frac{d^n}{dx^n}(x^2-1)^n \tag{6.12}$$

We investigate the highest power of x for each member of (6.12). We recall that C_n is

$$\frac{(2n)!}{2^n(n!)^2}$$

and

$$\frac{(2n)!}{2^n(n!)^2}x^n = K\frac{d^nx^{2n}}{dx^n} = K(2n)(2n-1)\dots(2n-n+1)x^n$$

$$= K\frac{(2n)!}{n!}x^n$$

Therefore,

$$\frac{(2n)!}{2^n(n!)^2} = K\frac{(2n)!}{n!}$$

and

$$K = \frac{1}{2^n n!}$$

As a result,

$$P_n(x) = \frac{1}{2^n n!} \frac{d^n}{dx^n} (x^2 - 1)^n \qquad (6.13)$$

Equation (6.13) is known as *Rodrigues' formula* for generating Legendre polynomials.

Example 6.1. Show that

$$(2n + 1)P_n(x) = P'_{n+1}(x) - P'_{n-1}(x)$$

From (6.13) we obtain

$$P'_n(x) = \frac{2n}{2^n n!} \frac{d^n}{dx^n} [x(x^2 - 1)^{n-1}]$$

$$= \frac{2n}{2^n n!} \frac{d^{n-1}}{dx^{n-1}} [(2n - 2)x^2(x^2 - 1)^{n-2} + (x^2 - 1)(x^2 - 1)^{n-2}]$$

$$= \frac{1}{2^{n-1}(n - 1)!} \frac{d^{n-1}}{dx^{n-1}} [((2n - 1)x^2 - 1)(x^2 - 1)^{n-2}] \qquad (6.14)$$

Replacing n with $n + 1$ in (6.14), we write

$$P'_{n+1}(x) = \frac{1}{2^n n!} \frac{d^n}{dx^n} [((2n + 1)x^2 - 1)(x^2 - 1)^{n-1}] \qquad (6.15)$$

From (6.13),

$$P_{n-1}(x) = \frac{1}{2^{n-1}(n - 1)!} \frac{d^{n-1}}{dx^{n-1}} (x^2 - 1)^{n-1}$$

The derivative of $P_{n-1}(x)$ is

$$P'_{n-1}(x) = \frac{2n}{2^n n!} \frac{d^n}{dx^n} (x^2 - 1)^{n-1} \qquad (6.16)$$

The difference (6.15) − (6.16) is

$$P'_{n+1}(x) - P'_{n-1}(x) = \frac{1}{2^n n!} \frac{d^n}{dx^n} [((2n + 1)x^2 - 1)(x^2 - 1)^{n-1} - 2n(x^2 - 1)^{n-1}]$$

$$= \frac{2n + 1}{2^n n!} \frac{d^n}{dx^n} (x^2 - 1)^n$$

$$= (2n + 1)P_n(x)$$

or

$$(2n + 1)P_n(x) = P'_{n+1}(x) - P'_{n-1}(x) \qquad (6.17)$$

6.3. A GENERATING FUNCTION FOR $P_n(x)$

It can be shown that the coefficient of t^n in the expansion of

$$[1 - 2xt + t^2]^{-1/2}$$

is the Legendre polynomial $P_n(x)$. We write a few terms of the expansion

$$[1 - t(2x - 1)]^{-1/2} = 1 + \frac{1}{2} t(2x - t) + \frac{1 \cdot 3}{2^2 2!} t^2(2x - t)^2$$

$$+ \frac{1 \cdot 3 \cdot 5}{2^3 3!} t^3(2x - t)^3 + \cdots + \frac{1 \cdot 3 \ldots (2n - 1)}{2^n n!} t^n(2x - t)^n + \cdots$$

The term t^n appears in the term $t^n(2x - t)^n$ and in preceding terms. The coefficient of t^n is a finite series which we display:

$$\frac{1 \cdot 3 \ldots (2n - 1)}{2^n n!} (2x)^n - \frac{1 \cdot 3 \ldots (2n - 3)}{2^{n-1}(n - 1)!} \cdot \frac{n - 1}{1!} (2x)^{n-2}$$

$$+ \frac{1 \cdot 3 \ldots (2n - 5)}{2^{n-2}(n - 2)!} \cdot \frac{(n - 2)(n - 3)}{2!} (2x)^{n-4}$$

$$- \frac{1 \cdot 3 \ldots (2n - 7)}{2^{n-3}(n - 3)!} \cdot \frac{(n - 3)(n - 4)(n - 5)}{3!} (2x)^{n-6} + \cdots$$

By appropriate factorial arithmetic and other simplifications this series has the form:

$$\frac{(2n)!}{2^n n! n!} x^n - \frac{(2n - 2)!}{2^n (n - 1)!(n - 2)!} x^{n-2} + \frac{(2n - 4)!}{2^n 2!(n - 2)!(n - 4)!} x^{n-4}$$

$$- \frac{(2n - 6)!}{2^n 3!(n - 3)!(n - 6)!} x^{n-6} + \cdots \tag{6.18}$$

However, (6.18) represents the first few terms of $P_n(x)$ in (6.8). It is suggestive that

$$[1 - 2xt + t^2]^{-1/2} = \sum_{n=0}^{\infty} P_n(x)t^n \tag{6.19}$$

The expression

$$[1 - 2xt + t^2]^{-1/2}$$

is referred to as a *generating function* for the Legendre polynomial $P_n(x)$.

Example 6.2. Show that

$$(2n + 1)xP_n(x) = (n + 1)P_{n+1}(x) + nP_{n-1}(x)$$

Differentiating (6.19) relative to t, we have

$$(x - t)(1 - 2xt + t^2)^{-3/2} = \sum_{n=0}^{\infty} nP_n(x)t^{n-1} \tag{6.20}$$

If we multiply (6.20) by $1 - 2xt + t^2$, then

$$\sum_{n=0}^{\infty} (x - t)P_n(x)t^n = \sum_{n=0}^{\infty} (1 - 2xt + t^2)nP_n(x)t^{n-1}$$

or

$$\sum_{n=0}^{\infty} xP_n(x)t^n - \sum_{n=0}^{\infty} P_n(x)t^{n+1} = \sum_{n=0}^{\infty} nP_n(x)t^{n-1}$$

$$- \sum_{n=0}^{\infty} 2nxP_n(x)t^n + \sum_{n=0}^{\infty} nP_n(x)t^{n+1}$$

If we equate coefficients of t^n, then

$$xP_n(x) - P_{n-1}(x) = (n + 1)P_{n+1}(x) - 2nxP_n(x) + (n - 1)P_{n-1}(x)$$

Collecting coefficients of $P_n(x)$ and $P_{n-1}(x)$, one obtains

$$(2n + 1)xP_n(x) = (n + 1)P_{n+1}(x) + nP_{n-1}(x) \tag{6.21}$$

Example 6.3. Show that $P_n(1) = 1$.
   From (6.19), if $x = 1$, then

$$(1 - 2t + t^2)^{-1/2} = \sum_{n=0}^{\infty} P_n(1)t^n$$

$$(1 - t)^{-1} = \sum_{n=0}^{\infty} P_n(1)t^n$$

and

$$1 + t + t^2 + t^3 + \cdots + t^n + \cdots = \sum_{n=0}^{\infty} P_n(1)t^n$$

Thus, for all coefficients of t^n,

$$P_n(1) = 1$$

The coefficients (6.5) were assigned so that $P_n(1) = 1$. It can be shown that for $-1 \leqslant x \leqslant 1$.

$$|P_n(x)| \leqslant 1$$

6.4. THE LEGENDRE POLYNOMIAL $P_n(\cos\theta)$

If we replace the independent variable x with θ in (6.1) using the substitution

$$x = \cos\theta$$

then we obtain

$$\frac{dy}{dx} = -\csc\theta\,\frac{dy}{d\theta}$$

and

$$\frac{d^2y}{dx^2} = \csc^2\theta\left[\frac{d^2y}{d\theta^2} - \frac{\cos\theta}{\sin\theta}\frac{dy}{d\theta}\right]$$

The new equation becomes

$$\sin\theta\,\frac{d^2y}{d\theta^2} + \cos\theta\,\frac{dy}{d\theta} + n(n+1)\sin\theta\ y = 0 \tag{6.22}$$

Equation (6.22) has a solution

$$y = P_n(\cos\theta) \tag{6.23}$$

The form of the Legendre differential equation (6.22) with the solution (6.23) is frequently useful for solving BVPs.

Exercises 6.2

1. Solve for $P_{n+1}(x)$ in (6.21) and then determine
   (a) $P_2(x)$, (b) $P_3(x)$, (c) $P_4(x)$.
2. Using Rodrigues' formula, verify $P_0(x)$, $P_1(x)$, and $P_2(x)$.
3. Show that

$$(n+1)P_n(x) = P'_{n+1}(x) - xP'_n(x)$$

4. Establish the formula

$$xP'_n(x) = nP_n(x) + P'_{n-1}(x)$$

5. Show that

$$P_n(-1) = (-1)^n$$

6. Determine that

(a) $P_2(\cos \theta) = \dfrac{3 \cos 2\theta + 1}{4}$,

(b) $P_3(\cos \theta) = \dfrac{5 \cos 3\theta + 3 \cos \theta}{8}$.

7. Find the coefficients so that

$$x^2 = B_0 P_0(x) + B_1 P_1(x) + B_2 P_2(x) + B_3 P_3(x) + \cdots$$

8. Represent x^3 as a linear combination of Legendre polynomials.

9. Show that

$$\int_1^x P_n(t) \, dt = \frac{1}{2n+1} \, [P_{n+1}(x) - P_{n-1}(x)], \quad n \geqslant 1$$

10. Find

(a) $\displaystyle\int_{-1}^1 P_2^2(x) \, dx$,

(b) $\displaystyle\int_{-1}^1 P_1(x) P_3(x) \, dx$.

11. Show that

$$[(1 - x^2) P_n'(x)]' + n(n+1) P_n(x) = 0$$

6.5. ORTHOGONALITY AND NORMS OF $P_n(x)$

The Legendre differential equation (6.1) may be restated in the form

$$[(1 - x^2) y']' + \lambda y = 0 \qquad (6.24)$$

where $\lambda = n(n+1)$. Comparing (6.24) with (2.9) one observes that $p(x) = 1 - x^2$, $q(x) = 0$, and $r(x) = 1$. The λ has already been assigned $n(n+1)$. Therefore, (6.24) is a SLDE. The differential equation is singular at $x = \pm 1$. If we consider an interval $[-1, 1]$, then the SLP is the type discussed in Section 5.6 where $p(-1) = p(1) = 0$. In this type no end condition needs to accompany the differential equation to show orthogonality. Since a solution set of (6.24) is $\{P_n(x)\}$ it is an orthogonal set relative to the weight function $r(x) = 1$. As a result ordinary orthogonality is implied, and

$$\int_{-1}^{1} P_n(x)P_m(x)\, dx = 0 \quad \text{if } m \neq n \tag{6.25}$$

To find the norm of $P_n(x)$, we employ the generating function of Section 6.3. We begin by squaring both members of (6.19), so that

$$[1 - 2xt + t^2]^{-1} = \left[\sum_{n=0}^{\infty} P_n(x)t^n\right]^2 \tag{6.26}$$

By integrating both members of (6.26) relative to x, we obtain

$$\int_{-1}^{1} \frac{dx}{1 - 2xt + t^2} = \sum_{n=0}^{\infty} t^{2n} \int_{-1}^{1} P_n^2(x)\, dx$$

$$-\frac{1}{2t}\left[\ln|1 - 2xt + t^2|\right]_{x=-1}^{1} = \sum_{n=0}^{\infty} t^{2n} \int_{-1}^{1} P_n^2(x)\, dx$$

$$\frac{1}{t} \ln\left|\frac{1+t}{1-t}\right| = \sum_{n=0}^{\infty} t^{2n} \int_{-1}^{1} P_n^2(x)\, dx$$

$$2\left(1 + \frac{t^2}{3} + \frac{t^4}{5} + \cdots + \frac{t^{2n}}{2n+1} + \cdots\right) = \sum_{n=0}^{\infty} t^{2n} \int_{-1}^{1} P_n^2(x)\, dx \tag{6.27}$$

Equating coefficients of t^{2n} in (6.27) one finds that

$$\|P_n(x)\|^2 = \int_{-1}^{1} P_n^2(x)\, dx = \frac{2}{2n+1} \tag{6.28}$$

From (6.28), the norm of $P_n(x)$ is

$$\|P_n(x)\| = \sqrt{\frac{2}{2n+1}}$$

Exercises 6.3

1. Determine

(a) $\displaystyle\int_{-1}^{1} xP_6(x)\, dx$,

(b) $\displaystyle\int_{-1}^{1} P_3(x)P_7(x)\, dx$.

2. Show that

$$\int_{-1}^{1} [AP_0(x) + BP_1(x)]P_n(x)\, dx = 0, \quad \text{if } n = 2, 3, 4, \ldots$$

3. Show that

$$\int_{-1}^{1} P_n(x)\, dx = 0, \quad \text{if } n \in \mathbf{N}$$

4. establish that

$$\int_{-1}^{1} x^2 P_n(x)\, dx = 0, \quad \text{if } n = 3, 4, 5, \ldots$$

5. Show that

$$\int_{0}^{\pi} \sin \theta P_n(\cos \theta) P_m(\cos \theta)\, d\theta = 0, \quad \text{if } m \neq n$$

6. Demonstrate that

$$\|P_n(\cos \theta)\|^2 = \int_{0}^{\pi} \sin \theta P_n^2(\cos \theta)\, d\theta = \frac{2}{2n + 1}$$

6.6. LEGENDRE SERIES

Since we have shown that the set of functions $\{P_n(x)\}$, $-1 \leq x \leq 1$, $n \in \mathbf{N}_0$, is orthogonal, we may construct a series based on the set in the same manner used in Section 2.7. Following the pattern of the previous section, $P_n(x)$ replaces $g_n(x)$, and

$$f(x) \sim \sum_{n=0}^{\infty} C_n P_n(x), \quad -1 \leq x \leq 1 \tag{6.29}$$

where

$$C_n = \frac{1}{\|P_n(x)\|^2} \int_{-1}^{1} f(x) P_n(x)\, dx$$

According to (6.28), we may write

$$C_n = \frac{2n + 1}{2} \int_{-1}^{1} f(x) P_n(x)\, dx \tag{6.30}$$

Thus (6.29) is the Legendre series representation for a function f, and (6.30) is the formula for the coefficients of the series.

A convergence theorem for the Legendre series is discussed by Jackson [27, pp. 65–68]. A similar theorem is included here without proof.

Theorem 6.1. If f is sectionally smooth on the interval $(-1, 1)$ the series (6.29) with its appropriate coefficients (6.30) converges to

$$\frac{f(x+) + f(x-)}{2} \tag{6.31}$$

Example 6.4. (a) Find the Legendre series for

$$f(x) = \begin{cases} 1 & \text{when } -1 < x < 0 \\ 0 & \text{when } 0 < x < 1 \end{cases}$$

(b) Determine the convergence of the series when $x = 0$.
In part (a), the coefficients are

$$C_n = \frac{2n+1}{2} \int_{-1}^{1} f(x) P_n(x)\, dx$$

$$= \frac{2n+1}{2} \int_{-1}^{0} 1 \cdot P_n(x)\, dx + \frac{2n+1}{2} \int_{0}^{1} 0 \cdot P_n(x)\, dx$$

Therefore,

$$C_n = \frac{2n+1}{2} \int_{-1}^{0} P_n(x)\, dx$$

and we compute several coefficients which follow:

$$C_0 = \tfrac{1}{2} \int_{-1}^{0} P_0(x)\, dx = \tfrac{1}{2} \int_{-1}^{0} 1 \cdot dx = \tfrac{1}{2}$$

$$C_1 = \tfrac{3}{2} \int_{-1}^{0} x\, dx = -\tfrac{3}{4}$$

$$C_2 = \tfrac{5}{2} \int_{-1}^{0} \tfrac{1}{2}(3x^2 - 1)\, dx = 0$$

$$C_3 = \tfrac{7}{2} \int_{-1}^{0} \tfrac{1}{2}(5x^3 - 3x)\, dx = \tfrac{7}{16}$$

$$C_4 = \tfrac{9}{2} \int_{-1}^{0} \tfrac{1}{8}(35x^4 - 30x^2 + 3)\, dx = 0$$

$$C_5 = \tfrac{11}{2} \int_{-1}^{0} \tfrac{1}{8}(63x^5 - 70x^3 + 15x)\, dx = -\tfrac{11}{32}$$

Using the computed coefficients, we display a few terms of the Legendre series

$$f(x) \sim \tfrac{1}{2} P_0(x) - \tfrac{3}{4} P_1(x) + \tfrac{7}{16} P_3(x) - \tfrac{11}{32} P_5(x) + \cdots \tag{6.32}$$

In part (b), according to Theorem 6.1, the series converges to the average of the right and left hand limits at 0. Thus, the convergence is $\frac{1}{2}$.

Example 6.5. If $f(x) = x^2$, find the series so that

$$x^2 = \sum_{n=0}^{\infty} C_n P_n(x)$$

According to (6.30),

$$C_n = \frac{2n+1}{2} \int_{-1}^{1} x^2 P_n(x) \, dx$$

Therefore,

$$C_0 = \frac{1}{2} \int_{-1}^{1} x^2 P_0(x) \, dx = \frac{1}{2} \int_{-1}^{1} x^2 \, dx = \frac{1}{3}$$

$$C_1 = \frac{3}{2} \int_{-1}^{1} x^2 \cdot x \, dx = 0$$

$$C_2 = \frac{5}{2} \int_{-1}^{1} \frac{x^2}{2} (3x^2 - 1) \, dx = \frac{2}{3}$$

Using the result of No. 4, Exercises 6.3

$$\int_{-1}^{1} x^2 P_n(x) \, dx = 0, \quad \text{if } n = 3, 4, 5, \ldots$$

Thus the series is written

$$x^2 = \frac{1}{3} P_0(x) + \frac{2}{3} P_2(x)$$

and all other coefficients are zeros.

In this example the function is a polynomial on the interval $-1 \leqslant x \leqslant 1$ and the representation is a truncated series of Legendre polynomials. We observe that x^2 is an even function on the symmetric interval. Only even functions $P_0(x)$ and $P_2(x)$ appear in the expansion. More general ideas concerning even and odd functions follow later in No. 6 of Exercises 6.4.

Exercises 6.4

1. Obtain the Legendre series for $f(x) = 1$ when $-1 < x < 1$. Is the expansion valid for all real x?

2. If $f(x) = |x|$, $-1 < x < 1$, determine the first three nonzero terms of the Legendre series.

3. Determine the Legendre series for x^3.

4. Find the Legendre expansion for

$$f(x) = \begin{cases} 1 & \text{when } -1 < x < 0 \\ x & \text{when } 0 < x < 1 \end{cases}$$

Express the first three nonzero terms of the series.

5. Write the first three nonzero terms for the Legendre series representing

$$f(x) = \begin{cases} 2 & \text{when } -1 < x < 0 \\ 1 & \text{when } 0 < x < 1 \end{cases}$$

6. (a) If f on $(0, 1)$ has an even extension, then show that the Legendre series for f may be expressed

$$f(x) \sim \sum_{n=0}^{\infty} C_{2n} P_{2n}(x), \quad 0 < x < 1$$

where

$$C_{2n} = (4n + 1) \int_0^1 f(x) P_{2n}(x) \, dx$$

(b) When f on $(0, 1)$ has an odd extension, show that the Legendre series representing f may be written

$$f(x) \sim \sum_{n=0}^{\infty} C_{2n+1} P_{2n+1}(x), \quad 0 < x < 1$$

where

$$C_{2n+1} = (4n + 3) \int_0^1 f(x) P_{2n+1}(x) \, dx$$

7. Determine the Legendre series for the function

$$f(x) = \begin{cases} -1 & \text{when } -1 < x < 0 \\ 1 & \text{when } 0 < x < 1 \end{cases}$$

using exercise 6. Also employ in the problem

$$\int_x^1 P_n(t) \, dt = \frac{1}{2n + 1} [P_{n-1}(x) - P_{n+1}(x)]$$

when x is 0 and n is replaced by $2n + 1$. Show that

$$C_{2n+1} = \frac{4n + 3}{2n + 2} \frac{(-1)^n (2n)!}{2^{2n} (n!)^2}$$

8. Using Exercise 6(a), write the first three nonzero terms of the Legendre series if $f(x) = x$, $0 < x < 1$. What function is represented by the same series on $-1 < x < 0$?

6.7. LEGENDRE POLYNOMIALS AND SPHERICAL GEOMETRY

Rectangular, cylindrical, and spherical coordinates of a point P are shown in Figure 6.2. In Section 5.9, we considered cylindrical-rectangular coordinate relations. Formulas relating spherical-rectangular coordinates follow:

$$x = \rho \sin \phi \cos \theta, \quad y = \rho \sin \phi \sin \theta, \quad z = \rho \cos \phi$$

Spherical-cylindrical coordinates are related by

$$r = \rho \sin \phi, \quad \theta = \theta, \quad z = \rho \cos \phi \qquad (6.33)$$

Even though rectangular coordinates seem the most common and fundamental of the three systems, we shall develop the Laplacian in spherical coordinates from spherical-cylindrical coordinate relations. The θ coordinate is unchanged in both cylindrical and spherical systems. In cylindrical coordinates

$$\nabla^2 u = u_{rr} + \frac{1}{r} u_r + \frac{1}{r^2} u_{\theta\theta} + u_{zz} \qquad (6.34)$$

as listed in (5.67). To determine $\nabla^2 u$ in spherical coordinates we must find u_r, u_{rr}, and u_{zz}. The term $u_{\theta\theta}$ will remain unchanged in the spherical system. From (6.33),

$$\rho^2 = z^2 + r^2, \quad \tan \phi = \frac{r}{z}, \quad \theta = \theta$$

$$u_r = u_\rho \rho_r + u_\phi \phi_r$$

If ϕ is a function of r and z, then

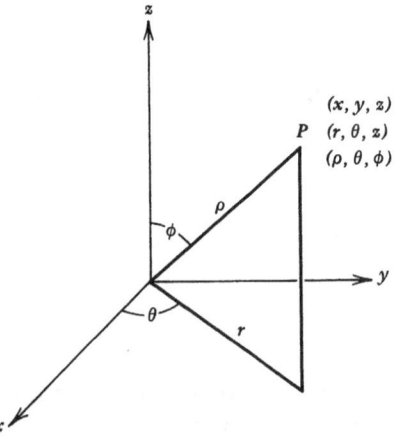

$$(x, y, z)$$
$$P \quad (r, \theta, z)$$
$$(\rho, \theta, \phi)$$

Figure 6.2. Cylindrical and spherical coordinates related to a rectangular system.

$$\frac{1}{z} = \sec^2\phi\,\phi_r$$

$$\phi_r = \frac{\cos^2\phi}{z} = \frac{\cos\phi}{\rho}$$

If ρ is a function of r and z, then

$$2\rho\rho_r = 2r$$

and

$$\rho_r = \frac{r}{\rho} = \sin\phi$$

Therefore,

$$u_r = u_\rho \sin\phi + u_\phi\left(\frac{\cos\phi}{\rho}\right) \tag{6.35}$$

$$u_{rr} = \left[u_\rho \sin\phi + u_\phi\,\frac{\cos\phi}{\rho}\right]_\rho \rho_r + \left[u_\rho \sin\phi + u_\phi\left(\frac{\cos\phi}{\rho}\right)\right]_\phi \phi_r$$

$$= u_{\rho\rho} \sin^2\phi + 2u_{\rho\phi}\,\frac{\cos\phi\sin\phi}{\rho} - 2u_\phi\,\frac{\cos\phi\sin\phi}{\rho^2}$$

$$+ u_\rho\,\frac{\cos^2\phi}{\rho} + u_{\phi\phi}\,\frac{\cos^2\phi}{\rho^2} \tag{6.36}$$

Next,

$$u_z = u_\rho \rho_z + u_\phi \phi_z$$

However,

$$2\rho\rho_z = 2z$$

and

$$\rho_z = \frac{\rho\cos\phi}{\rho} = \cos\phi$$

Now,

$$\sec^2\phi\,\phi_z = -\frac{r}{z^2}$$

and

$$\phi_z = -\frac{r}{z^2} \cos^2\phi = -\frac{\sin\phi}{\rho}$$

Therefore,

$$u_z = u_\rho \cos\phi + u_\phi\left(-\frac{\sin\phi}{\rho}\right)$$

$$u_{zz} = \left[u_\rho \cos\phi + u_\phi\left(-\frac{\sin\phi}{\rho}\right)\right]_\rho \rho_z + \left[u_\rho \cos\phi + u_\phi\left(-\frac{\sin\phi}{\rho}\right)\right]_\phi \phi_z$$

$$= u_{\rho\rho} \cos^2\phi - 2u_{\rho\phi} \frac{\sin\phi \cos\phi}{\rho} + 2u_\phi \frac{\sin\phi \cos\phi}{\rho^2}$$

$$+ u_\rho \frac{\sin^2\phi}{\rho} + u_{\phi\phi} \frac{\sin^2\phi}{\rho^2} \tag{6.37}$$

Adding (6.36) and (6.37), we obtain

$$u_{rr} + u_{zz} = u_{\rho\rho} + \frac{1}{\rho} u_\rho + \frac{1}{\rho^2} u_{\phi\phi} \tag{6.38}$$

According to (6.35) and basic relations,

$$\frac{1}{r} u_r = \frac{1}{\rho \sin\phi}\left[u_\rho \sin\phi + u_\phi \frac{\cos\phi}{\rho}\right] = \frac{1}{\rho} u_\rho + \frac{\cot\phi}{\rho^2} u_\phi \tag{6.39}$$

$$\frac{1}{r^2} u_{\theta\theta} = \frac{1}{\rho^2 \sin^2\phi} u_{\theta\theta} \tag{6.40}$$

From (6.38), (6.39), and (6.40),

$$\nabla^2 u = u_{\rho\rho} + \frac{2}{\rho} u_\rho + \frac{1}{\rho^2 \sin^2\phi} u_{\theta\theta} + \frac{1}{\rho^2} u_{\phi\phi} + \frac{\cot\phi}{\rho^2} u_\phi \tag{6.41}$$

If $\nabla^2 u$ depends on ρ alone, then

$$\nabla^2 u = u_{\rho\rho} + \frac{2}{\rho} u_\rho$$

From (6.41), the Laplace differential equation in spherical coordinates may be expressed as

$$u_{\rho\rho} + \frac{2}{\rho} u_\rho + \frac{1}{\rho^2} u_{\phi\phi} + \frac{\cot\phi}{\rho^2} u_\phi + \frac{1}{\rho^2 \sin^2\phi} u_{\theta\theta} = 0 \tag{6.42}$$

As an exercise it is left for the reader to show the equivalence of (6.42) and

$$\rho(\rho u)_{\rho\rho} + \frac{1}{\sin \phi} (\sin \phi u_\phi)_\phi + \frac{1}{\sin^2\phi} u_{\theta\theta} = 0 \qquad (6.43)$$

or

$$(\rho^2 u_\rho)_\rho + \frac{1}{\sin \phi} (\sin \phi u_\phi)_\phi + \frac{1}{\sin^2\phi} u_{\theta\theta} = 0 \qquad (6.44)$$

6.8. SPHERICAL HARMONICS

If $f(rx, ry, rz) = r^n f(x, y, z)$ where r is a constant, then $f(x, y, z)$ is *homogeneous of degree n*. *Harmonics* are the solutions of the Laplace differential equation. A function $V_n(x, y, z)$ is referred to as a *solid spherical harmonic of degree n* if $\nabla^2 V_n = 0$ and $V_n(x, y, z)$ is homogeneous of degree n. If V_n is a solid spherical harmonic of degree n, then there is another function S_n, referred to as a *spherical surface harmonic*, such that

$$V_n = \rho^n S_n \qquad (6.45)$$

If (6.45) is substituted into (6.43) and the result divided by ρ^n, one obtains the differential equation for S_n, a function of ϕ and θ alone. Thus

$$\frac{1}{\sin^2\phi} \frac{\partial^2 S_n}{\partial \theta^2} + \frac{1}{\sin \phi} \frac{\partial}{\partial \phi} \left(\sin \phi \frac{\partial S_n}{\partial \phi} \right) + n(n+1)S_n = 0 \qquad (6.46)$$

A *solid zonal harmonic* is a spherical harmonic which is expressed as a function of ρ and z. A *surface zonal harmonic of degree n* is a solid zonal harmonic of degree n divided by ρ^n. If Z_n is a *surface zonal harmonic* in spherical coordinates it does not contain θ and (6.46) becomes

$$\frac{1}{\sin \phi} \frac{\partial}{\partial \phi} \left(\sin \phi \frac{\partial Z_n}{\partial \phi} \right) + n(n+1)Z_n = 0 \qquad (6.47)$$

If $x = \cos \phi$, then (6.47) may be expressed as an ODE

$$(1 - x^2) \frac{d^2 Z_n}{dx^2} - 2x \frac{dZ_n}{dx} + n(n+1)Z_n = 0 \qquad (6.48)$$

This transformation is considered in the exercises. In (6.48) Z_n is the Legendre polynomial $P_n(x)$.

Exercises 6.5

1. Find the degree of homogeneity for
   (a) $ax + by + cz,$
   (b) $xyz,$
   (c) $x^2 + y^2 + z^2.$
2. After multiplying (6.42) by ρ^2 and recognizing certain partial derivative forms, show that the Laplace differential equation may be expressed as (6.43) or (6.44).
3. If V_n is a solution for (6.43) show that the differentiation of (6.45) with respect to ρ becomes

$$\rho(\rho V_n)_{\rho\rho} = n(n+1)\rho^n S_n \qquad (6.49)$$

4. Show that the differential equation for S_n is (6.46).
5. Show that if $x = \cos \phi$, (6.47) becomes (6.48) which is a Legendre differential equation with a solution $P_n(x)$.

6.9. THE GENERALIZED LEGENDRE EQUATION

By differentiating the equation

$$(1 - x^2)y''(x) - 2xy'(x) + n(n+1)y(x) = 0 \qquad (6.50)$$

m times relative to x and then replacing $y^{(m)}(x)$ with a new variable u, one obtains

$$(1 - x^2)u''(x) - 2x(m+1)u'(x) + (n-m)(n+m+1)u(x) = 0 \qquad (6.51)$$

Apparently,

$$u(x) = \frac{d^m P_n(x)}{dx^m}$$

satisfies (6.51). Letting $v = u(1 - x^2)^{m/2}$, we find that (6.51) becomes

$$(1 - x^2)v''(x) - 2xv'(x) + \left[n(n+1) - \frac{m^2}{1 - x^2} \right]v(x) = 0 \qquad (6.52)$$

This equation (6.52) is called the *associated Legendre differential equation.* Its solution may be written

$$v(x) = (1 - x^2)^{m/2} \frac{d^m P_n(x)}{dx^m}$$

which is known as an *associated Legendre polynomial* and is designated by $P_n^m(x)$. If $m > n$, then

$$P_n^m(x) = 0$$

If $m = 0$, then the differential equation (6.52) becomes (6.50) again. The functions $P_n^m(x)$ are said to be of *degree n and order m*.

It is possible to match the equation (6.52) with the SLDE since it can be written

$$[(1 - x^2)v']' + \left[n(n + 1) - \frac{m^2}{1 - x^2} \right] v = 0 \qquad (6.53)$$

with $p(x) = 1 - x^2$, $q(x) = m^2/(1 - x^2)$, $r(x) = 1$, and $\lambda = n(n + 1)$. We observe that (6.53) has singular points at $x = \pm 1$ and $p(1) = p(-1) = 0$. The SLP is the type discussed in Section 5.6. No end condition is required to show orthogonality. Consequently,

$$\int_{-1}^{1} P_n^m(x) P_k^m(x) \, dx = 0, \quad \text{if } n \neq k \qquad (6.54)$$

The norm squared is

$$\int_{-1}^{1} [P_n^m(x)]^2 \, dx = \frac{2}{2n + 1} \frac{(n + m)!}{(n - m)!} \qquad (6.55)$$

as shown in Whitaker and Watson [53, pp. 324–325].

Relations (6.54) and (6.55) allow us to write expansions for certain functions f. If

$$f(x) = \sum_{n=0}^{\infty} C_n P_n^m(x), \quad -1 < x < 1$$

then

$$\int_{-1}^{1} f(x) P_n^m(x) \, dx = C_n \int_{-1}^{1} [P_n^m(x)]^2 \, dx = \frac{2}{2n + 1} \frac{(n + m)!}{(n - m)!} C_n$$

Therefore,

$$C_n = \frac{(2n + 1)(n - m)!}{2(n + m)!} \int_{-1}^{1} f(x) P_n^m(x) \, dx$$

Without asserting conditions for convergence, we express the representation,

$$f(x) \sim \sum_{n=0}^{\infty} C_n P_n^m(x), \quad -1 < x < 1$$

where

$$C_n = \frac{(2n+1)(n-m)!}{2(n+m)!} \int_{-1}^{1} f(x) P_n^m(x) \, dx$$

The associated Legendre functions can be generated from the Legendre functions $P_n(x)$ and $Q_n(x)$. We state two relations:

$$P_n^m(x) = (1 - x^2)^{m/2} \frac{d^m}{dx^m} P_n(x)$$

$$Q_n^m(x) = (1 - x^2)^{m/2} \frac{d^m}{dx^m} Q_n(x)$$

where $Q_n^m(x)$ is called the *associated Legendre function of the second kind*.

Example 6.6. Determine the associated Legendre function $P_3^2(x)$.
From Section 6.1,

$$P_3(x) = \tfrac{1}{2}(5x^3 - 3x)$$

$$P_3^2(x) = (1 - x^2)^{2/2} \frac{d^2}{dx^2} P_3(x) = (1 - x^2) \frac{d}{dx} \left[\frac{1}{2}(15x^2 - 3) \right]$$

$$= (1 - x^2)\left(\frac{1}{2}\right)(30x) = 15x(1 - x^2)$$

Example 6.7. Determine $P_2^4(x)$.
According to Section 6.1,

$$P_2(x) = \tfrac{1}{2}(3x^2 - 1)$$

$$P_2^4(x) = (1 - x^2)^{4/2} \frac{d^4}{dx^4} P_2(x) = (1 - x^2)^2 \frac{d^4}{dx^4} \left[\tfrac{1}{2}(3x^2 - 1) \right] = 0$$

This result could have been written immediately from the statement that if $m > n$, $P_n^m(x) = 0$.

Example 6.8. Investigate a solution for the Laplace equation in spherical form

$$\nabla^2 u = u_{\rho\rho} + \frac{2}{\rho} u_\rho + \frac{1}{\rho^2 \sin \phi} (u_\phi \sin \phi)_\phi + \frac{1}{\rho^2 \sin^2 \phi} u_{\theta\theta} = 0$$

1. *Separation of Variables.* Let $u(\rho, \theta, \phi) = R(\rho)\Theta(\theta)\Phi(\phi)$. Then

$$R''\Theta\Phi + \frac{2}{\rho} R'\Theta\Phi + \frac{1}{\rho^2 \sin \phi} (R\Theta\Phi' \sin \phi)_\phi + \frac{1}{\rho^2 \sin^2\phi} R\Theta''\Phi = 0$$

Multiplying by $(\rho^2 \sin^2\phi)/R\Theta\Phi$, we obtain

$$\frac{R'' + \frac{2}{\rho} R'}{R} \rho^2 \sin^2\phi + \frac{(\Phi' \sin \phi)' \sin \phi}{\Phi} = -\frac{\Theta''}{\Theta}$$

and assign

$$\frac{\Theta''}{\Theta} = -\alpha^2$$

Then,

$$\Theta'' + \alpha^2\Theta = 0$$

and

$$\frac{\rho^2 R'' + 2\rho R'}{R} \sin^2\phi + \frac{(\Phi' \sin \phi)' \sin \phi}{\Phi} = \alpha^2$$

If we let

$$\frac{\rho^2 R'' + 2\rho R'}{R} = n(n+1)$$

then

$$\rho^2 R'' + 2\rho R' - n(n+1)R = 0$$

Finally,

$$n(n+1) \sin^2\phi + \frac{(\Phi' \sin \phi)' \sin \phi}{\Phi} = \alpha^2$$

and

$$(\Phi' \sin \phi)' \sin \phi + [n(n+1) \sin^2\phi - \alpha^2]\Phi = 0$$

2. *The Related ODEs:*

$$\Theta'' + \alpha^2\Theta = 0$$
$$\rho^2 R'' + 2\rho R' - n(n+1)R = 0$$
$$\sin \phi(\Phi' \sin \phi)' + [n(n+1) \sin^2\phi - \alpha^2]\Phi = 0$$

3. *The Θ Equation.* The solution of

$$\Theta'' + \alpha^2\Theta = 0$$

is

$$\Theta(\theta) = A_1 \cos \alpha\theta + A_2 \sin \alpha\theta$$

For $\Theta(\theta)$ to be a periodic function we need

$$\Theta(\theta + 2\pi) = \Theta(\theta)$$

or

$$A_1 \cos \alpha(\theta + 2\pi) + A_2 \sin \alpha(\theta + 2\pi) = A_1 \cos \alpha\theta + A_2 \sin \alpha\theta$$

This condition is true if

$$A_1 \cos (\alpha\theta + 2\pi\alpha) = A_1 \cos \alpha\theta$$
$$A_2 \sin (\alpha\theta + 2\pi\alpha) = A_2 \sin \alpha\theta \tag{6.56}$$

The statements of (6.56) are valid if $\alpha = m$ and $\alpha^2 = m^2$, $m \in \mathbf{N}_0$. Therefore,

$$\Theta_m(\theta) = A_1 \cos m\theta + A_2 \sin m\theta , \quad m \in \mathbf{N}_0 \tag{6.57}$$

4. *The R Equation.* The solution of

$$\rho^2 R'' + 2\rho R' - n(n+1)R = 0$$

will be given in (8.85). We duplicate it for reference here:

$$R(\rho) = B_1\rho^n + \frac{B_2}{\rho^{n+1}} , \quad \rho \neq 0 \tag{6.58}$$

5. *The Φ Equation.* In the equation with constants inserted,

$$\sin \phi(\Phi' \sin \phi)' + [n(n+1) \sin^2\phi - m^2]\Phi = 0 \tag{6.59}$$

We let $\gamma = \cos \phi$, so that (6.59) becomes

$$(1 - \gamma^2) \frac{d}{d\gamma} \left[(1 - \gamma^2) \frac{d\Phi}{d\gamma}\right] + [n(n+1)(1 - \gamma^2) - m^2]\Phi = 0$$

or

$$\frac{d}{d\gamma} \left[(1 - \gamma^2) \frac{d\Phi}{d\gamma}\right] + \left[n(n+1) - \frac{m^2}{1 - \gamma^2}\right]\Phi = 0 \tag{6.60}$$

Equation (6.60) is Legendre's associated differential equation

$$(1 - \gamma^2) \frac{d^2\Phi}{d\gamma^2} - 2\gamma \frac{d\Phi}{d\gamma} + \left[n(n+1) - \frac{m^2}{1 - \gamma^2} \right] \Phi = 0 \quad (6.61)$$

A solution for (6.61) is

$$\Phi(\gamma) = P_n^m(\gamma) = (1 - \gamma^2)^{m/2} \frac{d^m}{d\gamma^m} P_n(\gamma)$$

or

$$\Phi(\phi) = C_1 P_n^m(\cos \phi) \quad (6.62)$$

Therefore a solution of $\nabla^2 u = 0$ may be expressed as

$$u(\rho, \theta, \phi) = R(\rho)\Theta(\theta)\Phi(\phi)$$

where according to (6.57), (6.58), and (6.62)

$$R(\rho) = B_1 \rho^n + \frac{B_2}{\rho^{n+1}}, \quad \rho \neq 0$$

$$\Theta(\theta) = A_1 \cos m\theta + A_2 \sin m\theta$$

$$\Phi(\phi) = C_1 P_n^m(\cos \phi)$$

Exercises 6.6

1. Find the associated Legendre functions
   (a) $P_3^1(x)$, (b) $P_4^2(x)$, (c) $Q_0^2(x)$, (d) $P_3^5(x)$.
2. Determine a series for the function $x(1 - x^2)^{1/2}$ in the form $\sum_{n=0}^{\infty} C_n P_n^m(x)$ where $m = 1$.
3. If we let $x = \cos \phi$, so that $\partial x/\partial \phi = -\sin \phi$, show that the differential equation for S_n, (6.46) can be transformed into the equation

$$\frac{1}{1 - x^2} \frac{\partial^2 S_n}{\partial \theta^2} + \frac{\partial}{\partial x} \left[(1 - x^2) \frac{\partial S_n}{\partial x} \right] + n(n+1)S_n = 0 \quad (6.63)$$

4. Using separation of variables to solve (6.63) if $S_n(x, \theta) = X(x)\Theta(\theta)$ and $\Theta''/\Theta = -m^2$, then the ODEs become

$$\Theta'' + m^2\Theta = 0 \quad \text{and} \quad [(1 - x^2)X']' + \left[n(n+1) - \frac{m^2}{1 - x^2} \right]X = 0$$

   Notice that the second ODE is the equivalent of (6.52).
5. Show that
   (a) $X(\cos \phi) = CP_n^m(\cos \phi)$.
   (b) $S_n(\phi, \theta) = (A \cos m\theta + B \sin m\theta)P_n^m(\cos \phi)$ is the solution for the spherical surface harmonics.
   (c) $V_n(\rho, \phi, \theta) = \rho^n(A \cos m\theta + B \sin m\theta)P_n^m(\cos \phi)$ is the solution for the corresponding solid spherical harmonics of degree n.

7

INTEGRAL TRANSFORMS

We have encountered the idea of an operator or a transformation applied to a function to produce another function in Section 1.1. The introduction of integral transforms can sometimes simplify manipulations. For example, certain linear differential equations can be converted to algebraic equations by means of transforms. An integral transformation is defined by

$$T\{f(t)\} = \int_a^b K(s, t)f(t)\, dt = F(s) \tag{7.1}$$

A function f is transformed into another function F by this integral. The definition of a particular integral transformation is determined once the *kernel K* and the limits of integration are given. For example, if the kernel is e^{-st} with limits $a = 0$ and $b = \infty$ then $F(s)$ is the *Laplace transformation* of $f(t)$ which we will discuss in a moment. Usually the function being transformed is represented by a lower case letter and the transformed function by the same capital letter or by an operator symbol such as $\mathscr{L}\{\ \}$ which is used for the Laplace transform. The transformations defined by (7.1) satisfy the condition (1.1) for a *linear operator*. That is,

$$T\{c_1 f(t) + c_2 g(t)\} = c_1 T\{f(t)\} + c_2 T\{g(t)\} \tag{7.2}$$

This property is sometimes convenient for termwise operations with transforms.

7.1. LAPLACE TRANSFORMS

The *Laplace transform* is defined by the equation

$$\mathscr{L}\{f(t)\} = \int_0^\infty e^{-st}f(t)\, dt = F(s) \tag{7.3}$$

Example 7.1. Determine $F(s)$ if $f(t) = e^{kt}$. Substituting into the definition (7.3), we have

$$F(s) = \int_0^\infty e^{-st}e^{kt}\,dt = \int_0^\infty e^{-(s-k)t}\,dt = \frac{e^{-(s-k)t}}{k-s}\Big|_0^\infty = \frac{1}{s-k}, \quad \text{if } s-k>0$$

Although the improper integral (7.3) converges for the example above, this is not always the case. Consider the following.

Example 7.2. Attempt to find $\mathcal{L}\{1/t^2\}$.

$$\mathcal{L}\left\{\frac{1}{t^2}\right\} = \int_0^\infty \frac{e^{-st}}{t^2}\,dt = \int_0^1 \frac{e^{-st}}{t^2}\,dt + \int_1^\infty \frac{e^{-st}}{t^2}\,dt$$

When $0 \le t \le 1$, $e^{-st} \ge e^{-s}$ if $s > 0$. Therefore,

$$\int_0^\infty \frac{e^{-st}}{t^2}\,dt \ge \int_0^1 \frac{e^{-s}}{t^2}\,dt + \int_1^\infty \frac{e^{-st}}{t^2}\,dt$$

$$\int_0^\infty \frac{e^{-st}}{t^2}\,dt \ge^{-s} \int_0^1 \frac{dt}{t^2} + \int_1^\infty \frac{e^{-st}}{t^2}\,dt$$

But $\int_0^1 1/t^2\,dt$ diverges and hence $\mathcal{L}\{1/t^2\}$ fails to converge. As a result $1/t^2$ fails to have a Laplace transform.

7.2. EXISTENCE OF THE TRANSFORM

PWC functions are discussed at the beginning of Chapter 3. A function $f(t)$ is of *exponential order as* $t \to \infty$ if nonnegative constants k, M, and T exist so that $|f(t)| \le Me^{kt}$ for all $t \ge T$. Another way of expressing this condition is to say $f(t)$ *is of the order* e^{kt}, written $f(t) = O(e^{kt})$. The Laplace transform exists if $f(t)$ is PWC when $t \ge 0$ and $f(t) = O(e^{kt})$. To show this is true, we observe that if $f(t)$ is PWC, then $e^{-st}f(t)$ is integrable over the interval $[0, T]$ for any $T > 0$, and

$$\int_T^\infty |e^{-st}f(t)|\,dt \le M \int_T^\infty e^{-st}e^{kt}\,dt = \frac{Me^{-T(s-k)}}{s-k}$$

When $s > k$, the last expression approaches zero as $t \to \infty$. Therefore $\int_T^\infty |e^{-st}f(t)|\,dt$ is convergent if $s > k$ and hence $\mathcal{L}\{f(t)\}$ exists.

If a constant k exists so that the limit of $e^{-kt}|f(t)|$ exists as $t \to \infty$, then $f(t) = O(e^{kt})$. If this limit is infinite, then $f(t)$ is not of exponential order. All bounded functions are of exponential order with $k = 0$. However e^{t^2} is not of exponential order.

We emphasize that the existence statement is one of *sufficiency* and not one of *necessity*. Consider the following.

Example 7.3. Is the function $f(t) = t^{-1/2}$ PWC? Determine $\mathscr{L}\{t^{-1/2}\}$ if the transform exists.

This function is not PWC for all $t \geqslant 0$ since $t^{-1/2} \to \infty$ as $t \to 0^+$. Investigate

$$\mathscr{L}\{t^{-1/2}\} = \int_0^\infty e^{-st} t^{-1/2} \, dt$$

Let us make the change of variable $t = y^2$ followed by a second change $\sqrt{s}y = x$. Then

$$\mathscr{L}\{t^{-1/2}\} = 2 \int_0^\infty e^{-sy^2} \, dy = 2s^{-1/2} \int_0^\infty e^{-x^2} \, dx$$

Using (9.67) with $b = 1$, we obtain $\int_0^\infty e^{-x^2} \, dx = \frac{1}{2}\sqrt{\pi}$. Therefore,

$$\mathscr{L}\{t^{-1/2}\} = \sqrt{\frac{\pi}{s}}$$

Even though the condition of PWC is violated, the Laplace transform of $t^{-1/2}$ exists. Thus PWC is not a necessary condition for existence of the transform.

7.3. THE GAMMA FUNCTION AND LAPLACE TRANSFORMS

Using the definition for the *gamma function* (5.2), we have

$$\Gamma(\alpha + 1) = \int_0^\infty e^{-\beta} \beta^\alpha \, d\beta$$

If $\beta = st$, $s > 0$, then

$$\Gamma(\alpha + 1) = \int_0^\infty e^{-st} s^{\alpha+1} t^\alpha \, dt = s^{\alpha+1} \int_0^\infty e^{-st} t^\alpha \, dt$$

Therefore if $\alpha > -1$ and $s > 0$,

$$\frac{\Gamma(\alpha + 1)}{s^{\alpha+1}} = \int_0^\infty e^{-st} t^\alpha \, dt$$

and

$$\mathscr{L}\{t^\alpha\} = \frac{\Gamma(\alpha + 1)}{s^{\alpha+1}}$$

In Section 5.1, it is pointed out that the integral (5.2) and the factorial have common values when $\alpha \in N$. It follows that

$$\mathcal{L}\{t^n\} = \frac{n!}{s^{n+1}}$$

for $n \in N$ and $s > 0$.

Exercises 7.1

1. As $t \to \infty$ test $f(t)$ for $O(e^{kt})$ if (a) $f(t) = t^2$, (b) $f(t) = e^{t^2}$.
2. Find $\mathcal{L}\{1\}$.
3. Determine $\mathcal{L}\{\sin kt\}$ using result (14) in the Selected Integrals in Appendix 1.
4. Obtain $\mathcal{L}\{\sin kt\}$ using the complex variable definition given in Section 2.3.
5. Determine $\mathcal{L}\{\cos kt\}$ using result (15) in the Selected Integrals in Appendix 1.
6. Obtain (a) $\mathcal{L}\{\sinh kt\}$, (b) $\mathcal{L}\{\cosh kt\}$. Use the exponential definitions for $\sinh kt$ and $\cosh kt$.
7. (a) Without using the gamma function, show that if $s > 0$, then

$$\mathcal{L}\{t^n\} = \frac{n!}{s^{n+1}}$$

   (b) Find $\mathcal{L}\{t^5\}$.
8. Compute (a) $\mathcal{L}\{t^{1/2}\}$, (b) $\mathcal{L}\{t^{3/2}\}$, (c) $\mathcal{L}\{t^{5/2}\}$. See (5.3).
9. (a) Assume that $F(s) = \mathcal{L}\{f(t)\}$ exists for $s > k$. Show that if $k > 0$, then

$$\mathcal{L}\{f(kt)\} = \frac{1}{k} F\left(\frac{s}{k}\right)$$

   (b) Show that $F(\alpha s) = \frac{1}{\alpha} \mathcal{L}\left\{f\left(\frac{t}{\alpha}\right)\right\}$.

   (c) Find $\mathcal{L}\left\{\sin \frac{t}{k}\right\}$.

10. Does $\mathcal{L}\{1/t^3\}$ exist? From the definition of a Laplace transform, attempt to resolve the problem.
11. Using the properties for integrals, establish that a Laplace transform is a linear operator. That is,

$$\mathcal{L}\{c_1 f(t) + c_2 g(t)\} = c_1 \mathcal{L}\{f(t)\} + c_2 \mathcal{L}\{g(t)\}$$

7.4. TRANSFORMS OF DERIVATIVES

If a function $f(t)$ is PWC and of exponential order, there is no assurance that $f'(t)$ will meet the two conditions. When we seek $\mathcal{L}\{f'(t)\}$ we shall specify that $f'(t)$ be PWC and $f'(t) = O(e^{kt})$, as well as $f(t)$ continuous and $f(t) = O(e^{kt})$. Using integration by parts,

$$\mathcal{L}\{f'(t)\} = \int_0^\infty e^{-st} f'(t) \, dt = [e^{-st} f(t)]_0^\infty + s \int_0^\infty e^{-st} f(t) \, dt$$

$$\mathcal{L}\{f'(t)\} = s\mathcal{L}\{f(t)\} - f(0) \tag{7.4}$$

If $f''(t)$ is now restricted as $f'(t)$ was in (7.4) with $f'(t)$ and $f(t)$ continuous and of exponential order, then

$$\mathcal{L}\{f''(t)\} = s\mathcal{L}\{f'(t)\} - f'(0) = s[s\mathcal{L}\{f(t)\} - f(0)] - f'(0)$$

$$\mathcal{L}\{f''(t)\} = s^2\mathcal{L}\{f(t)\} - sf(0) - f'(0) \tag{7.5}$$

Assume $f(t), f'(t), \ldots, f^{(n-1)}(t)$ are continuous when $t > 0$ and of exponential order as $t \to \infty$. Also assume that $f^{(n)}(t)$ is PWC and of exponential order. Repeated use of the above process allows us to state,

$$\mathcal{L}\{f^{(n)}(t)\} = s^n\mathcal{L}\{f(t)\} - s^{n-1}f(0) - s^{n-2}f'(0) - \cdots - f^{(n-1)}(0) \tag{7.6}$$

Example 7.4. Use (7.6) to find $\mathcal{L}\{t^3\}$.
    Let $f(t) = t^3$, $f'(t) = 3t^2$, $f''(t) = 6t$, $f'''(t) = 6$, and apply (7.6) to obtain

$$\mathcal{L}\{6\} = s^3\mathcal{L}\{t^3\} - s^2 \cdot 0 - s \cdot 0 - 0$$

$$6\mathcal{L}\{1\} = s^3\mathcal{L}\{t^3\}$$

$$\mathcal{L}\{t^3\} = \frac{3!}{s^4}$$

7.5. DERIVATIVES OF TRANSFORMS

If $F(s) = \int_0^\infty e^{-st} f(t) \, dt$ and differentiation under the integral is permissible, then

$$F'(s) = \int_0^\infty (-t)e^{-st} f(t) \, dt = \int_0^\infty e^{-st} [-tf(t)] \, dt$$

Therefore,

$$\mathcal{L}\{-tf(t)\} = F'(s) \quad \text{and} \quad \mathcal{L}\{t^2f(t)\} = F''(s)$$

Continuing this process, we have

$$\mathcal{L}\{(-t)^n f(t)\} = F^{(n)}(s)$$

Example 7.5. Obtain $\mathcal{L}\{t^2 e^{3t}\}$.

$$\mathcal{L}\{t^2 e^{3t}\} = \frac{d^2}{ds^2}\left[\frac{1}{s-3}\right] = \frac{d}{ds}\left[-\frac{1}{(s-3)^2}\right] = \frac{2}{(s-3)^3}$$

7.6. THE INVERSE LAPLACE TRANSFORM

For certain procedures it is desirable to find $f(t)$ when $F(s)$ is known. This is known as the *inversion problem* for the Laplace transform. If $\mathcal{L}\{f(t)\} = F(s)$, we indicate that $f(t)$ is an *inverse Laplace transform* of $F(s)$ by the notation

$$f(t) = \mathcal{L}^{-1}\{F(s)\}$$

The complex integral,

$$f(t) = \frac{1}{2\pi i}\int_{c-i\infty}^{c+i\infty} e^{st}F(s)\,ds$$

is an inverse of $F(s)$. For those wishing to use this representation, see Churchill [14, pp. 193–195].

A transform does not in general have a unique inverse, since, for example, functions which differ at only one point will have the same transform. However, any transform can have at most one continuous inverse, and in general inverses of the same transform can differ only at their points of discontinuity (see Churchill [14, pp. 201–202]). For our purposes an effective method for determining inverse transforms is to use a table of transforms such as Appendix 2.

7.7. SOLUTIONS OF ODEs AND IVPs

Linear ODEs with constant coefficients and their corresponding IVPs can be converted into algebraic equations by taking the Laplace transform of both sides of the differential equation. Since transforms of derivatives involve initial values of the original function and its derivatives, the procedure is a natural one for solving IVPs. The *inverse transform* of the solution of the algebraic equation is the solution of the ODE or IVP. We illustrate the process with the following example.

Example 7.6. Find the solution for the initial value problem

$$y''(t) + y'(t) - 6y(t) = e^t \quad \text{with } y(0) = 1, \quad y'(0) = 0$$

Using Laplace transforms of both sides of the equation, we have

$$s^2 Y(s) - s + sY(s) - 1 - 6Y(s) = \frac{1}{s-1}$$

$$(s^2 + s - 6)Y(s) = s + 1 + \frac{1}{s-1} = \frac{(s+1)(s-1)+1}{s-1}$$

$$Y(s) = \frac{s^2}{(s-1)(s-2)(s+3)} = \frac{A}{s-1} + \frac{B}{s-2} + \frac{C}{s+3}$$

Using partial fraction expansions, one obtains

$$Y(s) = -\frac{-1}{4(s-1)} + \frac{4}{5(s-2)} + \frac{9}{20(s+3)}$$

Using the table of Laplace transforms from Appendix 2, we have

$$y(t) = -\frac{1}{4} e^t + \frac{4}{5} e^{2t} + \frac{9}{20} e^{-3t}$$

We assumed initially that the solution has a Laplace transform. Once we find what we believe is the solution our assumption should be justified by inserting it in the differential equation and the initial conditions. This is left as an exercise.

7.8. PARTIAL FRACTIONS

In our brief exposure to solving an ODE using Laplace transforms we have seen the need to find the inverse transformation of a rational fraction $F(s)/G(s)$. The numerator of the rational fraction contains a polynomial of degree smaller that the degree of the polynomial denominator. The fraction has the partial fraction expansion as used in calculus. We are able to replace a complicated expression in the transform with a sum of simpler transforms and then use the linearity of the inverse. These simpler transforms may have inverses that are recognizable. Formulas are available for partial fraction expansions of rational fractions. Except for the exercises below we will not discuss any other formulas.

Exercises 7.2

1. Find (a) $\mathcal{L}\{t \cosh 3t\}$, (b) $\mathcal{L}\{t^2 \sin 3t\}$.
2. Determine $\mathcal{L}\{\cos kt\}$ using (7.5).

3. Obtain (a) $\mathcal{L}\{t^2 e^{3t}\}$, (b) $\mathcal{L}\{t^2 \sinh t\}$.

4. Find

(a) $\mathcal{L}^{-1}\left\{ \dfrac{s^2 - 8s + 15}{(s-1)(s-2)(s-3)} \right\}$,

(b) $\mathcal{L}^{-1}\left\{ \dfrac{s^2 - 7s + 4}{(s-1)(s^2 + 4)} \right\}$.

Using Laplace transforms solve the following ODEs and IVPs.

5. $y''(t) + 3y'(t) + 2y(t) = 0$, $y(0) = 0$, $y'(0) = 1$.
6. $y''(t) - y'(t) - 6y(t) = 0$. For a general solution, let $y(0) = A$, $y'(0) = B$.
7. $y''(t) + y(t) = \sin t$, $y(0) = 0$, $y'(0) = 1$.
8. $y''(t) - 3y'(t) + 2y(t) = e^{-t}$ (see Exercise 6).
9. $y^{(4)}(t) - y(t) = 0$, $y(0) = 1$, $y'(0) = 0$, $y''(0) = 0$, $y'''(0) = 1$.
10. $y''(t) - y'(t) + 4y(t) = 0$, $y(0) = 1$, $y'(0) = 1$.
11. $y'''(t) + 3y''(t) + 3y'(t) + y(t) = 1$, $y(0) = 1$, $y'(0) = 0$, $y''(0) = -2$.
12. $y''(t) + y(t) = 0$, $y(\pi/2) = 2$, $y(0) = 0$ (let $y'(0) = c$ and find c).
13. Suppose $F(s)$ and $G(s)$ are polynomials in s with no common factors, the degree of $F(s)$ is less than the degree of $G(s)$, and $G(s)$ has distinct zeros $r_1, r_2, \ldots, r_n$. Prove the *Heaviside expansion formula*

$$\mathcal{L}^{-1}\left\{ \frac{F(s)}{G(s)} \right\} = \sum_{k=1}^{n} e^{r_k t} \frac{F(r_k)}{G'(r_k)}$$

Hint: Expand $F(s)/G(s)$ in partial fractions and multiply both sides of the corresponding equation by $s - r_k$. Then use l'Hospital's rule to show that

$$\frac{s - r_k}{G(s)} \to \frac{1}{G'(r_k)} \quad \text{as } s \to r_k$$

Hence

$$\frac{s - r_k}{G(s)} F(s) \to \frac{F(r_k)}{G'(r_k)} \quad \text{as } s \to r_k$$

Use this to determine the coefficients of the partial fraction of $F(s)/G(s)$.

7.9. THE UNIT STEP FUNCTION

The function U defined by $U(t - k) = 0$ if $t < k$ and $U(t - k) = 1$ if $t \geq k$ is referred to as the *unit step function of Heaviside* (Figure 7.1). The function is PWC and bounded. Therefore, it has a Laplace transform.

Figure 7.1. Unit step function $U(t-k)$.

$$\mathcal{L}\{U(t-k)\} = \int_0^\infty e^{-st}U(t-k)\,dt = \int_0^k e^{-st}\cdot 0\,dt + \int_k^\infty e^{-st}\cdot 1\,dt = \frac{1}{s}\,e^{-ks}$$

If $k = 0$, then $\mathcal{L}\{U(t)\} = 1/s$.

7.10. SHIFTING PROPERTIES

Suppose that $f(t)$ is such that its Laplace transform integral converges when $s > k$. Then

$$\mathcal{L}\{e^{kt}f(t)\} = \int_0^\infty e^{-st}e^{kt}f(t)\,dt = \int_0^\infty e^{-(s-k)t}f(t)\,dt$$

$$\mathcal{L}\{e^{kt}f(t)\} = F(s-k) \quad \text{or} \quad \mathcal{L}^{-1}\{F(s-k)\} = e^{kt}f(t) \qquad (7.7)$$

Also if $k > 0$

$$\mathcal{L}\{f(t-k)U(t-k)\} = \int_0^\infty e^{-st}f(t-k)U(t-k)\,dt = \int_k^\infty e^{-st}f(t-k)\,dt$$

If we let $r = t - k$, then

$$\mathcal{L}\{f(t-k)U(t-k)\} = \int_0^\infty e^{-s(r+k)}f(r)\,dr = e^{-sk}\mathcal{L}\{f(t)\}$$

Thus

$$\mathcal{L}\{f(t-k)U(t-k)\} = e^{-sk}\mathcal{L}\{f(t)\} \quad \text{and} \quad \mathcal{L}^{-1}\{e^{-sk}F(s)\}$$
$$= f(t-k)U(t-k) \qquad (7.8)$$

The two results (7.7) and (7.8) are frequently referred to as *shifting properties*. In the first the *shift* or *translation* occurs in the transform F. In the second the translation is in the function f, although the new function is zero until $t \geq k$.

Example 7.7. (a) Find $\mathscr{L}^{-1}\{F(s)\}$ if $F(s) = 1/(s^2 + 2s + 5)$.

Rearrange the fraction in $F(s)$ by completing the square:

$$F(s) = \frac{1}{(s+1)^2 + 2^2}$$

We recognize this as a form $F^*(s+1) = 1/((s+1)^2 + 2^2)$, where

$$f(t) = \frac{1}{2}\sin 2t$$

Therefore

$$\mathscr{L}^{-1}\{F(s)\} = \frac{1}{2}e^{-t}\sin 2t$$

(b) Determine $\mathscr{L}\{\cos(t - \pi)U(t - \pi)\}$.

$$\mathscr{L}\{\cos(t - \pi)U(t - \pi)\} = e^{-\pi s}\mathscr{L}\{\cos t\} = \frac{se^{-\pi s}}{s^2 + 1}$$

7.11. THE DIRAC DELTA FUNCTION

The *unit impulse function* or *Dirac delta function* $\delta(t - k)$ is defined so that

$$\delta(t - k) = 0 \quad \text{when } t \neq k \tag{7.9}$$

$$\int_{-\infty}^{\infty} \delta(t - k)\, dt = 1 \tag{7.10}$$

$$\int_{-\infty}^{\infty} f(t)\delta(t - k)\, dt = f(k) \tag{7.11}$$

Although other definitions are possible, for any $\varepsilon > 0$, we define

$$\delta_\varepsilon(t - k) = \begin{cases} 1/2\varepsilon & \text{if } k - \varepsilon < t < k + \varepsilon \\ 0 & \text{if } t \leq k - \varepsilon \text{ and } t \geq k + \varepsilon \end{cases}$$

and observe that $\delta_\varepsilon(t - k)$ has properties which approximate properties (7.9)–(7.11) of the *Dirac delta function* (or *symbol*, as some prefer) (Figure 7.2). Also note that $\delta_\varepsilon(t - k)$ is PWC. To obtain the infinite spike or surge, we define

$$\delta(t - k) = \lim_{\varepsilon \to 0} \delta_\varepsilon(t - k)$$

No ordinary function has all the conditions of (7.9)–(7.11).

Figure 7.2. The function $\delta_\varepsilon(t-k)$

The delta function fails to satisfy the sufficient conditions for existence of the Laplace transform (given earlier). Since we have defined $\delta(t-k)$ as a limit of $\delta_\varepsilon(t-k)$ as $\varepsilon \to 0$, we define $\mathscr{L}\{\delta(t-k)\}$ as a similar limit, so that

$$\mathscr{L}\{\delta(t-k)\} = \lim_{\varepsilon \to 0} \mathscr{L}\{\delta_\varepsilon(t-k)\}$$

First, we determine

$$\mathscr{L}\{\delta_\varepsilon(t-k)\} = \frac{1}{2\varepsilon} \int_{k-\varepsilon}^{k+\varepsilon} e^{-st}\, dt = -\frac{1}{2\varepsilon s} [e^{-st}]_{k-\varepsilon}^{k+\varepsilon} = \frac{e^{-sk}}{2\varepsilon s} [e^{s\varepsilon} - e^{-s\varepsilon}]$$

$$= e^{-sk}\, \frac{\sinh s\varepsilon}{s\varepsilon}$$

Therefore,

$$\mathscr{L}\{\delta(t-k)\} = \lim_{\varepsilon \to 0} e^{-sk}\, \frac{\sinh s\varepsilon}{s\varepsilon} = e^{-sk} \lim_{\varepsilon \to 0} s\, \frac{\cosh s\varepsilon}{s} = e^{-sk} \qquad (7.12)$$

It is possible to define the integral of the product of the delta function and any continuous function $f(t)$ in a similar fashion. That is, define

$$\int_{-\infty}^{\infty} \delta(t-k)f(t)\, dt = \lim_{\varepsilon \to 0} \int_{-\infty}^{\infty} \delta_\varepsilon(t-k)f(t)\, dt$$

Using the mean value theorem for integrals, we see that

$$\int_{-\infty}^{\infty} \delta_{\varepsilon}(t-k)f(t)\,dt = \frac{1}{2\varepsilon}\int_{k-\varepsilon}^{k+\varepsilon} f(t)\,dt = \frac{1}{2\varepsilon}\cdot 2\varepsilon\cdot f(t^*) = f(t^*)$$

where $k-\varepsilon < t^* < k+\varepsilon$. But as $\varepsilon \to 0$, $t^* \to k$. Therefore,

$$\int_{-\infty}^{\infty} \delta(t-k)f(t)\,dt = f(k) \tag{7.13}$$

The delta function is a useful tool to deal with applications involving impulsive phenomena such as voltages or forces of extreme magnitude acting for an instant of time. We have carried out formal operations using limits and integrals as though the delta function is mathematically legitimate. To justify these procedures rigorous mathematics exists, but it is beyond the scope of our discussion. These "functions" are sometimes referred to as *generalized functions* or *distributions*. (See Danese [17, Vol. 2, Chapter 24].)

Exercises 7.3

1. Find (a) $\mathcal{L}\{te^{-3t}\}$, (b) $\mathcal{L}\{e^{2t}\cos 3t\}$.
2. Determine (a) $\mathcal{L}\{U(t-3)\sinh(t-3)\}$, (b) $\mathcal{L}\{U(t-2)\cosh(t-2)\}$.
3. Sketch the graphs of the functions and find their Laplace transforms:

   (a) $f(t) = U(t-1) + U(t-2) - 2U(t-3)$;

   (b) $f(t) = \begin{cases} 0 & \text{if } t < 2 \\ t-2 & \text{if } t \geq 2; \end{cases}$

   (c) $\sin(t-\pi)U(t-\pi)$;

   (d) $f(t) = 1 - U(t-1)$.

4. Obtain $\mathcal{L}\{U(t-2)e^{t-2}\}$.
5. Determine

   (a) $\mathcal{L}^{-1}\left\{\dfrac{3e^{-s}}{s^2}\right\}$,

   (b) $\mathcal{L}^{-1}\left\{\dfrac{2e^{-3s}}{s^2+4}\right\}$,

   (c) $\mathcal{L}^{-1}\left\{\dfrac{e^{-3s}}{s^2+4}\right\}$.

6. If $f(t+p) = f(t)$, then $f(t)$ is periodic. Show that if $f(t)$ is periodic of period p, then

$$\mathcal{L}\{f(t)\} = \frac{1}{1-e^{-sp}}\int_0^p e^{-st}f(t)\,dt$$

   Hint: Begin by observing that

$$\int_0^\infty e^{-st}f(t)\,dt = \sum_{m=0}^\infty \int_{mp}^{(m+1)p} e^{-st}f(t)\,dt$$

Then let $t = mp + r$ and notice that the integral is over a single period of f. Therefore, the sum becomes

$$\sum_{m=0}^\infty e^{-smp} \int_0^p e^{-sr}f(r)\,dr$$

The sum

$$\sum_{m=0}^\infty (e^{-sp})^m$$

is a geometric series converging to $1/(1 - e^{-sp})$ under suitable conditions.

7. Use the result of Exercise 6 to determine the Laplace transforms of

(a) $f(t) = \begin{cases} 1 & \text{if } 0 \leqslant t < 1 \\ t & \text{if } 1 \leqslant t < 2 \end{cases}$, $f(t + 2) = f(t)$,

(b) $f(t) = \begin{cases} 0 & \text{if } 0 \leqslant t < \pi \\ \cos t & \text{if } \pi \leqslant t < 2\pi \end{cases}$, $f(t + 2\pi) = f(t)$,

(c) $f(t) = e^t$ if $0 \leqslant t < 1$, $f(t + 1) = f(t)$.

8. Solve the following initial value problems:
   (a) $y''(t) + y(t) = U(t - \pi)$, $y(0) = 0$, $y'(0) = 1$;
   (b) $y''(t) + 3y'(t) + 2y(t) = U(t - 2)$, $y(0) = y'(0) = 0$;
   (c) $y''(t) - 3y'(t) + 2y(t) = U(t - 2)$, $y(0) = 1$, $y'(0) = 0$;
   (d) $y''(t) + y(t) = U(t - \pi) \sin (t - \pi)$, $y(0) = y'(0) = 0$.

9. Solve the IVPs:
   (a) $y'(t) + 2y(t) = 3\delta(t - 1)$, $y(0) = 1$;
   (b) $y''(t) + 4y(t) = \delta(t - \pi)$, $y(0) = 1$, $y'(0) = 0$;
   (c) $y''(t) + 2y'(t) + 2y(t) = 4\delta(t - 1)$, $y(0) = 1$, $y'(0) = 0$;
   (d) $y''(t) + y(t) = \delta(t - \pi) \cos t$, $y(0) = 1$, $y'(0) = 0$.

7.12. CONVOLUTION

There are situations where we recognize a Laplace transform as the product of two other transforms. It is our problem to determine the nature of this product of two Laplace transforms and the inverse of this product.

We define the *convolution* h of two functions f and g by the equation

$$h(t) = \int_0^t f(t - \tau)g(\tau)\,d\tau$$

The convolution is often denoted by $f*g$. Let us assume that we have sufficient conditions on f and g for the existence of the Laplace transforms of f, g, and $h = f*g$. Then

$$\mathcal{L}\{h(t)\} = H(s) = \int_0^\infty e^{-st}h(t)\ dt = \int_0^\infty e^{-st}\left[\int_0^t f(t-\tau)g(\tau)\ d\tau\right] dt$$

$$= \int_0^\infty \int_0^t e^{-st}f(t-\tau)g(\tau)\cdot d\tau\ dt$$

Assuming that the order of integration can be changed we have

$$H(s) = \int_0^\infty \int_\tau^\infty e^{-st}f(t-\tau)g(\tau)\ dt\ d\tau$$

$$= \int_0^\infty g(\tau)\left[\int_\tau^\infty e^{-st}f(t-\tau)\ dt\right] d\tau \qquad (7.14)$$

Let $z = t - \tau$ in the inner integral of (7.14) to obtain

$$H(s) = \int_0^\infty g(\tau)\left[\int_0^\infty e^{-s(z+\tau)}f(z)\ dz\right] d\tau$$

$$= \int_0^\infty g(\tau)e^{-s\tau}\left[\int_0^\infty e^{-sz}f(z)\ dz\right] d\tau$$

$$= F(s)\int_0^\infty g(\tau)e^{-s\tau}\ d\tau = F(s)G(s)$$

We have shown that if $h = f*g$, then

$$H(s) = F(s)G(s)$$

or

$$\mathcal{L}^{-1}\{F(s)G(s)\} = h(t) = \int_0^t f(t-\tau)g(\tau)\ d\tau$$

Let us also note that the convolution satisfies the commutative law so that $f*g = g*f$. The proof is left to the reader.

Exercises 7.4

1. Find the Laplace transform of $f(t)$ if

(a) $f(t) = \displaystyle\int_0^t (t-\tau)\cos \tau\ d\tau$,

(b) $f(t) = \displaystyle\int_0^t e^{-(t-\tau)}\cos \tau\ d\tau$

2. Using the convolution method, determine

(a) $\mathscr{L}^{-1}\left\{\dfrac{1}{(s-1)(s-2)}\right\}$,

(b) $\mathscr{L}^{-1}\left\{\dfrac{2}{s(s^2+4)}\right\}$,

(c) $\mathscr{L}^{-1}\left\{\dfrac{2}{s^2(s^2+4)}\right\}$,

(d) $\mathscr{L}^{-1}\left\{\dfrac{1}{(s-2)(s^2+1)}\right\}$,

(e) $\mathscr{L}^{-1}\left\{\dfrac{e^{-2s}}{s(s-2)}\right\}$,

(f) $\mathscr{L}^{-1}\left\{\dfrac{1}{(s^2+1)^2}\right\}$.

3. (a) Using the convolution of 1 and $f(t)$, show that

$$\mathscr{L}^{-1}\left\{\frac{F(s)}{s}\right\} = \int_0^t f(\tau)\,d\tau$$

(b) Find $\mathscr{L}\left\{\displaystyle\int_0^t \cos\tau\,d\tau\right\}$.

(c) Find $\mathscr{L}\left\{\displaystyle\int_0^t e^\tau\,d\tau\right\}$.

(d) Determine

$$\mathscr{L}^{-1}\left\{\frac{1}{s(s^2-9)}\right\}$$

(e) Obtain

$$\mathscr{L}^{-1}\left\{\frac{e^{-s}}{s(s-2)}\right\}$$

4. (a) Show that

$$\mathscr{L}^{-1}\left\{\frac{F(s)}{s^2}\right\} = \int_0^t \int_0^\tau f(\alpha)\,d\alpha\,d\tau$$

(b) Obtain

$$\mathscr{L}^{-1}\left\{\frac{1}{s^2(s-k)}\right\}$$

(c) Using (a) check Exercise 2(c).

5. (a) Show that if $\mathscr{L}\{f(t)\} = F(s)$ and if $f(t)/t = O(e^{kt})$ and $f(t)/t$ is PWC, then the integral of the transform $\int_s^\infty F(\alpha)\, d\alpha$ is given by

$$\int_s^\infty F(\alpha)\, d\alpha = \mathscr{L}\left\{\frac{f(t)}{t}\right\}$$

Hint: Observe that

$$\int_s^\infty F(\alpha)\, d\alpha = \int_s^\infty \int_0^\infty e^{-\alpha t} f(t)\, dt\, d\alpha$$

Change the order of integration and compute the inside integral. Then

$$\int_s^\infty F(\alpha)\, d\alpha = \int_0^\infty \int_s^\infty e^{-\alpha t} f(t)\, d\alpha\, dt = \int_0^\infty e^{-st}\, \frac{f(t)}{t}\, dt$$

(b) Show that

$$\mathscr{L}\left\{\frac{\sin kt}{t}\right\} = \frac{\pi}{2} - \arctan\frac{s}{k} = \arctan\frac{k}{s}\,, \quad s > 0$$

(c) Find

$$\mathscr{L}\left\{\frac{e^{-kt} - e^{-mt}}{t}\right\}$$

6. The error function, erf (t) is defined by $\text{erf}(t) = \dfrac{2}{\sqrt{\pi}} \displaystyle\int_0^t e^{-\alpha^2}\, d\alpha.$

Show that

$$\mathscr{L}\left\{\frac{1}{\sqrt{s}(s-1)}\right\} = \frac{1}{\sqrt{\pi t}} * e^t = \frac{2e^t}{\sqrt{\pi}}\int_0^t \frac{e^{-\tau}\, d\tau}{2\sqrt{\tau}}$$

Hint: Let $\alpha = \sqrt{\tau}$ in the integral. Then we have $\mathscr{L}^{-1}\left\{\dfrac{1}{\sqrt{s}(s-1)}\right\} = e^t \, \text{erf}\,\sqrt{t}.$

7. (a) Show that

$$\frac{1}{\sqrt{s}-1} = \frac{1}{\sqrt{s}}\left(1 + \frac{1}{s-1}\right) + \frac{1}{s-1}$$

(b) Show that

$$\mathscr{L}^{-1}\left\{\frac{1}{\sqrt{s}-1}\right\} = e^t + \frac{1}{\sqrt{\pi t}} + \frac{1}{\sqrt{\pi t}} * e^t = e^t + \frac{1}{\sqrt{\pi t}} + e^t \, \text{erf}\,\sqrt{t}$$

8. Solve the IVPs:
  (a) $y''(t) + 4y(t) = e^t$, $y(0) = y'(0) = 0$;
  (b) $y''(t) - 2y'(t) + y(t) = f(t)$, $y(0) = y'(0) = 0$;
  (c) $y'''(t) + y'(t) = f(t)$, $y(0) = y'(0) = y''(0) = 0$.

9. An equation of the type

$$y(t) = f(t) + \int_0^t y(\tau)\beta(t - \tau)\, d\tau \qquad (7.15)$$

in which the unknown function appears inside the integral is referred to as an *integral equation*. The example given here is called a *Volterra integral equation*. Since the integral is a convolution integral, we can express its Laplace transform under suitable conditions.
  (a) Solve for $y(t)$ and check your solution if

$$y(t) = \sin t + \int_0^t y(t - \tau)\sin \tau\, d\tau$$

  (b) Solve the integral equation and verify the solution

$$y(t) = 1 + \int_0^t e^\tau y(t - \tau)\, d\tau$$

  (c) Solve for $y(t)$ in the integro-differential equation and check the solution

$$\int_0^t y(\tau)\, d\tau - y'(t) = 1, \quad y(0) = 1$$

  (d) Solve the integral equation

$$y(t) = t + \int_0^t y(\tau)e^{t-\tau}\, d\tau$$

  (e) Find the Laplace transforms for (7.15) and solve for $Y(s)$.

10. The Taylor series for

$$\cos t = \sum_{k=0}^\infty \frac{(-1)^k t^{2k}}{(2k)!} = 1 - \frac{t^2}{2!} + \frac{t^4}{4!} - \cdots$$

Assuming that the Laplace transform of this series can be computed termwise, show that

$$\mathcal{L}\{\cos t\} = \frac{s}{s^2 + 1}$$

Hint: Find a geometric series.

11. Use the Taylor series method to show that $\mathcal{L}\{\sin t\} = 1/(s^2 + 1)$.

12. From (1.45) if $n = 0$,

$$J_0(t) = \sum_{k=0}^{\infty} \frac{(-1)^k}{(k!)^2} \left(\frac{t}{2}\right)^{2k} = 1 - \frac{t^2}{2^2} + \frac{t^4}{2^2(4^2)} - \cdots$$

Assuming that the Laplace transform of this Taylor series can be computed termwise, show that

$$\mathcal{L}\{J_0(t)\} = \frac{1}{(\sqrt{s^2 + 1})}, \quad s > 1$$

13. In (1.39) if the index $n = 0$, the Bessel differential equation is expressed as

$$t^2 y''(t) + t y'(t) + t^2 y(t) = 0$$

If $y(0) = 1$, the transformed equation is

$$-\frac{d}{ds}[s^2 Y(s) - s - y'(0)] + s Y(s) - 1 - \frac{d}{ds} Y(s) = 0$$

or

$$(s^2 + 1) Y'(s) + s Y(s) = 0$$

From this we obtain

$$Y(s) = \frac{C}{\sqrt{s^2 + 1}}$$

If $y(0) = 1$, then $C = 1$. But $y(t) = J_0(t)$. Therefore,

$$\mathcal{L}\{J_0(t)\} = \frac{1}{\sqrt{s^2 + 1}}$$

14. The Laplace transform method is useful when applied to certain systems of differential equations. Find a solution for each of the systems given below.

(a) $\left.\begin{array}{l} 2x'(t) + 2x(t) + y'(t) - y(t) = 1 \\ x'(t) + x(t) + y'(t) + y(t) = 1 \end{array}\right\}$, $x(0) = 1$, $y(0) = 0$

(b) $\left.\begin{array}{l} x'(t) = y(t) + z(t) \\ y'(t) = x(t) + y(t) \\ z'(t) = x(t) - y(t) \end{array}\right\}$, $x(0) = 0$, $y(0) = 1$, $z(0) = 0$

(c) $\left.\begin{array}{l} x'(t) + y'(t) = 2 \cosh t \\ x'(t) - y'(t) = 2 \sinh t \end{array}\right\}$, $x(0) = 2$, $y(0) = 1$

7.13. LAPLACE TRANSFORM METHOD FOR PDEs

The Laplace transform method is useful for certain BVPs which contain linear PDEs with constant coefficients. The process involves transforming the PDE relative to one of the independent variables and reducing the equation to an ODE or one with a smaller number of independent variables. So far as the Laplace transform is concerned the remaining independent variables are treated as constants. The transformed equation must be solved considering boundary conditions, and the inverse transform of this result is the required solution. One should be aware that there are problems for which the Laplace transform method is not advised. Instead of attempting a lengthy discussion of the process, we consider the solution of an example accompanied by a complete verification that the solution is a proper one which satisfies the conditions of the BVP.

Example 7.8. Solve the BVP

$$y_{xx} = 4y_{tt}, \quad t>0, \quad x>0 \tag{7.16}$$

$$y(x, 0) = 0, \quad x>0 \tag{7.17}$$

$$y_t(x, 0) = -1, \quad x>0 \tag{7.18}$$

$$y(0, t) = t, \quad x>0 \tag{7.19}$$

$$\lim_{x \to \infty} y(x, t) \text{ exists for a fixed } t>0 \tag{7.20}$$

Some of the features of this BVP, which suggest that the Laplace transform method may be desirable, follow. The PDE is linear with constant coefficients. One independent variable (at least) has a range $(0, \infty)$. Initial conditions $(t = 0)$ involving the independent variable are assumed. The independent variable x also has a range $(0, \infty)$, but only (7.19) has a condition at $(x = 0)$. For this equation two conditions are necessary for the transformation of a second order derivative. We use the Laplace transform relative to t first, so that

$$\mathcal{L}\{y(x, t)\} = Y(x, s) = \int_0^\infty e^{-st} y(x, t) \, dt \tag{7.21}$$

and x is treated as a constant in the transform. The PDE (7.16) will be transformed into an ODE with the independent variable x and the parameter s. Applying (7.21) to (7.16) we have

$$\mathcal{L}\{y_{xx}\} = \frac{\partial^2 \mathcal{L}\{y\}}{\partial x^2} = 4\mathcal{L}\{y_{tt}\} \tag{7.22}$$

Notice that in (7.22) we have exchanged the operation of transforming and differentiating. This interchange may not be valid, but if the process produces a solution to the problem, we will offer no further explanation of its validity. Thus, it is *essential to verify the result* when this process is used. Proceeding with (7.22) along with (7.17) and (7.18), one obtains

$$\frac{d^2Y}{dx^2} = 4(s^2Y + 1) \quad \text{or} \quad \frac{d^2Y}{dx^2} - 4s^2Y = 4 \tag{7.23}$$

It would be possible to solve (7.23) using the Laplace transform method, but for this ODE we select the process described in Section 1.3. The solution of (7.23) may be expressed

$$Y(x, s) = C_1(s)e^{-2sx} + C_2(s)e^{2sx} - \frac{1}{s^2} \tag{7.24}$$

With another note of caution concerning (7.20), we write

$$\int_0^\infty e^{-st}[\lim_{x\to\infty} y(x, t)]\, dt = \lim_{x\to\infty} \int_0^\infty e^{-st}y(x, t)\, dt = \lim_{x\to\infty} Y(x, s) \tag{7.25}$$

It follows from (7.25) that $\lim_{x\to\infty} Y(x, s)$ must exist. Therefore $C_2(s)$ must be zero, and

$$Y(x, s) = C_1(s)e^{-2sx} - \frac{1}{s^2}$$

From (7.19) using (7.21), the transform of $y(0, t)$ is

$$\mathcal{L}\{y(0, t)\} = Y(0, s) = \int_0^\infty e^{-st}t\, dt = \frac{1}{s^2} \tag{7.26}$$

Therefore,

$$Y(0, s) = C_1(s) - \frac{1}{s^2} = \frac{1}{s^2} \quad \text{or} \quad C_1(s) = \frac{2}{s^2}$$

and

$$Y(x, s) = \frac{2}{s^2} e^{-2sx} - \frac{1}{s^2} \tag{7.27}$$

The solution is the inverse transform of (7.27),

$$y(x, t) = -t + 2(t - 2x)U(t - 2x) \tag{7.28}$$

To see that (7.28) is indeed a solution of the original BVP, we check each of the conditions (7.16)–(7.20) beginning with (7.17). Note that $y(x, 0) =$

$0 + 2(-2x)U(-2x)$. But $U(-2x) = 0$; so $y(x, 0) = 0$ for all $x > 0$. Next let us compute y_t. For $x > 0$, $t > 0$ and $t \neq 2x$,

$$y_t(x, t) = -1 + 2U(t - 2x)$$

There is a discontinuity in the derivative along $t = 2x$ in the first quadrant of an xt plane. The solution (7.28) does not satisfy (7.16) along the line $t = 2x$ since the second order derivative fails to exist there. Substituting $t = 0$, we obtain for $x > 0$

$$y_t(x, 0) = -1 + 2U(-2x) = -1$$

Checking (7.19) we see that for $t > 0$

$$y(0, t) = -t + 2tU(t) = -t + 2t = t$$

For the final boundary condition note that

$$\lim_{x \to \infty} y(x, t) = -t + 2 \lim_{x \to \infty} (t - 2x)U(t - 2x)$$

For a fixed t, $t - 2x$ is negative for all adequately large x, and $U(t - 2x) = 0$. Therefore,

$$\lim_{x \to \infty} y(x, t) = -t ,$$

and (7.20) is satisfied. The verification is complete except for the PDE (7.16). We have discussed the difficulty along $t = 2x$. The derivative $y_t(x, t)$ has been computed, and

$$y_{tt}(x, t) = 0 \quad \text{for } x > 0, \quad t > 0, \quad t \neq 2x$$

$$y_x(x, t) = 0 + 2(-2)U(t - 2x) = -4U(t - 2x), \quad x > 0, \quad t > 0, \quad t \neq 2x$$

$$y_{xx}(x, t) = 0 \quad \text{for } x > 0, \quad t > 0, \quad t \neq 2x$$

Therefore, $y_{xx} = 4y_{tt}$ is satisfied in the xt region for $x > 0$, $t > 0$, $t \neq 2x$, and the verification is complete.

Exercises 7.5

Solve the following BVPs and check each solution completely.
1. $y_x + 2y_t = -4t$, $t > 0$, $x > 0$; $y(x, 0) = 0$, $x > 0$; $y(0, t) = t^2$, $t > 0$.
2. Solve Exercise 1 with the condition $y(x, 0) = x$, $x > 0$, replacing $y(x, 0) = 0$.

3. $y_{xx} = 4y_{tt}$, $t > 0$, $x > 0$; $y(x, 0) = 0$, $x > 0$; $y(0, t) = t^2$, $t > 0$; $y_t(x, 0) = -2$, $x > 0$; $\lim_{x \to \infty} y(x, t)$ exists for a fixed $t > 0$.

4. $y_{xx} = 16y_{tt}$, $t > 0$, $x > 0$; $y(x, 0) = 0$, $x > 0$; $y_t(x, 0) = -2$, $x > 0$; $y(0, t) = t^2 - t$, $t > 0$; $\lim_{x \to \infty} y(x, t)$ exists for a fixed $t > 0$.

Of all the integral transforms it seems that the Laplace transform is the most important. Although there certainly are examples where other integral transforms have definite advantages, the Laplace transform is perhaps the most powerful for its use with IVPs generally. In our work we will introduce other integral transforms, but not with the same intensity as we have devoted to the Laplace transform. Even our discussion of the Laplace transform has been brief, with emphasis on the development of tools for solving IVPs and BVPs. For a more extensive coverage see such references as Churchill [14].

7.14. FINITE FOURIER TRANSFORMS

Finite sine and cosine transforms are defined from corresponding sine and cosine Fourier series. Assume that f is a PWC function on an interval $(0, L)$. Then we define the *finite Fourier sine transform* by

$$S_n\{f\} = \int_0^L f(t) \sin\left(\frac{n\pi t}{L}\right) dt = F_s(n), \quad n \in \mathbb{N} \qquad (7.29)$$

The *inverse* of the transform is the Fourier series with the factor $2/L$.

$$f(t) = \frac{2}{L} \sum_{n=1}^{\infty} F_s(n) \sin\left(\frac{n\pi t}{L}\right) \qquad (7.30)$$

The *finite Fourier cosine transform* is defined in a similar way by

$$C_n\{f\} = \int_0^L f(t) \cos\left(\frac{n\pi t}{L}\right) dt = F_c(n), \quad n \in \mathbb{N}_0 \qquad (7.31)$$

where the *inverse* is

$$f(t) = \frac{F_c(0)}{L} + \frac{2}{L} \sum_{n=1}^{\infty} F_c(n) \cos\left(\frac{n\pi t}{L}\right) \qquad (7.32)$$

The factor $2/L$ may be associated with either the transform or the inverse of the transform or the factor $\sqrt{2/L}$ may be associated with both the transform and the inverse. We have elected to associate $2/L$ with the inverse of the transform in this discussion, even though $2/L$ is the factor in the

Fourier sine series coefficients. The two types of notation employed to represent the transforms are both useful. $S_n\{f\}$ specifies the actual function being transformed, while $F_s(n)$ indicates the index of the accompanying series. Both notations contain an S, indicating the "sine" transform.

If we choose the exponential representation for the Fourier series (3.13) and the coefficients (3.14), then our definition for the *finite Fourier exponential transform* may be written

$$E_n\{f\} = \int_{-\pi}^{\pi} fe^{-int} \, dt = F_e(n), \quad n \in \mathbf{Z} \tag{7.33}$$

and the *inverse* is

$$f(t) = \frac{1}{2\pi} \sum_{n=-\infty}^{\infty} F_e(n)e^{int}, \quad -\pi < x < \pi \tag{7.34}$$

Example 7.9. If f' is continuous and f'' is PWC on $[0, L]$ show that

$$S_n\{f''\} = \frac{n\pi}{L} [f(0) - (-1)^n f(L)] - \frac{n^2\pi^2}{L^2} F_s(n) \tag{7.35}$$

To show this relation, we replace f in (7.29) by f'' and integrate by parts two times.

$$S_n\{f''\} = \int_0^L f'' \sin\left(\frac{n\pi t}{L}\right) dt$$

$$= -\frac{n\pi}{L} \int_0^L f' \cos\left(\frac{n\pi t}{L}\right) dt$$

$$= \frac{n\pi}{L} [f(0) - (-1)^n f(L)] - \frac{n^2\pi^2}{L^2} F_s(n)$$

Exercises 7.6

1. If f' is continuous and f'' PWC on $[0, L]$ show that

$$C_n\{f''\} = (-1)^n f'(L) - f'(0) - \frac{n^2\pi^2}{L^2} F_c(n)$$

2. Show that if f' is continuous on $[0, L]$ then

(a) $S_n\{f'\} = -\dfrac{n\pi}{L} F_c(n),$

(b) $C_n\{f'\} = (-1)^n f(L) - f(0) + \dfrac{n\pi}{L} F_s(n).$

3. Find

    (a) $S_n\{1\}$,

    (b) $S_n\{t\}$,

    (c) $C_n\{1\}$,

    (d) $C_n\{t\}$,

    (e) $C_n\{e^{kt}\}$,

    (f) $S_n\{e^{kt}\}$,

    (g) $E_n\{t\}$,

    (h) $E_n\{\sin t\}$.

4. If f' is continuous and f'' PWC on $[0, \pi]$ show that

$$E_n\{f''\} = [f'(\pi) - f'(-\pi)](-1)^n + in[f(\pi) - f(-\pi)](-1)^n - n^2 F_e(n)$$

5. If f' is continuous show that

$$E_n\{f'\} = [f(\pi) - f(-\pi)](-1)^n + inF_e(n)$$

7.15. FOURIER TRANSFORMS

Just as finite Fourier transforms are related to Fourier series and coefficients, the Fourier transforms follow from Fourier integrals and corresponding coefficients. Using the pattern of (4.18) with coefficients (4.19), we define the *Fourier cosine transform* $C_\alpha\{f\}$ of the function f as

$$C_\alpha\{f\} = \int_0^\infty f(t) \cos \alpha t \, dt = F_c(\alpha) , \quad \alpha > 0 \qquad (7.36)$$

The *inverse of the transform* F_c is given by the Fourier cosine integral as

$$f(t) = \frac{2}{\pi} \int_0^\infty F_c(\alpha) \cos \alpha t \, d\alpha , \quad t > 0 \qquad (7.37)$$

The factor $2/\pi$ found in (7.37) is sometimes split into two factors $\sqrt{2/\pi}$ and associated with both the F_c integral and the f integral. The situation is the same as we discussed concerning finite Fourier transforms. The user of Fourier transforms needs to be familiar with the definitions assumed.

We define the *Fourier sine transform* $S_\alpha\{f\}$ of the function f by

$$S_\alpha\{f\} = \int_0^\infty f(t) \sin \alpha t \, dt = F_s(\alpha) , \quad \alpha > 0 \qquad (7.38)$$

The *inverse of* F_s is defined by

$$f(t) = \frac{2}{\pi} \int_0^\infty F_s(\alpha) \sin \alpha t \, d\alpha , \quad t > 0 \qquad (7.39)$$

From the definition of the Fourier cosine transform (7.36), we write the transform of f'' and integrate by parts twice to obtain an operational formula. We assume that f and its first and second order derivatives are continuous and AI on $(0, \infty)$.

$$C_\alpha\{f''\} = \int_0^\infty f''(t) \cos \alpha t \, dt = [f'(t) \cos \alpha t]_0^\infty + \alpha \int_0^\infty f'(t) \sin \alpha t \, dt$$

If $f'(t) \to 0$ as $t \to \infty$, then

$$C_\alpha\{f''\} = -f'(0) + \alpha\left\{[f(t) \sin \alpha t]_0^\infty - \alpha \int_0^\infty f(t) \cos \alpha t \, dt\right\}$$

If $f(t) \to 0$ as $t \to \infty$, then

$$C_\alpha\{f''\} = -f'(0) - \alpha^2 C_\alpha\{f\} \tag{7.40}$$

For a *Fourier exponential transform*, we investigate the form of the Fourier integral (4.31) and its coefficients (4.32). We let

$$E_\alpha\{f\} = \int_{-\infty}^\infty f(t)e^{-i\alpha t} \, dt = F_e(\alpha), \quad -\infty < \alpha < \infty$$

be the definition of the transform. The inverse of the transform of F_e is defined by

$$f(t) = \frac{1}{2\pi} \int_{-\infty}^\infty F_e(\alpha)e^{i\alpha t} \, d\alpha, \quad -\infty < t < \infty$$

Assume that f and f' are continuous and AI on $(-\infty, \infty)$,

$$E_\alpha\{f'\} = \int_{-\infty}^\infty f'(t)e^{-i\alpha t} \, dt = [f(t)e^{-i\alpha t}]_{-\infty}^\infty + i\alpha \int_{-\infty}^\infty f(t)e^{-i\alpha t} \, dt$$

If $f(t) \to 0$ as $|t| \to \infty$, then

$$E_\alpha\{f'\} = i\alpha E_\alpha\{f\} \tag{7.41}$$

Other operational formulas will be found in the exercises.

We assume that the two functions f and g are each PWC and AI on the real axis. In the context of Fourier transforms, the *convolution of f and g* defined by

$$f * g = \int_{-\infty}^\infty f(t - \tau)g(\tau) \, d\tau \tag{7.42}$$

To formally compute the Fourier transform of the convolution (7.42), we note that

$$E_\alpha\{f*g\} = \int_{-\infty}^{\infty}\left[\int_{-\infty}^{\infty} f(t-\tau)g(\tau)\,d\tau\right]e^{-i\alpha t}\,dt$$

$$= \int_{-\infty}^{\infty}\int_{-\infty}^{\infty} f(t-\tau)g(\tau)e^{-i\alpha t}\,dt\,d\tau$$

$$= \int_{-\infty}^{\infty} g(\tau)e^{-i\alpha\tau}\int_{-\infty}^{\infty} f(t-\tau)e^{-i\alpha(t-\tau)}\,dt\,d\tau$$

If we let $z = t - \tau$ in the inner integral, then

$$E_\alpha\{f*g\} = \int_{-\infty}^{\infty} g(\tau)e^{-i\alpha\tau}\left[\int_{-\infty}^{\infty} f(z)e^{-i\alpha z}\,dz\right]d\tau$$

$$= E_\alpha\{f\}\int_{-\infty}^{\infty} g(\tau)e^{-i\alpha\tau}\,d\tau = E_\alpha\{f\}\cdot E_\alpha\{g\}$$

Thus

$$E_\alpha\{f*g\} = E_\alpha\{f\}\cdot E_\alpha\{g\} \tag{7.43}$$

Formula (7.43) is the conclusion of the *convolution theorem*.

Example 7.10. (a) Find the Fourier sine transform for the function $f(t) = e^{-kt}$, $k>0$, from the definition.

(b) Find the Fourier cosine transform for $f(t) = e^{-kt}$, $k>0$, using the operational formula (7.40).

(a) $\quad S_\alpha\{e^{-kt}\} = \int_0^{\infty} e^{-kt}\sin \alpha t\,dt = \left[\frac{e^{-kt}}{k^2+\alpha^2}(-k\sin \alpha t - \alpha\cos \alpha t)\right]_0^{\infty}$

$$= \frac{\alpha}{k^2+\alpha^2}, \quad k>0$$

(b) $C_\alpha\{e^{-kt}\}$ is to be determined by the formula

$$C_\alpha\{f''\} = -f'(0) - \alpha^2 C_\alpha\{f\}$$

If

$$f = e^{-kt}$$
$$f' = -ke^{-kt}, \quad f'(0) = -k$$
$$f'' = k^2 e^{-kt}$$
$$C_\alpha\{k^2 e^{-kt}\} = -(-k) - \alpha^2 C_\alpha\{e^{-kt}\} \tag{7.44}$$

$C_\alpha\{f\}$ is a linear operator. Therefore,

$$C_\alpha\{k^2 e^{-kt}\} = k^2 C_\alpha\{e^{-kt}\}$$

We can write (7.44) as

$$k^2 C_\alpha\{e^{-kt}\} = k - \alpha^2 C_\alpha\{e^{-kt}\}$$
$$(k^2 + \alpha^2)C_\alpha\{e^{-kt}\} = k$$

and

$$C_\alpha\{e^{-kt}\} = \frac{k}{k^2 + \alpha^2}, \quad k > 0$$

Example 7.11. Solve the integral equation

$$\int_0^\infty f(t) \cos \alpha t \, dt = \begin{cases} 1 & \text{when } 0 < \alpha < \pi \\ 0 & \text{when } \pi < \alpha < \infty \end{cases}$$

An equation of the form

$$\int_0^\infty f(t) \cos \alpha t \, dt = g(\alpha)$$

is an *integral equation*. To solve the equation one needs to determine f. Therefore,

$$f(t) = \frac{2}{\pi} \int_0^\infty g(\alpha) \cos \alpha t \, d\alpha$$

In our problem,

$$f(t) = \frac{2}{\pi} \int_0^\pi 1 \cdot \cos \alpha t \, d\alpha = \frac{2}{\pi} \left[\frac{1}{t} \sin \alpha t \right]_0^\pi$$

$$f(t) = \frac{2}{\pi t} \sin \pi t$$

Example 7.12. Show that

$$E_\alpha\{f(t - c)\} = e^{-i\alpha c} E_\alpha\{f(t)\}$$

if c is a real constant.

$$E_\alpha\{f(t - c)\} = \int_{-\infty}^\infty f(t - c) e^{-i\alpha t} \, dt$$

If $z = t - c$ or $t = z + c$, then

$$E_\alpha\{f(t-c)\} = \int_{-\infty}^{\infty} f(z)e^{-i\alpha(z+c)}\, dz$$

$$= e^{-i\alpha c}\int_{-\infty}^{\infty} f(z)e^{-i\alpha z}\, dz = e^{-i\alpha c}E_\alpha\{f(t)\}$$

Hence

$$E_\alpha\{f(t-c)\} = e^{-i\alpha c}E_\alpha\{f(t)\} \tag{7.45}$$

This result is sometimes referred to as *shifting* or *translating*.

Example 7.13. Determine a solution for the integral equation

$$f(x) = g(x) + \int_{-\infty}^{\infty} f(t)h(x-t)\, dt$$

We assume that the Fourier transforms of f, g, and h exist and are represented by $F_e(\alpha)$, $G_e(\alpha)$, and $H_e(\alpha)$, respectively. If we write the transforms of the equation, using the convolution theorem, we have

$$F_e(\alpha) = G_e(\alpha) + F_e(\alpha)\cdot H_e(\alpha) \tag{7.46}$$

From (7.46) we obtain

$$F_e(\alpha)[1 - H_e(\alpha)] = G_e(\alpha)$$

or

$$F_e(\alpha) = \frac{G_e(\alpha)}{1 - H_e(\alpha)}$$

If $F_e(\alpha)$ has an inverse, then

$$f(x) = \frac{1}{2\pi}\int_{-\infty}^{\infty} \frac{G_e(\alpha)}{1 - H_e(\alpha)}\, e^{i\alpha x}\, d\alpha$$

Exercises 7.7

1. If

$$f(t) = \begin{cases} 1 & \text{when } 0 < x < L \\ 0 & \text{when } L < x < \infty \\ \frac{1}{2} & \text{when } x = L \end{cases}$$

show that

$$f(t) = \frac{2}{\pi}\int_0^{\infty} \frac{1 - \cos L\alpha}{\alpha}\, \sin \alpha t\, d\alpha, \quad 0 < t < \infty$$

2. Find the Fourier sine transform for

$$f(t) = te^{-t}, \quad t \geq 0$$

3. Determine the Fourier cosine transform for

$$f(t) = e^{-t} \cos t, \quad t \geq 0$$

and show that

$$e^{-t} \cos t = \frac{2}{\pi} \int_0^\infty \frac{(\alpha^2 + 2) \cos \alpha t}{\alpha^4 + 4} \, d\alpha, \quad 0 \leq t$$

4. Even though $f(t) = \sin t$ is continuous and has continuous derivatives, explain why it fails to have a Fourier sine transform.

5. Show that $C_\alpha\{f\}$ is a linear transform.

6. Show that

$$E_\alpha\{f(ct)\} = \frac{1}{|c|} F_e\left(\frac{\alpha}{c}\right), \quad c \neq 0$$

This is a *scaling* formula.

7. (a) For appropriate conditions on f, show that

$$S_\alpha\{f''\} = \alpha f(0) - \alpha^2 S_\alpha\{f\}$$

State sufficient conditions on f.

(b) Derive the formula

$$E_\alpha\{f''\} = -\alpha^2 E_\alpha\{f\}$$

Do the conditions on f in (a) need to be altered for this derivation?

(c) Generalize (b) and show that

$$E_\alpha\{f^{(n)}\} = (i\alpha)^n E_\alpha\{f\}, \quad n \in \mathbf{N}$$

8. Solve the integral equation

$$\int_0^\infty f(t) \sin \alpha t \, dt = \begin{cases} -\alpha & \text{when } 0 < \alpha < 1 \\ 0 & \text{when } 1 < \alpha < \infty \end{cases}$$

9. Determine f in the integral equation

$$\int_0^\infty f(t) \sin \alpha t \, dt = e^{-\alpha}$$

10. Show that

    (a) $f * g = g * f$,

    (b) $f * (g + h) = f * g + f * h$.

11. Verify the convolution theorem if $f(t)$ and $g(t)$ are both defined as

$$f(t) = g(t) = \begin{cases} 1 & \text{when } -1 < t < 1 \\ 0 & \text{when } t < -1 \text{ or } t > 1 \end{cases}$$

12. Show that $E_\alpha\{f(-t)\} = F_e(-\alpha)$

13. Find $E_\alpha\{e^{ict}f(t)\}$.

14. Determine $E_\alpha\{e^{-|t|}\}$.

15. Show that

$$E_\alpha\{e^{-ct}U(t)\} = \frac{1}{c + i\alpha}, \quad c > 0$$

16. Find $E_\alpha\{e^t U(-t)\}$.

17. Show that:

    (a) $E_\alpha\{tf(t)\} = iF_e'(\alpha)$;

    (b) $E_\alpha\{t^2 f(t)\} = -F_e''(\alpha)$;

    (c) $E_\alpha\{t^n f(t)\} = i^n F_e^{(n)}(\alpha)$;

    (d) $E_\alpha\{tf'(t)\} = -\dfrac{d}{d\alpha}[\alpha F_e(\alpha)]$;

    (e) $E_\alpha\{tf''(t)\} = -i\dfrac{d}{d\alpha}[\alpha^2 F_e(\alpha)]$;

    (f) $E_\alpha\{t^2 f''(t)\} = \dfrac{d^2}{d\alpha^2}[\alpha^2 F_e(\alpha)]$.

18. Solve the differential equation, $y''(t) + ty'(t) + y(t) = 0$. Use Fourier transforms.

19. Using Fourier transforms, solve the differential equation, $y''(t) + ty'(t) + ty(t) = 0$. Check the solution.

7.16. THE DISCRETE FOURIER TRANSFORM

The discrete Fourier transform is convenient for the harmonic analysis of discrete data such as those obtained from experimental measurements or by sampling a function at a finite set of points. The utility of the discrete transform has been markedly increased by the development of an algorithm, known as the fast Fourier transform, which permits rapid computer evaluation of the transform at a large number of points. We will first discuss the discrete transform and then we will outline the steps in the fast Fourier transform algorithm.

Let $T = \{t_0, t_1, \ldots, t_{n-1}\}$ be a finite set of uniformly spaced real numbers and let $f(t)$ and $g(t)$ be complex-valued functions defined on T. We will define a discrete norm and inner product for such functions by the equations

$$\| f(t) \|^2 = \sum_{j=0}^{n-1} |f(t_j)|^2 \tag{7.47}$$

$$(f, g) = \sum_{j=0}^{n-1} f(t_j)\overline{g(t_j)} \tag{7.48}$$

where $\overline{g(t_j)}$ represents the complex conjugate of $g(t_j)$. The norm defined by (7.47) is sometimes called a pseudo-norm because in the context of an interval $[a, b]$ which contains T, it fails to have the property that $\| f \| = 0$ implies $f(t)$ is identically zero.

For notational simplicity in defining the discrete transform we will take $t_j = j, j = 0, 1, \ldots, n - 1$. (One can always convert the problem to this form by a linear change of variable when the points are uniformly spaced.) The discrete Fourier transform (DFT) is defined by

$$J_f(\omega) = \frac{1}{n} \sum_{j=0}^{n-1} f(t_j)e^{-i\omega t_j} = F_d(\omega) \tag{7.49}$$

The inverse transform is defined by

$$f(t) = \sum_{j=0}^{n-1} F_d(\omega_j)e^{i\omega_j t} , \tag{7.50}$$

where $\omega_j = 2\pi j/n$. The inverse transform agrees with the original function $f(t)$ at points of T but not necessarily at all points of $[a, b]$ if $T \subset [a, b]$.

We would now like to develop some properties of the DFT and point out its connection to the discrete least squares approximation. First, let us show that the inverse transform defined by (7.50) actually agrees with the original function at points in T.

Theorem 7.1. For $k = 0, 1, \ldots, n - 1$,

$$\sum_{j=0}^{n-1} F_d(\omega_j)e^{i\omega_j t_k} = f(t_k) \tag{7.51}$$

To avoid confusion replace the dummy variable j by ν in (7.49). Then replace $F_d(\omega_j)$ by the middle member of (7.49) to obtain

$$\sum_{j=0}^{n-1} F_d(\omega_j)e^{i\omega_j t_k} = \sum_{j=0}^{n-1} \frac{1}{n} \sum_{\nu=0}^{n-1} f(t_\nu)e^{-i\omega_j t_\nu}e^{i\omega_j t_k}$$

$$= \frac{1}{n} \sum_{\nu=0}^{n-1} f(t_\nu) \sum_{j=0}^{n-1} e^{i(t_k-t_\nu)\omega_j} \tag{7.52}$$

Now let us examine the inner sum in (7.52). Recall that $\omega_j = 2\pi j/n$ and let $z = e^{i2\pi(t_k-t_\nu)/n}$. Note that $z = 1$ if $k = \nu$. Since $0 \leqslant k,\ \nu < n$, it follows that $z \neq 1$ when $k \neq \nu$. Note that $z^n = 1$ in either case. Using the formula for the sum of a geometric progression we obtain for $z \neq 1$,

$$\sum_{j=0}^{n-1} z^j = \frac{z^n - 1}{z - 1} = 0 \tag{7.53}$$

Obviously the left member of (7.53) is n when $z = 1$. It follows that

$$\sum_{j=0}^{n-1} e^{i(t_k-t_\nu)\omega_j} = \sum_{j=0}^{n-1} z^j = \begin{cases} n & \text{if } k = \nu \\ 0 & \text{otherwise} \end{cases}$$

Thus the inner sum in (7.52) is zero except when $\nu = k$ and the result follows.

The following corollary will facilitate the comparison of discrete and continuous least squares approximations.

Corollary 7.2. Let M be any set of n consecutive integers. Then for $k = 0, 1, \ldots, n - 1$,

$$\sum_{j\in M} F_d(\omega_j)e^{i\omega_j t_k} = f(t_k)$$

The proof is essentially the same as that for Theorem 7.1.

Next, we note without proof that the DFT is a linear operator.

Theorem 7.3. Let the DFTs of $f(t)$, $g(t)$, and $h(t)$ be $F_d(\omega)$, $G_d(\omega)$, and $H_d(\omega)$, respectively. If $h(t) = c_1 f(t) + c_2 g(t)$, then $H_d(\omega) = c_1 F_d(\omega) + c_2 G_d(\omega)$.

Next, let us show that the DFT is a periodic function with period 2π.

Theorem 7.4. If $F_d(\omega)$ is the DFT of $f(t)$, then

(i) $F_d(\omega + 2\pi) = F_d(\omega)$,
(ii) $F_d(2\pi - \omega) = F_d(-\omega)$.

Substitute $\omega + 2\pi$ for ω in (7.49) to obtain

$$F_d(\omega + 2\pi) = \frac{1}{n} \sum_{j=0}^{n-1} f(t_j)e^{-i(\omega+2\pi)t_j} = \frac{1}{n} \sum_{j=0}^{n-1} f(t_j)e^{-i\omega t_j}e^{-i2\pi t_j}$$

But $t_j = j$; so

$$e^{-i2\pi t_j} = \cos 2\pi j - i \sin 2\pi j = 1$$

and (i) follows.

Part (ii) is obtained from (i) by replacing ω in (i) by $-\omega$.

We will define a *forward difference operator* Δ by the equations

$$\Delta(f(t_j)) = f(t_{j+1}) - f(t_j)$$

$$\Delta F(\omega_j) = F(\omega_{j+1}) - F(\omega_j)$$

The following theorem lists some periodicity and shifting properties for the DFT along with formulas analogous to those given for continuous transforms which relate the transform of the derivative to the transform of the function and so forth.

Theorem 7.5. Suppose $f(t)$ is periodic with period n. Then

(i) $J_{\Delta f}(\omega_k) = (e^{i\omega_k} - 1)J_f(\omega_k)$,

(ii) $\Delta J_f(\omega_k) = J_g(\omega_k)$, where $g(t) = (e^{-i\omega_1 t} - 1)f(t)$

Substituting in (7.49) we obtain

$$J_{\Delta f}(\omega_k) = \frac{1}{n} \sum_{j=0}^{n-1} [f(t_{j+1}) - f(t_j)]e^{-i\omega_k t_j}$$

$$= \frac{1}{n} \sum_{j=0}^{n-1} f(t_{j+1})e^{-i\omega_k t_j} - J_f(\omega_k) \qquad (7.54)$$

Next, let $\nu = j + 1$ in the right sum in (7.54) to obtain

$$\frac{1}{n} \sum_{j=0}^{n-1} f(t_{j+1})e^{-i\omega_k t_j} = \frac{1}{n} \sum_{\nu=1}^{n} f(t_\nu)e^{-i\omega_k(\nu-1)}$$

Now $f(n) = f(0)$ because $f(t)$ is periodic, and $e^{-i\omega_k t_n} = e^{-i2\pi k} = 1$ because $t_n = n$ and $\omega_k = 2\pi k/n$. Thus $f(t_n)e^{-i\omega_k t_n} = f(t_0)e^{-i\omega_k t_0}$. It follows that

$$\frac{1}{n} \sum_{\nu=1}^{n} f(t_\nu)e^{-i\omega_k(\nu-1)} = \frac{e^{i\omega_k}}{n} \sum_{\nu=0}^{n-1} f(t_\nu)e^{-i\omega_k t_\nu} = e^{i\omega_k}J_f(\omega_k) \qquad (7.55)$$

Part (i) now follows from (7.54) and (7.55).

To prove (ii), note that

$$\Delta J_f(\omega_k) = J_f(\omega_{k+1}) - J_f(\omega_k) = \frac{1}{n} \sum_{j=0}^{n-1} f(t_j)[e^{-i\omega_{k+1}t_j} - e^{-i\omega_k t_j}] \quad (7.56)$$

Now $\omega_{k+1} = 2(k+1)\pi/n = \omega_k + \omega_1$; hence

$$e^{-i\omega_{k+1}t_j} = e^{-i\omega_k t_j}e^{-i\omega_1 t_j}$$

Substituting this in (7.56) we have

$$\Delta J_f(\omega_k) = \frac{1}{n} \sum_{j=0}^{n-1} f(t_j)(e^{-i\omega_1 t_j} - 1)e^{-i\omega_k t_j} = J_g(\omega_k)$$

This completes the proof of Theorem 7.5.

Let us now show the connection between the DFT and the discrete least squares approximation. This connection comes about because the complex exponential functions $e^{i\omega_k t}$ are orthogonal with respect to the discrete inner product defined by (7.48). Define the sequence $\{\phi_k(t)\}$ by the equation

$$\phi_k(t) = e^{i\omega_k t}, \quad k = 0, 1, \dots, n-1$$

Theorem 7.6. The functions $\{\phi_k(t)\}$ are orthogonal with respect to the discrete inner product (7.48).

Note that

$$(\phi_k, \phi_\nu) = \sum_{j=0}^{n-1} e^{i\omega_k j}\overline{e^{i\omega_\nu j}} = \sum_{j=0}^{n-1} e^{i(\omega_k - \omega_\nu)j}$$

Letting $z = e^{i(\omega_k - \omega_\nu)} = e^{i2\pi(k-\nu)/n}$, we proceed as in the proof of Theorem 7.1 to conclude:

$$(\phi_k, \phi_\nu) = \sum_{j=0}^{n-1} z^j = \begin{cases} n & \text{if } k = \nu \\ 0 & \text{otherwise} \end{cases}$$

and the proof is complete.

Now consider the problem of approximating the function $f(t)$ on the set T by a linear combination of the complex exponential functions $\{\phi_k\}_{k=0}^{m}$, $m \leq n-1$. The least squares criterion requires that the sum of the squares of the deviations be minimized. That is, $\alpha_0, \alpha_1, \dots, \alpha_m$ should be chosen so that

$$S = \sum_{j=0}^{n-1} \left[f(t_j) - \sum_{j=0}^{m} \alpha_k \phi_k(t_j) \right]^2$$

is minimized. It can be shown that the discrete analog of Theorem 2.7 holds; so that S is minimized when

$$\alpha_k = \frac{(f, \phi_k)}{\|\phi_k\|^2} = \frac{1}{n} \sum_{j=0}^{n-1} f(t_j)\overline{\phi_k(t_j)} = \frac{1}{n} \sum_{j=0}^{n-1} f(t_j)e^{-i\omega_k t_j} = F_d(\omega_k)$$

Theorem 7.1 shows that when $m = n - 1$, the function

$$\Phi(t) = \sum_{j=0}^{n-1} \alpha_j e^{i\omega_j t}, \quad \alpha_j = F_d(\omega_j)$$

approximates $f(t)$ exactly on T. In this case the function $\Phi(t)$ is said to *interpolate* $f(t)$ on the set T.

Example 7.14. Let $T = \{0, 1, \ldots, 20\}$ and let $G_d(\omega)$ be the DFT of $g(t)$, where

$$g(t) = \begin{cases} 0 & 0 < t < 10 \\ \pi(t-10)/10 & 10 \leqslant t < 20 \\ \pi/2 & t = 0 \text{ or } 20 \end{cases}$$

Then

$$G_d(\omega) = \frac{1}{21} \sum_{j=0}^{20} g(j)e^{-i\omega j} = \frac{\pi}{42}(1 + e^{-20i\omega}) + \frac{1}{21} \sum_{j=11}^{19} \frac{\pi}{10}(j-10)e^{-i\omega j}$$

Now let $\nu = j - 10$ to obtain

$$G_d(\omega) = \frac{\pi}{42}(1 + e^{-20i\omega}) + \frac{\pi}{210} \sum_{\nu=1}^{9} \nu e^{-i\omega(\nu+10)}$$

$$= \frac{\pi}{42}(1 + e^{-20i\omega}) + \frac{e^{-10i\omega}}{210} \pi \sum_{\nu=1}^{9} \nu e^{-i\omega\nu}$$

Let $z = e^{-i\omega}$ and use the result of Problem 2, Exercises 7.8 to obtain

$$G_d(\omega) = \frac{\pi}{42}(1 + e^{-20i\omega}) + \frac{\pi e^{-10i\omega}}{210} \sum_{\nu=1}^{9} \nu z^\nu$$

$$= \frac{\pi}{42}(1 + e^{-20i\omega}) + \frac{z\pi e^{-10i\omega} z}{210(z-1)^2} [9z^{10} - 10z^9 + 1]$$

or

$$G_d(\omega) = \frac{\pi}{42}(1 + e^{-20i\omega}) + \frac{e^{-11i\omega} \pi}{210(e^{-i\omega} - 1)^2} [9e^{-10i\omega} - 10e^{-9i\omega} + 1]$$

$$(7.57)$$

Further details on discrete Fourier transforms are given by Weaver [51], Bloomfield [3], and Brigham [9]. Applications of DFTs to PDEs are discussed by Weaver [51], Vichnevetsky and Bowles [49], and Pickering [38].

Exercises 7.8

1. Let $f(t) = t^2$ and $T = \{0, 1, 2. 3, 4\}$. Write out the terms for the DFT. Then separate this expression into real and imaginary parts.
2. Derive a formula for $\sum_{j=1}^{n-1} jz^j$. Hint: Differentiate both sides of the formula for the sum of a geometric progression.
3. Let $G_d(\omega)$ be given by (7.57) and define

$$\Phi(t) = \sum_{j=0}^{20} \alpha_j e^{i\omega t}, \quad \alpha_j = G_d(\omega_j)$$

Write a computer program to compute $\Phi(t)$ for $t = 0, 0.5, 1.0, \ldots, 20$. Then compare the graphs of $g(t)$ and $\Phi(t)$.

7.17. THE FAST FOURIER TRANSFORM

The fast Fourier transform (FFT) is an efficient algorithm for computing the discrete Fourier transform. This algorithm requires roughly $2n \log_2 n$ arithmetic operations to compute the complete DFT for the Fourier frequencies $\omega_j = 2\pi j/n, j = 0, 1, \ldots, n-1$, associated with n data points. This contrasts with the roughly $2n^2$ operations which would be required for an algorithm based directly on (7.49). Our intention here is to present the key idea behind this algorithm, but not to discuss the details involved in a computer implementation of the FFT. These details (including a program listing) are discussed by Bloomfield [3]. Other references on the FFT include Weaver [51] and Brigham [9].

The central idea behind the FFT is to exploit the periodicity of the terms $e^{-i\omega_k t_j}$ in (7.49) which results in a duplication of terms in the expressions for $F_d(\omega_k)$. The FFT algorithm saves these duplicate terms so that they need only be computed once. Let $z = e^{-i\omega_1} = e^{-i2\pi/n}$. Then z is a so-called *root of unity* because $z^n = e^{-2\pi i} = 1$. Note that

$$z^{n+1} = z^n z = z, \quad z^{n+2} = z^n z^2 = z^2$$

and so forth. This illustrates the fact that the integral powers of z are periodic with period n. In fact,

$$z^j = z^k \quad \text{if } k \equiv j \pmod{n} \tag{7.58}$$

To see how the FFT works, let us examine the computation of $F_d(\omega_k)$, $k = 0, 1, \ldots, 7$ associated with 8 data points $y_j = f(t_j)$, $j = 0, 1, \ldots, 7$. As before assume that $x_j = j$. Note that $e^{-i\omega_k t_j} = z^{kj}$. Then (7.49) becomes

$$F_d(\omega_k) = \frac{1}{8} \sum_{j=0}^{7} y_j z^{kj}, \quad k = 0, 1, \ldots, 7 \tag{7.59}$$

We will split this sum into two parts, one involving terms with even indices and the other involving terms with odd indices. For $k = 0, 1, 2, 3$ let

$$S_e(k) = \sum_{j=0}^{3} y_{2j} z^{2jk}, \quad S_o(k) = \sum_{j=0}^{3} y_{2j+1} z^{(2j+1)k} \tag{7.60}$$

It is understood in (7.60) that the exponents of z are reduced to numbers between 0 and 7 by means of (7.58). Note, for example, that

$$8F_d(\omega_0) = y_0 + y_1 + \cdots + y_7 = S_e(0) + S_o(0)$$

while

$$8F_d(\omega_4) = y_0 + y_1 z^4 + y_2 z^8 + \cdots + y_7 z^{28}$$
$$= y_0 + y_1 z^4 + y_2 + y_3 z^4 + y_4 + y_5 z^4 + y_6 + y_7 z^4$$
$$= S_e(0) + z^4 S_o(0)$$

Having saved $S_e(0)$ and $S_o(0)$ from the calculation of $F_d(\omega_0)$ we use fewer additions and multiplications in the calculation of $F_d(\omega_4)$ than would be required by direct application of (7.59). This is the crux of the FFT. In a similar manner one can use the fact that $z^4 = e^{i\pi} = -1$ to show that

$$8F_d(\omega_k) = S_e(k) + S_o(k), \quad k = 0, 1, 2, 3$$
$$8F_d(\omega_k) = S_e(k-4) - S_o(k-4), \quad k = 4, 5, 6, 7 \tag{7.61}$$

Now let us do a careful count of the number of arithmetic operations required to compute

$$F_d(\omega_0), \ldots, F_d(\omega_7)$$

using (7.61) and compare that to the number required for a direct application of (7.59). Such operation counts actually depend on the exact details of the computer program used to perform the calculations. However, in order to simplify the analysis, we envision a program based directly on the formulas which, for example, does not take into account of the fact that the term $y_0 z^0$ would not actually require a multiplication.

For both methods let us assume that $z^0, z^1, \ldots, z^7$ are computed and stored in a table. This requires 7 multiplications. Note that (7.60) requires 3 additions and 4 multiplications to compute each term $S_e(k)$ or $S_o(k)$, for a subtotal of 24 additions and 32 multiplications. The transforms $F_d(\omega_k)$, $k = 0, 1, 2, \ldots, 7$ computed from (7.61) each require one addition and one division for a second subtotal of 8 additions and 8 divisions. Adding the subtotals and the 7 multiplications required for the powers of z we obtain the results shown in the second row of Table 7.1. A similar analysis applied to (7.59) was used to obtain the first row in this table. The savings obtained by storing $S_e(k)$ and $S_o(k)$, $k = 0, 1, 2, 3$ for later use amounted to 24 additions and 32 multiplications.

A further saving can be effected by applying the same type of reasoning to each of the terms $S_e(k)$ and $S_o(k)$. Let

$$S_{ee}(k) = y_0 + y_4 z^{4k}, \quad S_{eo}(k) = y_2 z^{2k} + y_6 z^{6k}, \quad k = 0, 1$$
$$S_{oe}(k) = y_1 z^k + y_5 z^{5k}, \quad S_{oo}(k) = y_3 z^{3k} + y_7 z^{7k}, \quad k = 0, 1 \quad (7.62)$$

Then

$$S_e(k) = S_{ee}(k) + S_{eo}(k), \quad S_o(k) = S_{oe}(k) + S_{oo}(k), \quad k = 0, 1$$
$$S_e(k) = S_{ee}(k) - S_{eo}(k), \quad S_o(k) = S_{oe}(k) - S_{oo}(k), \quad k = 2, 3$$
$$(7.63)$$

Now $S_e(k)$ and $S_o(k)$, $k = 0, 1, 2, 3$ can be computed by means of (7.62) and (7.63) with a total of 16 additions and 16 multiplications for an additional saving of 8 additions and 16 multiplications. The totals for this approach are shown on the third line of Table 7.1.

The fast Fourier transform algorithm proceeds along the lines illustrated in the above example. This is an example of the divide and conquer strategy which has wide application in computer science. The original sums required for the DFT are split into two sums, one involving terms with even index and the other involving terms with odd index. These new sums form two new DFT computations, each with half as many data points. The new DFT computations can be split in two again, and so on until the sums only have two terms. The next theorem shows that when n is a power of 2, the DFT calculations (7.59) can be accomplished with only $n \log_2 n$ multiplications and additions.

TABLE 7.1. Operation Counts for Formulas (7.59), (7.61), and (7.63)

	Additions	Multiplications	Divisions
Formula (7.59)	56	71	8
Formula (7.61)	32	39	8
Formula (7.63)	24	23	8

Theorem 7.7. Suppose that $n = 2^m$ for some integer $m \geq 1$, $z = e^{-2\pi i/n}$, and $y_0, y_1, \ldots, y_{n-1}$ are complex numbers. The sums

$$S(k) = \sum_{j=0}^{n-1} y_j z^{kj}, \quad k = 0, 1, \ldots, n-1 \tag{7.64}$$

can be evaluated using at most $n \log_2 n$ each of additions and multiplications.

This theorem will be proved using induction on m. When $m = 1$ ($n = 2$) we have

$$S(0) = y_0 + y_1 z^0, \quad S(1) = y_0 + y_1 z$$

These sums can be evaluated with at most 2 additions and 2 multiplications. Thus the theorem holds when $m = 1$ because in that case $n \log_2 n = 2$.

For the inductive step suppose the theorem holds for some $m \geq 1$. Let $n = 2^{m+1}$ and suppose $z = e^{2\pi i/n}$. Define

$$S_e(k) = \sum_{j=0}^{n/2-1} y_{2j} z^{2kj}, \quad k = 0, 1, \ldots, n-1$$

$$S_o(k) = \sum_{j=0}^{n/2-1} y_{2j+1} z^{(2j+1)k}, \quad k = 0, 1, \ldots, n-1 \tag{7.65}$$

For $k \geq (n/2)$, let $\lambda = k - (n/2)$. Then

$$S_e(k) = \sum_{j=0}^{n/2-1} y_{2j} z^{2(n/2+\lambda)j} = \sum_{j=0}^{n/2-1} y_{2j} z^{nj} z^{2\lambda j} = S_e(\lambda) = S_e\left(k - \frac{n}{2}\right)$$

because $z^{nj} = (z^n)^j = 1$. Similarly

$$S_o(k) = \sum_{j=0}^{n/2-1} y_{2j+1} z^{(n/2+\lambda)(2j+1)} = \sum_{j=0}^{n/2-1} y_{2j+1} z^{(n/2)(2j+1)} z^{(2j+1)\lambda}$$

$$= -S_o\left(k - \frac{n}{2}\right)$$

because $z^{(2j+1)n/2} = z^{nj} z^{n/2} = e^{\pi i} = -1$. Thus

$$S(k) = S_e(k) + S_o(k), \quad k = 0, 1, \ldots, \frac{n}{2} - 1$$

$$S(k) = S_e\left(k - \frac{n}{2}\right) - S_o\left(k - \frac{n}{2}\right), \quad k = \frac{n}{2}, \ldots, n-1 \tag{7.66}$$

Now let us determine the number of operations required to evaluate $S_e(k)$ and $S_o(k)$, $k = 0, 1, \ldots, n/2 - 1$. Let $v = n/2$, $w = z^2$, $u_j = y_{2j}$, and $v_j = y_{2j+1}$. Substituting these in (7.65) and simplifying, we obtain

$$S_e(k) = \sum_{j=0}^{\nu-1} u_j w^{kj}, \quad k = 0, 1, \ldots, \nu - 1 \tag{7.67}$$

$$z^{-k} S_o(k) = \sum_{j=0}^{\nu-1} v_j w^{kj}, \quad k = 0, 1, \ldots, \nu - 1 \tag{7.68}$$

Note that $w^\nu = z^n = 1$; so one can infer that (7.67) and (7.68) are both problems of the form (7.65) with n replaced by $\nu = n/2$. The inductive hypothesis then implies that (7.67) and (7.68) can each be evaluated with $\nu \log_2 \nu$ additions and multiplications. As before, we will assume that $z^0, z^1, \ldots, z^{\nu-1}$ are computed and stored. This leads to the following breakdown for the operations required to compute the $S(k)'s$, as shown in Table 7.2. The totals follow from the fact that $n = 2\nu$ and the equation

$$n + n \log_2 \frac{n}{2} = n + n[\log_2 n - 1] = n \log_2 n$$

This completes the inductive step and the proof is complete.

Corollary 7.8. Suppose that $n = 2^m$ for some integer $m \geq 1$, $\omega_j = 2\pi j/n$, and $y_0, \ldots, y_{n-1}$ are complex numbers. The discrete Fourier transform F_d can be evaluated at $\omega_0, \ldots, \omega_{n-1}$ using at most n divisions, and $n \log_2 n$ each of additions and multiplications.

This corollary follows from Theorem 7.7 and (7.49).

The fast Fourier transform can be used to approximate Fourier series and transforms, and to speed up the process of approximating solutions to PDEs. These applications are discussed by Weaver [51], Pickering [38], Vichnevetsky and Bowles [49], and Brigham [9].

Exercises 7.9

1. (a) Write out expressions for the sums $S(k)$ in (7.64) for $n = 6$.
   (b) Write out expressions for $S_o(k)$ and $S_e(k)$ in (7.65) for $n = 6$.

TABLE 7.2. Operation Counts for Fast Fourier Transform

	Additions	Multiplications
Compute $S(k)$ by (7.66)	n	
Compute $S_e(k)$ by (7.67)	$\nu \log_2 \nu$	$\nu \log_2 \nu$
Compute $z^{-k} S_o(k)$ by (7.68)	$\nu \log_2 \nu$	$\nu \log_2 \nu$
Compute $z^0, \ldots, z^{\nu-1}$		ν
Multiply $z^{-k} S_o(k)$ by z^k		ν
Total	$n \log_2 n$	$n \log_2 n$

(c) Show that the formulas (7.66) are valid for this example.

(d) Determine the number of operations required for calculating $F_d(\omega_k)$, $k = 0, 1, \ldots, 5$ using (7.49) and compare with those required for (7.66) for $n = 6$.

7.18. FOURIER TRANSFORMS OF FUNCTIONS OF TWO VARIABLES

Fourier transforms may be extended to functions of two or more variables. We consider the case for functions of two variables primarily to understand an origin for a Hankel transform. If $f(x, y)$ is a function of two independent variables x and y, let us consider f as a function of x alone. Then the Fourier exponential transform is

$$\hat{f}(\alpha, y) = \int_{-\infty}^{\infty} f(x, y)e^{-i\alpha x} \, dx \tag{7.69}$$

Considering $\hat{f}(\alpha, y)$ as a function of y, we have

$$F(\alpha, \beta) = \int_{-\infty}^{\infty} \hat{f}(\alpha, y)e^{-i\beta y} \, dy \tag{7.70}$$

Combining (7.69) and (7.70),

$$F(\alpha, \beta) = \int_{-\infty}^{\infty} \left[\int_{-\infty}^{\infty} f(x, y)e^{-i\alpha x} \, dx \right] e^{-i\beta y} \, dy$$

or

$$F(\alpha, \beta) = \int_{-\infty}^{\infty} \int_{-\infty}^{\infty} f(x, y)e^{-i(\alpha x + \beta y)} \, dx \, dy \tag{7.71}$$

This is the *exponential Fourier transform of $f(x, y)$*.

To consider the inversion of $F(\alpha, \beta)$, we begin by considering

$$f(x, y) = \frac{1}{2\pi} \int_{-\infty}^{\infty} \hat{f}(\alpha, y)e^{i\alpha x} \, d\alpha \tag{7.72}$$

In a similar manner,

$$\hat{f}(\alpha, y) = \frac{1}{2\pi} \int_{-\infty}^{\infty} F(\alpha, \beta)e^{i\beta y} \, d\beta \tag{7.73}$$

From (7.72) and (7.73), we obtain

$$f(x, y) = \frac{1}{2\pi} \int_{-\infty}^{\infty} \left[\frac{1}{2\pi} \int_{-\infty}^{\infty} F(\alpha, \beta)e^{i\beta y} \, d\beta \right] e^{i\alpha x} \, d\alpha$$

or

$$f(x, y) = \frac{1}{(2\pi)^2} \int_{-\infty}^{\infty} \int_{-\infty}^{\infty} F(\alpha, \beta) e^{i(\alpha x + \beta y)} \, d\alpha \, d\beta \qquad (7.74)$$

Thus (7.71) and (7.74) may be considered an *exponential Fourier transform pair*. From this pair we may obtain a new pair by transforming coordinates.

7.19. HANKEL TRANSFORMS

In (7.71) we introduce the polar coordinate substitutions,

$$x = r \cos \theta, \quad y = r \sin \theta, \quad \alpha = \rho \cos \phi, \quad \beta = \rho \sin \phi$$

$$dx \, dy = r \, dr \, d\theta, \quad r = \sqrt{x^2 + y^2}, \quad \rho = \sqrt{\alpha^2 + \beta^2},$$

$$\alpha x + \beta y = r\rho \cos(\theta - \phi),$$

$$d\alpha \, d\beta = \rho \, d\rho \, d\phi$$

Then,

$$F(\alpha, \beta) = \int_{-\infty}^{\infty} \int_{-\infty}^{\infty} f(\sqrt{x^2 + y^2}) e^{-i(\alpha x + \beta y)} \, dx \, dy$$

and this becomes,

$$F(\rho) = \int_0^{\infty} r f(r) \, dr \int_0^{2\pi} e^{-i\rho r \cos(\theta - \phi)} \, d\theta$$

Due to the periodic nature of the integrand,

$$\int_0^{2\pi} e^{-i\rho r \cos(\theta - \phi)} \, d\theta = \int_0^{2\pi} e^{-i\rho r \cos \theta} \, d\theta$$

From Problem 16 of Exercises 5.1,

$$\int_0^{2\pi} e^{i\rho r \cos \theta} \, d\theta = 2\pi J_0(\rho r)$$

and

$$\int_0^{2\pi} e^{-i\rho r \cos \theta} \, d\theta = 2\pi J_0(-\rho r)$$

From (5.6), $J_0(-\rho r) = J_0(\rho r)$. Therefore,

$$F(\rho) = \int_0^{\infty} r f(r) \, dr [2\pi J_0(\rho r)] \, dr$$

or

$$F(\rho) = 2\pi \int_0^\infty rf(r)J_0(\rho r)\, dr \qquad (7.75)$$

The inverse

$$f(x,\, y) = \frac{1}{(2\pi)^2} \int_{-\infty}^\infty \int_{-\infty}^\infty F(\sqrt{\alpha^2 + \beta^2})e^{i(\alpha x + \beta y)}\, d\alpha\, d\beta$$

becomes

$$f(r) = \frac{1}{(2\pi)^2} \int_0^\infty \int_0^{2\pi} F(\rho)e^{i\rho r \cos(\theta - \phi)}\rho\, d\rho\, d\theta$$

Then

$$f(r) = \frac{1}{(2\pi)^2} \int_0^\infty \rho F(\rho)[2\pi J_0(\rho r)]\, d\rho$$

It follows that

$$f(r) = \frac{1}{2\pi} \int_0^\infty \rho F(\rho)J_0(\rho r)\, d\rho \qquad (7.76)$$

The constant factors preceding the integrals of (7.75) and (7.76) come from the Fourier integral representation and could have been assigned differently (or neglected completely). It is important to know the exact statement of the transform pair when tabular entries are used. Except for the notation, Miles [34, p. 85] defines the *Hankel transform* or *Bessel transform* with the nth order Bessel function kernel by

$$F(\rho) = \int_0^\infty rf(r)J_n(\rho r)\, dr = B_n\{f(r)\} \qquad (7.77)$$

and the inverse

$$f(r) = \int_0^\infty \rho F(\rho)J_n(\rho r)\, d\rho \qquad (7.78)$$

In these definitions the transform and its inverse have identical integrals.

The other transforms we consider, involving Bessel functions are associated with the orthogonal series of Bessel functions. From (5.57) and (5.58) we define the *Finite Hankel transform* and its *inverse*, if the end condition $J_n(\lambda b) = 0$ is given.

$$H_n\{f(x)\} = \int_0^b xf(x)J_n(\lambda_k x)\, dx = F_{Hn}(\lambda_k) \qquad (7.79)$$

and

$$f(x) = \frac{2}{b^2} \sum_{k=1}^{\infty} \frac{F_{Hn}(\lambda_k) J_n(\lambda_k x)}{J_{n+1}^2(\lambda_k b)} \tag{7.80}$$

If λ_k are the eigenvalues from the end condition

$$h J_n(\lambda b) + \lambda b J_n'(\lambda b) = 0 , \quad h \geqslant 0 \tag{7.81}$$

then

$$H_n\{f(x)\} = \int_0^b x f(x) J_n(\lambda_k x) \, dx = F_{Hn}(\lambda_k) , \quad k \in \mathbf{N} \tag{7.82}$$

and

$$f(x) = 2 \sum_{k=1}^{\infty} \frac{\lambda_k^2 F_{Hn}(\lambda_k) J_n(\lambda_k x)}{[h^2 + (\lambda_k b)^2 - n^2] J_n^2(\lambda_k b)} \tag{7.83}$$

The case where the boundary condition is $J_0'(\lambda b) = 0$ is left for the problems. No special case is made for $J_n'(\lambda b) = 0$ if $n \neq 0$. In this situation transforms come from (7.82) and (7.83) with $h = 0$.

Example 7.15. Find the finite Hankel transform of $f'(x)$ if $J_n(\lambda_k b) = 0$.

$$H_n\{f'(x)\} = \int_0^b x f'(x) J_n(\lambda_k x) \, dx = [x f(x) J_n(\lambda_k x)]_0^b - \int_0^b f(x) [x J_n(\lambda_k x)]' \, dx$$

From (5.28)

$$J_n(\lambda_k x) = \frac{\lambda_k x}{2n} [J_{n-1}(\lambda_k x) + J_{n+1}(\lambda_k x)] \tag{7.84}$$

and from (5.27)

$$\lambda_k x J_n'(\lambda_k x) = \frac{\lambda_k x}{2} [J_{n-1}(\lambda_k x) - J_{n+1}(\lambda_k x)] \tag{7.85}$$

$$H_n\{f'(x)\} = b f(b) J_n(\lambda_k b) - 0 - \int_0^b f(x) [J_n(\lambda_k x) + \lambda_k x J_n(\lambda_k x)] \, dx$$

But $J_n(\lambda_k b) = 0$ since λ_k is an eigenvalue. Therefore,

$$H_n\{f'(x)\} = - \int_0^b f(x) \left[\frac{\lambda_k x}{2n} (J_{n-1}(\lambda_k x) + J_{n+1}(\lambda_k x)) \right.$$

$$\left. + \frac{n \lambda_k x}{2n} (J_{n-1}(\lambda_k x) - J_{n+1}(\lambda_k x)) \right] dx$$

$$H_n\{f'(x)\} = \int_0^b x f(x) \frac{\lambda_k}{2n} [(n-1) J_{n+1}(\lambda_k x) - (n+1) J_{n-1}(\lambda_k x)] \, dx$$

and

$$H_n\{f'(x)\} = \frac{\lambda_k}{2n} [(n-1)H_{n+1}\{f(x)\} - (n+1)H_{n-1}\{f(x)\}] \quad (7.86)$$

where H_{n+1} and H_{n-1} show (in the subscripts) the orders of the Bessel functions used in the transforms.

Example 7.16. If $J_n(\lambda_k b) = 0$, the Finite Hankel transform of $f(x)/x$ is

$$H_n\left\{\frac{f(x)}{x}\right\} = \int_0^b x\left(\frac{f(x)}{x}\right)J_n(\lambda_k x)\,dx$$

$$= \int_0^b f(x)\left[\frac{\lambda_k x}{2n}(J_{n-1}(\lambda_k x) + J_{n+1}(\lambda_k x))\right] dx$$

$$= \frac{\lambda_k}{2n}\int_0^b xf(x)[J_{n-1}(\lambda_k x) + J_{n+1}(\lambda_k x)]\,dx$$

or

$$H_n\left\{\frac{f(x)}{x}\right\} = \frac{\lambda_k}{2n}[H_{n-1}\{f(x)\} + H_{n+1}\{f(x)\}] \qquad (7.87)$$

Exercises 7.10

1. If $J_0'(\lambda b) = 0$, show that the finite Bessel transform pair is

$$H_0\{f(x)\} = \int_0^b xf(x)\,dx = F_{HO}(0)\,, \quad \text{if } \lambda_0 = 0, \quad k = 0\,,$$

$$H_0\{f(x)\} = \int_0^b xf(x)J_0(\lambda_k x)\,dx = F_{HO}(\lambda_k)\,, \quad \text{if } k \in \mathbf{N},$$

and

$$f(x) = \frac{2F_{HO}(0)}{b^2} + 2\sum_{k=1}^{\infty}\frac{F_{HO}(\lambda_k)J_0(\lambda_k x)}{b^2 J_0^2(\lambda_k b)}$$

2. Show that if $J_n(\lambda_k b) = 0$,

$$H_n\left\{f''(x) + \frac{1}{x}f'(x)\right\} = \frac{\lambda_k}{2}[H_{n+1}\{f'(x)\} - H_{n-1}\{f'(x)\}]\,,$$

where H_{n-1} and H_{n+1} show the order of the Bessel functions of the transforms.

3. (a) Using integration by parts show that

$$H_n\left\{f''(x) + \frac{1}{x}f'(x)\right\} = -\lambda_k \int_0^{\infty} xf'(x)J_n'(\lambda_k x)\,dx$$

if $J_n(\lambda_k b) = 0$.

(b) Determine the transform of $f''(x) + (1/x)f'(x) - (n^2/x^2)f(x)$. Show that it can be expressed by

$$H_n\left\{f''(x) + \frac{1}{x} f'(x) - \frac{n^2}{x^2} f(x)\right\} = -\lambda_k bf(b)J'_n(\lambda_k b) - \lambda_k^2 H_n\{f(x)\}$$

if $J_n(\lambda_k b) = 0$.

4. Show that if $J_1(\lambda_k b) = 0$,

$$H_1\{f'(x)\} = -\lambda_k H_0\{f(x)\}$$

5. (a) If $J_0(\lambda_k b) = 0$, show that

$$H_0\left\{f''(x) + \frac{f'(x)}{x}\right\} = \lambda_k bf(b)J_1(\lambda_k b) - \lambda_k^2 H_0\{f(x)\}$$

(b) If $f(b) = 0$, then

$$H_0\left\{f''(x) + \frac{f'(x)}{x}\right\} = -\lambda_k^2 H_0\{f(x)\}$$

6. Show that $\int_0^b x^{n+1}J_n(\lambda_k x) \, dx = (b^{n+1}/\lambda_k)J_{n+1}(\lambda_k b)$ and that $H_n\{x^n\} = (b^{n+1}/\lambda_k)J_{n+1}(\lambda_k b)$. For the following exercises use the transform pair (7.77) and (7.78).

7. Show that

$$B_n\{f(kr)\} = \frac{1}{k^2} F\left(\frac{\rho}{k}\right)$$

8. Show that

$$B_n\left\{\frac{f(r)}{r}\right\} = \frac{\rho}{2n} [B_{n+1}\{f(r)\} + B_{n-1}\{f(r)\}]$$

9. Assuming that $rf(r) \to 0$ as $r \to 0$ and $r \to \infty$, obtain the transform

$$B_n\{f'(r)\} = \frac{\rho}{2n} [(n-1)B_{n+1}\{f(r)\} - (n+1)B_{n-1}\{f(r)\}]$$

10. Assuming that $rf'(r) \to 0$ as $r \to 0$ and $r \to \infty$, show that

$$B_n\{f''(r)\} = \frac{\rho}{2n} [(n-1)B_{n+1}\{f'(r)\} - (n+1)B_{n-1}\{f'(r)\}]$$

11. (a) Suppose $rf'(r) \to 0$ as $r \to 0$ and $r \to \infty$. Using integration by parts, determine that

$$B_n\{f''(r)\} = -\int_0^\infty \frac{df}{dr} \frac{d}{dr} [rJ_n(\rho r)] \, dr$$

(b) If $rf(r) \to 0$ as $r \to 0$ and $r \to \infty$, show that

$$B_n\left\{f''(r) + \frac{1}{r} f'(r)\right\} = \rho \int_0^\infty f(r) \frac{d}{dr} [rJ_n'(\rho r)] \, dr$$

(c) Obtain the integral relationship

$$\int_0^\infty r\left[f''(r) + \frac{1}{r} f'(r) - \frac{n^2}{r^2} f(r)\right] J_n(\rho r) \, dr$$

$$= \rho^2 \int_0^\infty rf(r)\left[J_n''(\rho r) + \frac{J_n'(\rho r)}{\rho r} - \frac{n^2}{\rho^2 r^2} J_n(\rho r)\right] dr$$

(d) Finally, by employing the fact that $J_n(\rho r)$ satisfies the Bessel differential equation

$$y''(x) + \frac{1}{x} y'(x) + \left(1 - \frac{n^2}{x^2}\right) y(x) = 0$$

show that the following operational property holds:

$$B_n\left\{f''(r) + \frac{1}{r} f'(r) - \frac{n^2}{r^2} f(r)\right\} = -\rho^2 F(\rho)$$

7.20. LEGENDRE TRANSFORM

Finite Fourier transforms were defined from an observation of Fourier series and their coefficients. In much the same way, the finite Hankel transform was defined. Similar definitions are used here with the Legendre series from (6.29) and (6.30). The *Legendre transform of f(x)* is defined by the following equation:

$$L\{f(x)\} = \int_{-1}^1 f(x)P_n(x) \, dx = F_L(n) \tag{7.88}$$

and its *inverse*

$$f(x) = \frac{1}{2} \sum_{n=0}^\infty (2n + 1)F_L(n)P_n(x) \tag{7.89}$$

Using (6.24),

$$L\{[(1 - x^2)f'(x)]'\} = L\{-n(n + 1)f(x)\}$$

or we can express the *differentiation property*

$$L\{[(1-x^2)f'(x)]'\} = -n(n+1)F_L(n) \qquad (7.90)$$

Equation (6.22) can be written in the form,

$$\frac{d}{d\theta}\left(\sin\theta\,\frac{df}{d\theta}\right) + n(n+1)\sin\theta\,f = 0$$

Therefore,

$$\frac{1}{\sin\theta}\frac{d}{d\theta}\left(\sin\theta\,\frac{df}{d\theta}\right) = -n(n+1)f$$

and

$$L\left\{\frac{1}{\sin\theta}\frac{d}{d\theta}\left(\sin\theta\,\frac{df}{d\theta}\right)\right\} = -n(n+1)F_L(n) \qquad (7.91)$$

7.21. MELLIN TRANSFORM

We define the *Mellin transform* by selecting in the integral transform (7.1) the kernel $K(s, x) = x^{s-1}$, $a = 0$, and $b = \infty$. Therefore, the transform is displayed by

$$M_s\{f(x)\} = \int_0^\infty f(x)x^{s-1}\,dx = F_M(s) \qquad (7.92)$$

From definition (7.92) we begin our discussion by finding a transform for $xf'(x)$.

$$M_s\{xf'(x)\} = \int_0^\infty xf'(x)x^{s-1}\,dx$$

Integrating by parts, we have

$$M_s\{xf'(x)\} = [f(x)x^s]_0^\infty - s\int_0^\infty f(x)x^{s-1}\,dx$$

If we assume that $f(x)x^s \to 0$ as $x \to \infty$, then

$$M_s\{xf'(x)\} = -sF_M(s) \qquad (7.93)$$

By using the same procedure again,

$$M_s\{x(xf'(x))'\} = -sM_s\{xf'(x)\} = s^2F_M(s), \qquad (7.94)$$

assuming that both $f(x)x^s \to 0$ and $xf'(x)x^s \to 0$ as $x \to 0$. The function $x(xf'(x))' = x^2f''(x) + xf'(x)$, so that

$$M_s\{x^2f''(x) + xf'(x)\} = s^2F_M(s) \tag{7.95}$$

From (7.93) and (7.95), one obtains

$$M_s\{x^2f''(x)\} = (s^2 + s)F_M(s) \tag{7.96}$$

Using the definition (7.92) again, we determine a transform for $f'(x)$.

$$M_s\{f'(x)\} = \int_0^\infty f'(x)x^{s-1}\,dx$$

Using integration by parts,

$$M_s\{f'(x)\} = [f(x)x^{s-1}]_0^\infty - \int_0^\infty f(x)(s-1)x^{s-2}\,dx$$

If $f(x)x^{s-1} \to 0$ as $x \to \infty$, then

$$M_s\{f'(x)\} = -(s-1)F_M(s-1) \tag{7.97}$$

Continuing with transforms of derivatives of functions $f^{(n)}(x)$, we have

$$M_s\{f^{(n)}(x)\} = \int_0^\infty f^{(n)}(x)x^{s-1}\,dx$$
$$= [f^{(n-1)}(x)x^{s-1}]_0^\infty - \int_0^\infty f^{(n-1)}(x)(s-1)x^{s-2}\,dx$$

In our use of integration by parts, we assume that f is such that the content of the square brackets $[\ ]_0^\infty$ vanishes. Therefore

$$M_s\{f^{(n)}(x)\} = -\int_0^\infty f^{(n-1)}(x)(s-1)x^{s-2}\,dx$$
$$= -(s-1)\int_0^\infty f^{(n-2)}(x)(s-2)x^{s-3}\,dx$$
$$= (s-2)(s-1)(-1)\int_0^\infty f^{(n-3)}(x)(s-3)x^{s-4}\,dx$$
$$= -(s-3)(s-2)(s-1)\int_0^\infty f^{(n-3)}(x)x^{s-4}\,dx$$

Continuing this process we find that

$$M_s\{f^{(n)}(x)\} = (-1)^n(s-n)(s-(n-1))\ldots(s-2)(s-1)F_M(s-n) \tag{7.98}$$

This gives the Mellin transform of $f^{(n)}(x)$ in terms of the Mellin transform of the function with s replaced by $s - n$ in the transform.

Other transforms follow:

$$M_s\{f(ax)\} = \int_0^\infty f(ax)x^{s-1}\,dx$$

In this integral, let $u = ax$, so that

$$M_s\{f(ax)\} = \int_0^\infty f(u)\frac{u^{s-1}}{a^{s-1}}\frac{du}{a} = \frac{1}{a^s}\int_0^\infty f(u)u^{s-1}\,du$$

Therefore,

$$M_s\{f(ax)\} = a^{-s}F_M(s), \quad a > 0 \tag{7.99}$$

Also

$$M_s\{f(x^h)\} = \int_0^\infty f(x^h)x^{s-1}\,dx = \int_0^\infty f(u)u^{(s-1)/h}\,\frac{du}{hu^{(h-1)/h}}$$

if $u = x^h$, $du = hx^{h-1}\,dx$ and $x = u^{1/h}$. Then

$$M_s\{f(x^h)\} = \frac{1}{h}\int_0^\infty f(u)u^{(s-1-h+1)/h}\,du = \frac{1}{h}\int_0^\infty f(u)u^{(s/h)-1}\,du$$

Hence

$$M_s\{f(x^h)\} = \frac{1}{h}F_M\left(\frac{s}{h}\right), \quad h > 0 \tag{7.100}$$

The inversion formula, involving complex variables, may be stated as follows. Assume that $f(x)$ is PWS when $0 \leqslant x < \infty$ and $f(x) = O(e^{c_0 x})$, $x \geqslant 0$, $c_0 \geqslant 0$. If $F_M(s)$ is the transform of $f(x)$, then

$$f(x) = \frac{1}{2\pi i}\int_{c-i\infty}^{c+i\infty} F_M(s)x^{-s}\,c > c_0 \tag{7.101}$$

There is a convolution relation for the Mellin transform. Assume $F_M(s)$ is the transform of $f(x)$ when $a_1 < s < b_1$ and $G_M(s)$ is the transform of $g(x)$ when $a_2 < s < b_2$. If $\max(a_1, a_2) < s < \min(b_1, b_2)$, then

$$F_M(s)G_M(s) = M_s\left\{\int_0^\infty f(u)g\left(\frac{x}{u}\right)\frac{du}{u}\right\} \tag{7.102}$$

The Mellin transform becomes the two-sided or bilateral Laplace transform or a combination of the Laplace transform if the variable is changed so that

$$x = e^{-t}, \quad dx = -e^{-t}\,dt$$

Then

$$F_M(s) = \int_\infty^{-\infty} f(e^{-t}) e^{-t(s-1)} (-e^{-t} \, dt)$$

and

$$F_M(s) = \int_{-\infty}^{\infty} f(e^{-t}) e^{-st} \, dt \tag{7.103}$$

Assuming $s = p + iq$, a complex variable, then the Mellin transform may be expressed as an exponential Fourier transform. We assume also that the variable x is changed by the substitution $x = e^{-t}$. Then the Mellin transform

$$F_M(s) = \int_0^\infty f(x) x^{s-1} \, dx$$

becomes

$$F_M(p + iq) = \int_0^\infty f(e^{-t}) e^{-t(p+iq-1)} (-e^{-t} \, dt)$$

$$= \int_{-\infty}^\infty f(e^{-t}) e^{-pt} e^{-iqt} \, dt$$

Then let x become the integration variable. This gives

$$F_M(p + iq) = \int_{-\infty}^\infty [f(e^{-x}) e^{-px}] e^{-iqx} \, dx \tag{7.104}$$

It is possible to obtain Mellin transforms from the bilateral Laplace transforms and the exponential Fourier transform tables. Erdelyi [19] has extensive tables for the Mellin transform and its inverse. Some applications of the Mellin transform are given by Tranter [48]. Problems involving such forms as $xf'(x)$ and $x^2 f''(x)$ are likely candidates for the use of the Mellin transform. However, the conditions under which the transform exists are somewhat restrictive and this tends to limit its practical use.

Exercises 7.11

1. (a) Show that $M_s\{f(1/x)\} = F_M(-s)$.
   (b) Determine the Mellin transform for $x^a f(x)$.
   (c) Find $M_s\{f(x^{-h})\}$ if $h > 0$.
   (d) Show that $M_s\{x^a f(bx)\} = b^{-s-a} F_M(s + a)$, $a > 0$.
   (e) Demonstrate that $M_s\{e^{-x}\} = \Gamma(s)$.
   (f) Find $M_s\{e^{-ax}\}$.
   (g) Show that $M_s\{\sin x\} = \Gamma(s) \sin \pi s/2$.
   (h) Determine $M_s\{\cos x\}$.

8

APPLICATIONS OF BVPs

In this chapter we are concerned with the actual formulations and solutions of BVPs. Although some problems are stated without specific reference to physical situations, most of our models are related to "real world" physical or geometric concepts. Our idea of modeling is a process that progresses from a physical or ideological notion to a mathematical description of the concept in terms of IVPs, SLPs, BVPs, and so forth. Our primary objective is to bring together all of the ingredients of the *Fourier Method* for solving BVPs. This procedure involves (a) separation of variables, (b) SLPs, and (c) superposition for the homogeneous part of the problem. After superposition comes (d) evaluation of nonhomogeneous initial conditions. Proper sequencing of the various parts of the solution is very important. The methods discussed here are primarily suited for linear PDEs with homogeneous boundary conditions and one or more nonhomogeneous initial conditions. Most of the necessary mathematical tools have been discussed in previous chapters.

Although our main emphasis is centered on the systematic solution of BVPs, we have included several examples of existence and uniqueness demonstrations for BVPs. Numerical methods are illustrated by applying difference techniques to the heat and Laplace equations.

8.1. THE VIBRATING STRING

A vibrating string stretched between two points is our subject for modeling. Figure 8.1 is a representation of a small section or element of the string. We investigate displacements if the string satisfies the following characteristics:

1. Motion is entirely in the xy plane. Equilibrium points are positions along the x axis.
2. The string is completely flexible. Tensile forces $T(x)$ exerted on an element are tangent to the string midline at points of action (x, y) and $(x + \Delta x, y + \Delta y)$.

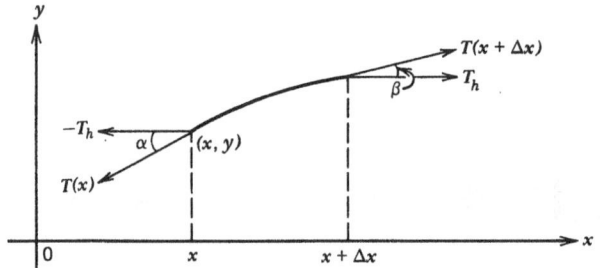

Figure 8.1. The element of a string.

3. Displacements $y(x, t)$ are small compared to the length of the string. A point moves only in the y direction.
4. Slopes $y_x(x, t)$ are small. Horizontal components T_h of $T(x)$ have equal magnitudes.
5. The string is a uniform substance which has a constant density ρ.

Equating forces in the vertical direction, we have

$$-T(x) \sin \alpha + T(x + \Delta x) \sin \beta - mg = my_{tt}(x, t) \qquad (8.1)$$

since Newton's second law describes the force as $my_{tt}(x, t)$. The force $mg = \rho \, \Delta x$ for the element.

According to assumption 4,

$$-T(x) \cos \alpha + T(x + \Delta x) \cos \beta = 0$$

and

$$T(x) \cos \alpha = T(x + \Delta x) \cos \beta = T_h \qquad (8.2)$$

Solving for $T(x)$ and $T(x + \Delta x)$ in (8.2) we find

$$T(x) = \frac{T_h}{\cos \alpha}, \quad T(x + \Delta x) = \frac{T_h}{\cos \beta} \qquad (8.3)$$

The results of (8.3) in (8.1) allow us to write

$$-T_h \tan \alpha + T_h \tan \beta - \rho \, \Delta x g = \rho \, \Delta x y_{tt}(x, t) \qquad (8.4)$$

The slopes at the ends (x, y) and $(x + \Delta x, y + \Delta y)$ are

$$y_x(x, t) = \tan \alpha, \quad y_x(x + \Delta x, t) = \tan \beta \qquad (8.5)$$

The results of (8.5) in (8.4) permit us to state that

$$-T_h y_x(x, t) + T_h y_x(x + \Delta x, t) - \rho g \, \Delta x = \rho y_{tt}(x, t) \, \Delta x$$

Dividing by Δx, we have

$$T_h\left[\frac{y_x(x+\Delta x, t)-y_x(x, t)}{\Delta x}\right]-\rho g = \rho y_{tt}$$

If $\Delta x \to 0$, then

$$T_h y_{xx}(x, t)-\rho g = \rho y_{tt}$$

This may be written

$$y_{tt}=\frac{T_h}{\rho}\ y_{xx}-g \qquad (8.6)$$

If the weight of the string is neglected in (8.6), then

$$y_{tt}=a^2 y_{xx} \qquad (8.7)$$

where $a^2 = T_h/\rho$. Equation (8.7) is the *wave equation*. If in place of *mg* in (8.1) we insert a damping force per unit length Δx proportional to the velocity y_t, we obtain

$$y_{tt}=a^2 y_{xx}-ky_t$$

where k includes the original proportionality constant and another factor.

From the description of the problem, we state an appropriate set of constraints

$$y(0, t) = y(L, t) = 0$$

The string is fastened at $x = 0$ and $x = L$. Displacement is zero at each end point.

$$y_t(x, 0) = 0$$

The string is at rest when we begin the experiment. Initial velocity is zero.

$$y(x, 0) = f(x)$$

Initially the string follows the curve $y = f(x)$ as suggested in Figure 8.2.

$$|y(x, t)| < M$$

Figure 8.2. Initial position of a string.

For all x and t in the domain, the displacement is bounded. We request a bounded solution to our problems generally, even though we may omit this restrictive notation. The formal statement of the BVP follows:

$$y_{tt}(x, t) = a^2 y_{xx}(x, t) , \quad 0 < x < L, \; t > 0$$
$$y(0, t) = y(L, t) = 0 , \quad t \geqslant 0$$
$$y_t(x, 0) = 0 , \quad y(x, 0) = f(x) , \quad 0 \leqslant x \leqslant L$$
$$|y(x, t)| < M , \quad 0 \leqslant x \leqslant L, \; t \geqslant 0 \tag{8.8}$$

To solve the BVP we proceed with the Fourier method.

1. *Separation of Variables.* Let $y(x, t) = X(x) T(t)$. We write the PDE of (8.8) in the form

$$XT'' = a^2 X'' T$$

and dividing by $a^2 XT$, one obtains

$$\frac{T''}{a^2 T} = \frac{X''}{X} \tag{8.9}$$

The ratios of (8.9) cannot be positive, since a bounded solution is required. Therefore

$$\frac{T''}{a^2 T} = \frac{X''}{X} = -\alpha^2$$

As a result we obtain two ODEs.

2. *Related ODEs*

$$X'' + \alpha^2 X = 0 , \quad T'' + \alpha^2 a^2 T = 0$$

To solve the PDE with its boundary conditions, we actually solve the related ODEs with their constraints. Therefore, we want boundary conditions associated with ODEs. We investigate the homogeneous constraints.

3. *Homogeneous Boundary Conditions.* Using the separation substitution of 1, we write

$$y(0, t) = X(0) T(t) = 0$$

Since the product $X(0) T(t)$ is zero, at least one factor is zero. If $T(t) = 0$, then the trivial solution follows, since $T(t)$ is a factor of $y(x, t)$. If $T(t) \neq 0$, then

$$X(0) = 0$$

In a similar argument with

$$y(L, t) = X(L)T(t) = 0$$

if $T(t) \neq 0$, then

$$X(L) = 0$$

The last homogeneous condition is

$$y_t(x, 0) = X(x)T'(0) = 0$$

If $X(x) \neq 0$, then

$$T'(0) = 0 \qquad (8.10)$$

From our related ODEs and newly discovered boundary conditions we write the following.

4. *A Related SLP*

$$X'' + \alpha^2 X = 0, \quad X(0) = X(L) = 0$$

The SLDE has a general solution

$$X = C_1 \cos \alpha x + C_2 \sin \alpha x$$

Using the first boundary condition of the SLP, we have

$$X(0) = C_1 + 0 = 0$$

The second condition along with $C_1 = 0$ permits us to write

$$X(L) = C_2 \sin \alpha L = 0$$

If C_2 and C_1 are both zero, then $X = 0$ is a trivial solution of the SLP. This implies that $y(x, t) = 0$. If $C_2 \neq 0$, then $\sin \alpha L = 0$. This means that $\alpha L = n\pi$ or $\alpha = n\pi/L$. Then

$$\alpha_n^2 = \frac{n^2 \pi^2}{L^2}, \quad n \in N \qquad (8.11)$$

is a set of eigenvalues for the SLP. Testing for $\alpha = 0$, we find only a trivial solution. Therefore, (8.11) is the complete set of eigenvalues. The matching eigenfunctions are

$$X_n(x) = \sin \frac{n\pi x}{L}, \quad n \in \mathbf{N} \tag{8.12}$$

In (8.12) we wrote the solution set with $C_2 = 1$. The general constant factor C_2 could have been included in (8.12). Because of the homogeneity of the linear problem, (8.12) is an adequate solution set or set of eigenfunctions. This is the only complete SLP, but we must solve the T equation with α^2 and the constraint (8.10) known.

5. *The New T Equation.* Now we have the equation and condition

$$T'' + \frac{n^2\pi^2 a^2}{L^2} T = 0, \quad T'(0) = 0$$

The general solution is

$$T = B_1 \cos \frac{n\pi a t}{L} + B_2 \sin \frac{n\pi a t}{L}$$

Since the constraint involves the derivative, we write

$$T' = \frac{n\pi a}{L} \left[-B_1 \sin \frac{n\pi a t}{L} + B_2 \cos \frac{n\pi a t}{L} \right]$$

$$T'(0) = 0 + \frac{n\pi a}{L} B_2 = 0$$

This implies that $B_2 = 0$. The T solution is

$$T_n(t) = \cos \frac{n\pi a t}{L} \tag{8.13}$$

We have employed all homogeneous conditions available in the problem. Now we write the following.

6. *Solution Set for Homogeneous Conditions.* Using the separation substitution along with the solutions (8.12) and (8.13) we write a solution set

$$y_n(x, t) = X_n(x) T_n(t) = \sin \frac{n\pi x}{L} \cos \frac{n\pi a t}{L}, \quad n \in \mathbf{N} \tag{8.14}$$

Since (8.14) is a solution set of a linear homogeneous differential equation accompanied by linear homogeneous boundary conditions, a linear combination of the set is also a solution. This is the *superposition principle* mentioned in Section 1.2. If the set is infinite, as it is here, the linear combination is an infinite series. Superposition of the solution set is still a solution if the series can be differentiated termwise enough times to account for all derivatives of the BVP and all the series are uniformly convergent.

7. *Superposition.* If we write the infinite linear combination of the solution set (8.14), then

$$y(x, t) = \sum_{n=1}^{\infty} C_n \sin \frac{n\pi x}{L} \cos \frac{n\pi at}{L} \qquad (8.15)$$

Finally, we attempt to determine the coefficients of (8.15) by using the remaining steps.

8. *Nonhomogeneous Initial Condition.* From the statement of the BVP (8.8) and (8.15), we observe that

$$y(x, 0) = f(x) = \sum_{n=1}^{\infty} C_n \sin \frac{n\pi x}{L}, \qquad 0 < x < L \qquad (8.16)$$

is the Fourier sine series for f. The coefficients are

$$C_n = \frac{2}{L} \int_0^L f(x) \sin \frac{n\pi x}{L} \, dx$$

9. *Solution for the Original BVP*

$$y(x, t) = \sum_{n=1}^{\infty} C_n \sin \frac{n\pi x}{L} \cos \frac{n\pi at}{L} \qquad (8.17)$$

where

$$C_n = \frac{2}{L} \int_0^L f(x) \sin \frac{n\pi x}{L} \, dx$$

Details of the formal solution are emphasized in items 1–9. The proper sequence of these items in the solution is important. For example, if one uses the nonhomogeneous condition $y(x, 0) = f(x)$ prematurely before superposition, $f(x)$ cannot be an arbitrary function. The items of the solution will vary in problems, but a logical outline of the procedure is recommended.

8.2. VERIFICATION AND UNIQUENESS OF THE SOLUTION OF THE VIBRATING STRING PROBLEM

If $y = f(x)$ is the initial pattern of the string, it is reasonable to assume that f is continuous. Since the string is fixed at $x = 0$ and $x = L$, $f(0) = f(L) = 0$. For convenience we assume that f is a smooth function. By a specialization of Theorem 3.4, the sine series of (8.16) converges to f absolutely and uniformly on $0 \le x \le L$.

Employing the identity

$$\sin \frac{n\pi x}{L} \cos \frac{n\pi at}{L} = \frac{1}{2}\left[\sin \frac{n\pi}{L}(x+at) + \sin \frac{n\pi}{L}(x-at)\right]$$

the solution (8.17) becomes

$$y(x, t) = \frac{1}{2}\left[\sum_{n=1}^{\infty} C_n \sin \frac{n\pi}{L}(x+at) + \sum_{n=1}^{\infty} C_n \sin \frac{n\pi}{L}(x-at)\right]$$

We define

$$F(x) = \sum_{n=1}^{\infty} C_n \sin \frac{n\pi x}{L}$$

for all real x. Then F is the odd periodic extension of f. F must satisfy the conditions

$$F(x) = f(x) \quad \text{for } 0 \leq x \leq L$$
$$F(-x) = -F(x) \quad \text{for all } x$$
$$F(x + 2L) = F(x) \quad \text{for all } x$$

Since

$$F(x + at) = \sum_{n=1}^{\infty} C_n \sin \frac{n\pi}{L}(x+at)$$

$$F(x - at) = \sum_{n=1}^{\infty} C_n \sin \frac{n\pi}{L}(x-at)$$

we have

$$y(x, t) = \tfrac{1}{2}[F(x+at) + F(x-at)] \tag{8.18}$$

To show the verification, we must check that boundary conditions of (8.8) are satisfied by (8.18). Since F is odd,

$$y(0, t) = \tfrac{1}{2}[F(at) + F(-at)] = 0$$
$$y(L, t) = \tfrac{1}{2}[F(L + at) + F(L - at)]$$
$$= \tfrac{1}{2}[F(L + at) + F(-L - at)] = 0$$

Next, we investigate the nonhomogeneous condition

$$y(x, 0) = \tfrac{1}{2}[F(x) + F(x)]$$
$$= F(x) = f(x), \quad 0 \leq x \leq L$$

The derivative f' on $0 \leqslant x \leqslant L$ is continuous, since f is smooth. Therefore our new function F' is continuous for all real x. Now we check

$$y_t(x, t) = \tfrac{1}{2}[aF'(x + at) - aF'(x - at)]$$

$$y_t(x, 0) = \tfrac{1}{2}[aF'(x) - aF'(x)] = 0$$

To check the PDE of the problem, we assume that, on $0 \leqslant x \leqslant L$, f'' is continuous and $f''(0) = f''(L) = 0$. As a result F'' is continuous for all x, and

$$u_{tt} = \tfrac{1}{2}[a^2F''(x + at) + a^2F''(x - at)]$$

$$u_{xx} = \tfrac{1}{2}[F''(x + at) + F''(x - at)]$$

This is adequate to show that

$$u_{tt} = a^2 u_{xx}$$

Since all conditions and the PDE are satisfied, we have shown that, under the conditions asserted, a solution exists.

To establish uniqueness for the solution, we let $y(x, t)$ and $z(x, t)$ be two solutions for the BVP (8.8). If $w(x, t) = y(x, t) - z(x, t)$, then $w(x, t)$ satisfies the BVP

$$
\begin{aligned}
w_{tt}(x, t) &= a^2 w_{xx}(x, t), \quad 0 < x < L, \, t > 0 \\
w(0, t) &= w(L, t) = 0, \quad t \geqslant 0 \\
w_t(x, 0) &= w(x, 0) = 0, \quad 0 \leqslant x \leqslant L
\end{aligned}
\tag{8.19}
$$

We need to show that $w(x, t) = 0$ throughout $(0 \leqslant x \leqslant L, \, t \geqslant 0)$. Physically, it seems that this is the case, since a string satisfying (8.19) is neither displaced initially nor dislodged from its rest position. The string is subjected to no external force except the tensile force. Certainly, our intuition would be that no displacement takes place in $w(x, t)$. To show this analytically we note from Section 1.11 that the PDE $w_{tt} = a^2 w_{xx}$ has the general solution

$$w(x, t) = F(x + at) + G(x - at) \tag{8.20}$$

where F and G are arbitrary (twice differentiable) functions. Thus

$$w_t(x, t) = a[F'(x + at) - G'(x - at)] \tag{8.21}$$

The initial condition $0 = w_t(x, 0)$ then becomes

$$0 = a[F'(x) - G'(x)]$$

It follows that $G(x) = F(x) + c$ for some constant c. The initial condition $w(x, 0) = 0$ can now be expressed as

$$0 = F(x) + G(x) = 2F(x) + c \qquad (8.22)$$

from which it follows that $F(x) = -c/2$ and $G(x) = c/2$. Thus

$$w(x, t) = F(x + at) + G(x - at) = -\frac{c}{2} + \frac{c}{2} = 0 \qquad (8.23)$$

Then w is identically zero and the solution to (8.8) is unique.

8.3. THE VIBRATING STRING WITH TWO NONHOMOGENEOUS CONDITIONS

This BVP is that of Section 8.1 with an arbitrary initial velocity. The problem follows.

$$y_{tt}(x, t) = a^2 y_{xx}(x, t), \quad 0 < x < L, \, t > 0$$
$$y(0, t) = y(L, t) = 0, \quad t \geq 0$$
$$y_t(x, 0) = g(x), \quad y(x, 0) = f(x), \quad 0 \leq x \leq L \qquad (8.24)$$
$$|y(x, t)| < M, \quad 0 \leq x \leq L, \, t \geq 0$$

Since the differential equation and the two linear homogeneous boundary conditions are the same as those in (8.8), the steps will coincide with those of Section 8.1 through result (8.12). The new T equation is the same as the one of Section 8.1, but there is no accompanying homogeneous boundary condition.

5. *The New T Equation.* The equation

$$T'' + \frac{n^2 \pi^2 a^2}{L^2} T = 0$$

has the general solution

$$T_n(t) = B_1 \cos \frac{n \pi a t}{L} + B_2 \sin \frac{n \pi a t}{L} \qquad (8.25)$$

At this point the procedure must be altered. We have exhausted all of the homogeneous boundary conditions.

6. *Solution Set for Homogeneous Conditions.* In this case the solution set from (8.12) and (8.25) is

$$y_n(x, t) = \sin \frac{n\pi x}{L} \left[B_1 \cos \frac{n\pi at}{L} + B_2 \sin \frac{n\pi at}{L} \right], \quad n \in \mathbf{N} \tag{8.26}$$

after considering all homogeneous conditions.

7. *Superposition.* The infinite linear combination

$$y(x, t) = \sum_{n=1}^{\infty} \sin \frac{n\pi x}{L} \left[K_n \cos \frac{n\pi at}{L} + M_n \sin \frac{n\pi at}{L} \right]$$

where the constants K_n and M_n absorb B_1 and B_2 of (8.26). Two nonhomogeneous conditions remain.

8. *Nonhomogeneous Initial Conditions.* The initial condition

$$y(x, 0) = f(x) = \sum_{n=1}^{\infty} K_n \sin \frac{n\pi x}{L}$$

is the same as (8.16) and therefore

$$K_n = \frac{2}{L} \int_0^L f(x) \sin \frac{n\pi x}{L} \, dx$$

Assuming that termwise differentiation is permitted, we write

$$y_t(x, t) = \sum_{n=1}^{\infty} \sin \frac{n\pi x}{L} \left[\frac{n\pi a}{L} \left(-K_n \sin \frac{n\pi at}{L} + M_n \cos \frac{n\pi at}{L} \right) \right]$$

The initial velocity condition

$$y_t(x, 0) = g(x) = \sum_{n=1}^{\infty} \frac{n\pi a}{L} M_n \sin \frac{n\pi x}{L}$$

gives us another sine series with

$$M_n = \frac{2}{n\pi a} \int_0^L g(x) \sin \frac{n\pi x}{L} \, dx$$

9. *Solution for the Original BVP*

$$y(x, t) = \sum_{n=1}^{\infty} \sin \frac{n\pi x}{L} \left[K_n \cos \frac{n\pi at}{L} + M_n \sin \frac{n\pi at}{L} \right]$$

where

$$K_n = \frac{2}{L} \int_0^L f(x) \sin \frac{n\pi x}{L} \, dx$$

$$M_n = \frac{2}{n\pi a} \int_0^L g(x) \sin \frac{n\pi x}{L} \, dx$$

8.4. LONGITUDINAL VIBRATIONS ALONG AN ELASTIC ROD

Assume that a rod has a natural length L and an axis is placed along its
length as in Figure 8.3. We include several idealizations in our definition of
the problem. First we assume that motion takes place only in a linear
direction parallel to the x axis. Longitudinal displacements along the rod at
any two positions x and $x + \Delta x$ and a common time t are given by $y(x, t)$ and
$y(x + \Delta x, t)$, respectively. The element of length Δx is stretched by an
amount $y(x + \Delta x, t) - y(x, t)$. From elementary physics, the modulus of
elasticity E is the ratio of tensile stress to tensile strain. Stress in this case is
the force per unit area or F/A. The strain is elongation per unit length or

$$\frac{y(x + \Delta x, t) - y(x, t)}{\Delta x}$$

Formalizing, we have

$$F = AE \frac{y(x + \Delta x, t) - y(x, t)}{\Delta x} \tag{8.27}$$

As $\Delta x \to 0$ in (8.27) the instantaneous force F becomes

$$F = AEy_x(x, t)$$

where A is a constant cross sectional area.

If ρ is the density factor, according to Newton's second law of motion, the
force for the element of length Δx is

$$\rho A \, \Delta x y_{tt}(x, t)$$

Equating forces for the element we have

$$\rho A \, \Delta x y_{tt}(x, t) = AEy_x(x + \Delta x, t) - AEy_x(x, t)$$

Solving for $y_{tt}(x, t)$ one obtains

Figure 8.3. An element of an elastic rod.

$$y_{tt}(x, t) = \frac{E}{\rho} \frac{y_x(x + \Delta x, t) - y_x(x, t)}{\Delta x} \qquad (8.28)$$

As $\Delta x \to 0$ in (8.28),

$$y_{tt}(x, t) = a^2 y_{xx}(x, t)$$

if $a^2 = E/\rho$. This is the *wave equation* which we determined originally with the vibrating string.

For our discussion, we are concerned with the case where both ends of the rod are free. We assume that the rate of change of longitudinal displacement relative to x is zero at both ends of the rod. The rod is initially at rest and $f(x)$ is the initial displacement for each x on $0 < x < L$. The boundary value problem is described as follows:

$$
\begin{aligned}
&y_{tt}(x, t) = a^2 y_{xx}(x, t) , \quad 0 < x < L, \ t > 0 \\
&y_x(0, t) = y_x(L, t) = 0 , \quad t > 0 \\
&y_t(x, 0) = 0 , \quad 0 < x < L \\
&y(x, 0) = f(x) , \quad 0 < x < L
\end{aligned}
\qquad (8.29)
$$

Details of the Fourier method accompany sequentially the BVP.

1. *Separation of Variables.* Let $y(x, t) = X(x)T(t)$. The PDE becomes

$$XT'' = a^2 X''T$$

from which we obtain

$$\frac{T''}{a^2 T} = \frac{X''}{X} = -\alpha^2$$

2. *Related ODEs*

$$X'' + \alpha^2 X = 0 , \quad T'' + \alpha^2 a^2 T = 0$$

3. *Homogeneous Boundary Conditions*

$$y_x(0, t) = X'(0)T(t) = 0$$

If $T(t) \neq 0$, then

$$X'(0) = 0$$
$$y_x(L, t) = X'(L)T(t) = 0$$

If $T(t) \neq 0$, then

$$X'(L) = 0$$

$$y_t(x, 0) = X(x)T'(0) = 0$$

If $X(x) \neq 0$, then

$$T'(0) = 0$$

4. *A Related SLP*

$$X'' + \alpha^2 X = 0, \quad X'(0) = X'(L) = 0$$

The general solution for the SLDE is

$$X = C_1 \cos \alpha x + C_2 \sin \alpha x$$

Differentiating, we have

$$X' = -\alpha C_1 \sin \alpha x + \alpha C_2 \cos \alpha x$$

$$X'(0) = 0 + \alpha C_2 = 0$$

Either $\alpha = 0$ or $C_2 = 0$. If $\alpha \neq 0$, $C_2 = 0$.

$$X'(L) = -\alpha C_1 \sin \alpha L = 0$$

If $\alpha \neq 0$, $C_1 \neq 0$, then $\sin \alpha L = 0$. This implies that $\alpha L = n\pi$, and

$$\alpha_n^2 = \frac{n^2 \pi^2}{L^2}$$

If $n = 0$, $\alpha = 0$, and

$$X'' = 0, \quad X' = K_1 \tag{8.30}$$

From the condition $X'(0) = 0$, we find $K_1 = 0$.

$$X = K_1 x + K_2$$

is the general solution of (8.30). We have found $K_1 = 0$, but K_2 is arbitrary. Therefore,

$$X_0(x) = 1$$

and

$$X_n(x) = \cos \frac{n\pi x}{L}, \quad n \in N_0 \tag{8.31}$$

5. *The New T Equation.* We do not have a complete SLP in $T(t)$, but

$$T'' + \frac{n^2\pi^2a^2}{L^2}\, T = 0, \quad T'(0) = 0$$

$$T = B_1 \cos \frac{n\pi at}{L} + B_2 \sin \frac{n\pi at}{L}$$

is the general solution. Differentiating, we obtain

$$T' = \frac{-n\pi a}{L}\, B_1 \sin \frac{n\pi at}{L} + \frac{n\pi a}{L}\, B_2 \cos \frac{n\pi at}{L}$$

$$T'(0) = \frac{n\pi a}{L}\, B_2 = 0$$

Therefore, $B_2 = 0$, and

$$T_n(t) = \cos \frac{n\pi at}{L}, \quad n \in \mathbf{N}_0 \tag{8.32}$$

6. *Solution Set for Homogeneous Conditions.* Employing the separation substitution and the functions (8.31) and (8.32), we have

$$y_n(x, t) = \cos \frac{n\pi x}{L} \cos \frac{n\pi at}{L}, \quad n \in \mathbf{N}_0 \tag{8.33}$$

7. *Superposition.* The infinite linear combination of (8.33) is the series

$$y(x, t) = \frac{A_0}{2} + \sum_{n=1}^{\infty} A_n \cos \frac{n\pi x}{L} \cos \frac{n\pi at}{L} \tag{8.34}$$

We observe that $A_0/2$ is the coefficient of $X_0 T_0 = 1$ in (8.34).

8. *Nonohomogeneous Boundary Condition*

$$y(x, 0) = f(x) = \frac{A_0}{2} + \sum_{n=1}^{\infty} A_n \cos \frac{n\pi x}{L}$$

contains a Fourier cosine series.

$$A_n = \frac{2}{L} \int_0^L f(x) \cos \frac{n\pi x}{L}\, dx, \quad n \in \mathbf{N}_0$$

9. *Solution for the Original BVP*

$$y(x, t) = \frac{A_0}{2} + \sum_{n=1}^{\infty} A_n \cos \frac{n\pi x}{L} \cos \frac{n\pi at}{L}$$

where

$$A_n = \frac{2}{L} \int_0^L f(x) \cos \frac{n\pi x}{L} \, dx, \quad n \in \mathbf{N}_0$$

Even though we omitted $|y(x, t)| < M$ in the BVP, we assume that a bounded solution is in order.

Exercises 8.1

1. A string is fastened at the end points $(0, 0)$ and $(2, 0)$, and initially has a velocity $g(x)$. It has an equilibrium position $y = 0$. Show that the BVP associated with these conditions is

$$y_{tt}(x, t) = a^2 y_{xx}(x, t), \quad 0 < x < 2, \ t > 0$$
$$y(0, t) = y(2, t) = 0, \quad t \geq 0$$
$$y(x, 0) = 0, \ y_t(x, 0) = g(x), \quad 0 < x < 2$$

   Solve the BVP.

2. Determine the solution for the BVP

$$y_{tt}(x, t) = a^2 y_{xx}(x, t), \quad 0 < x < \pi, \ t > 0$$
$$y_x(0, t) = y_x(\pi, t) = 0, \quad t > 0$$
$$y_t(x, 0) = 0, \ y(x, 0) = kx, \quad 0 < x < \pi, \ k \text{ constant}$$

3. A string is stretched between the points $(0, 0)$ and $(L, 0)$. The initial contour of the string is

$$y(x, 0) = \begin{cases} 0.02x & \text{if } 0 \leq x \leq L/2 \\ 0.02(L - x) & \text{if } L/2 \leq x \leq L \end{cases}$$

   The string is released from rest. Write an appropriate BVP for the string and solve it. This may be referred to as the *plucked string problem*.

4. A string is stretched between $(0, 0)$ and $(2, 0)$ and given an initial velocity

$$y_t(x, 0) = g(x) = \begin{cases} 0.05x & \text{if } 0 < x < 1 \\ 0.05(2 - x) & \text{if } 1 < x < 2 \end{cases}$$

   Initially the string is in an equilibrium position along $y = 0$. Construct a BVP matching the given conditions and solve the problem. This may be called the *struck string problem*.

5. Find the deflection for a string of length L if the two ends are attached at $(0, 0)$ and $(L, 0)$ and the initial deflection is $0.05 \sin(4\pi x/L)$. Initially the string is at rest.

6. Solve the BVP

$$y_{tt} = 4y_{xx}, \quad 0 < x < 5, \ t > 0$$

$$y(0, t) = y(5, t) = 0, \quad t \geqslant 0$$

$$y(x, 0) = 0, \ y_t(x, 0) = \sin 2\pi x, \quad 0 < x < 5$$

7. Verify that

$$y(x, t) = \frac{1}{2}[F(x + at) + F(x - at)] + \frac{1}{2a} \int_{x-at}^{x+at} G(\xi) \, d\xi$$

is a solution of the BVP (8.24) if F and G are the odd periodic extensions of f and g, respectively.

8. If the rod described in (8.29) has a fixed end at $x = 0$ and a free end at $x = L$, show that the BVP describing longitudinal displacement is

$$y_{tt}(x, t) = a^2 y_{xx}(x, t), \quad 0 < x < L, \ t > 0$$

$$y(0, t) = y_x(L, t) = 0, \quad t > 0$$

$$y_t(x, 0) = 0, \ y(x, 0) = f(x), \quad 0 < x < L$$

if all other information in the problem is unchanged. Solve the BVP.

9. It can be shown that if θ is the angular displacement in a torsionally vibrating shaft of length L, the wave equation

$$\theta_{tt} = a^2 \theta_{xx}, \quad 0 < x < L, \ t > 0$$

under suitable conditions describes the angular vibratory motion. We assume that the ends of the shaft are fixed, so that

$$\theta(0, t) = \theta(L, t) = 0$$

The shaft is twisted initially so that each cross section rotates through an angle proportional to x, or

$$\theta(x, 0) = kx, \quad 0 \leqslant x \leqslant L, \ k \text{ constant}$$

Finally, the shaft is released from rest in this position, and

$$\theta_t(x, 0) = 0, \quad 0 \leqslant x \leqslant L$$

Solve the BVP for $\theta(x, t)$.

10. The shaft in Exercise 9 has one end fixed and one end free, so that

$$\theta(0, t) = \theta_x(L, t) = 0$$

The remainder of the information is the same as in Exercise 9. Find the angular displacement $\theta(x, t)$ for the revised BVP.

8.5. HEAT CONDUCTION

Before attempting to model this problem, we state experimental observations concerning heat conduction.

1. Heat flows in the direction of the low temperature.
2. Heat flows through an area at a rate proportional to the area and to the temperature gradient normal to the area.
3. The amount of heat lost or gained by a substance because of a temperature change is proportional to the mass of the substance and the change in temperature.

The proportionality constant K in 2 is the *thermal conductivity* of the material. In 3, the proportionality constant is *specific heat C*.

We consider a rectangular parallelepiped element of a conducting solid with dimensions $\Delta x, \Delta y, \Delta z$ in Figure 8.4. The weight of the element is

$$\Delta w = \rho \, \Delta x \, \Delta y \, \Delta z = g \, \Delta m \qquad (8.35)$$

where ρ is the density factor, Δm is the mass of the element, and g is the gravitational constant. From (8.35)

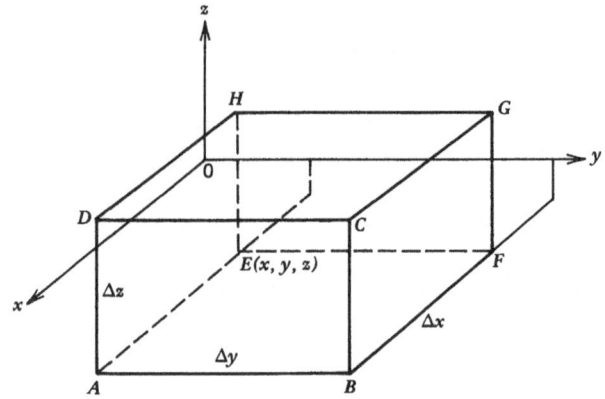

Figure 8.4. An element of a substance.

$$\Delta m = \frac{\rho \, \Delta x \, \Delta y \, \Delta z}{g} \qquad (8.36)$$

If the temperature change is Δu for a time interval Δt, the amount of heat ΔQ stored in the element according to 3 is

$$\Delta Q = C \, \Delta m \, \Delta u \qquad (8.37)$$

Inserting (8.36) in (8.37), one obtains

$$\Delta Q = \frac{C\rho}{g} \, \Delta x \, \Delta y \, \Delta z \, \Delta u \qquad (8.38)$$

Dividing (8.38) by Δt, we have

$$\frac{\Delta Q}{\Delta t} = \frac{C\rho}{g} \, \Delta x \, \Delta y \, \Delta z \, \frac{\Delta u}{\Delta t} \qquad (8.39)$$

According to 2 if heat flows into the element the rate of flow is

$$-K \, \Delta A u_x \qquad (8.40)$$

where ΔA is the area of a face and u_x is the gradient. If the flow is out of the element the sign in (8.40) is reversed. For the element, ΔA is $\Delta y \, \Delta z$ for the faces $ABCD$ and $EFGH$; ΔA is $\Delta x \, \Delta z$ for $BCGF$ and $ADHE$; and for $ABFE$ and $CDHG$ the area ΔA is $\Delta x \, \Delta y$.

Heat that causes the temperature change Δu comes from within the substance or the transfer of the heat through faces of the element. We let $q(x, y, z, t)$ represent the *amount of heat change internally per unit volume*.

The net rate of change of heat entering and leaving the element through faces $EFGH$ and $ABCD$ is

$$K \, \Delta y \, \Delta z [u_x|_{x+\Delta x} + u_x|_x]$$

We assume in this problem that $u_x|_{x+\Delta x}$ indicates the partial derivative relative to x evaluated at $x + \Delta x$ at the centroid of the face (y_c, z_c).

Equating the rate in (8.39) with the contribution in (8.40) for the total ΔA from the six faces of the element, we obtain

$$\frac{C\rho}{g} \, \Delta x \, \Delta y \, \Delta z \, \frac{\Delta u}{\Delta t} = K[\Delta y \, \Delta z (u_x|_{x+\Delta x} - u_x|_x) + \Delta x \, \Delta z (u_y|_{y+\Delta y} - u_y|_y)$$

$$+ \Delta x \, \Delta y (u_z|_{z+\Delta z} - u_z|_z)] + \Delta x \, \Delta y \, \Delta z q(x, y, z, t)$$

Dividing by $(C\rho/g) \, \Delta x \, \Delta y \, \Delta z$ and allowing Δx, Δy, Δz, and Δt to go to zero, we have

$$u_t = a^2(u_{xx} + u_{yy} + u_{zz}) + \frac{g}{c\rho} \, q(x, y, z, t)$$

where $a^2 = Kg/C\rho$. This is the *heat equation* for three dimensional conduction. If heat is neither generated nor lost within the substance, $q(x, y, z, t)$ is zero. Usually this will be our assumption.

If $u_t = 0$, and all changes relative to time have ceased, we have the *steady state* condition

$$u_{xx} + u_{yy} + u_{zz} = 0$$

The one dimensional heat equation is

$$u_t = a^2 u_{xx}$$

In this equation one assumes a heat transfer along a single x-axis. The case is illustrated by the temperature in a long rod or bar of homogeneous cross section with the material insulated laterally parallel to the x-axis. The equation

$$u_t = a^2 \nabla^2 u$$

is also the *diffusion equation* if a^2 is the coefficient of diffusion.

As an example of a heat conduction problem in one dimension, let us consider the temperature u at any point x and time t in a rod whose ends at $x = 0$ and $x = L$ are kept at zero temperature. The initial temperature in the bar is $f(x)$. The lateral surface of the rod is insulated.

Appropriate constraints follow.

$$u(0, t) = u(L, t) = 0$$

The ends of the rod of length L are kept at zero temperature.

$$u(x, 0) = f(x)$$

The initial temperature in the rod is dependent only on the location along the x-axis.

A formal statement of the BVP follows:

$$\begin{aligned}
&u_t(x, t) = a^2 u_{xx}(x, t), \quad 0 < x < L, \, t > 0 \\
&u(0, t) = u(L, t) = 0, \quad t \geq 0 \\
&u(x, 0) = f(x), \quad 0 < x < L
\end{aligned} \tag{8.41}$$

1. *Separation of Variables*. Let $u(x, t) = X(x)T(t)$. The PDE displayed in (8.41) becomes

$$XT' = a^2 X''T$$

Dividing by $a^2 XT$ and recognizing the constant ratios, we obtain

$$\frac{T'}{a^2 T} = \frac{X''}{X} = -\alpha^2$$

if a bounded solution is demanded. We obtain two related ODEs.

2. *Related ODEs*

$$X'' + \alpha^2 X = 0, \quad T' + \alpha^2 a^2 T = 0 \qquad (8.42)$$

From the BVP (8.41), we investigate the homogeneous boundary conditions.

3. *Homogeneous Boundary Conditions*. Using the separation substitution of 1, we have

$$u(0, t) = X(0)T(t) = 0$$

If $T(t) \neq 0$, then

$$X(0) = 0 \qquad (8.43)$$

In a similar manner

$$u(L, t) = X(L)T(t) = 0$$

If $T(t) \neq 0$, then

$$X(L) = 0 \qquad (8.44)$$

From (8.42), (8.43), and (8.44) we have a related SLP.

4. *A Related SLP*

$$X'' + \alpha^2 X = 0, \quad X(0) = X(L) = 0$$

The SLDE has a general solution

$$X = C_1 \cos \alpha x + C_2 \sin \alpha x$$

Employing the first boundary condition of the SLP, we have

$$X(0) = C_1 + 0 = 0$$

Using the second condition with $C_1 = 0$, we write

$$X(L) = 0 = C_2 \sin \alpha L$$

If $C_2 \neq 0$, then $\sin \alpha L = 0$. This implies that $\alpha L = n\pi$ or $\alpha = n\pi/L$. Then

$$\alpha_n^2 = \frac{n^2 \pi^2}{L^2}, \quad n \in \mathbf{N} \tag{8.45}$$

is a set of eigenvalues for the SLP. If $n = 0$ and $\alpha = 0$, we find a trivial solution. The complete set of eigenvalues is included in (8.45). The corresponding eigenfunctions are

$$X_n(x) = \sin \frac{n\pi x}{L}, \quad n \in \mathbf{N} \tag{8.46}$$

There are no other SLPs to be found in the data.

5. *The New T Equation.* We have obtained α^2, but there are not T constraints.

$$T' + \frac{n^2 \pi^2 a^2}{L^2} T = 0$$

is the new T equation. The solution is

$$T_n(t) = \exp\left(-\frac{n^2 \pi^2 a^2 t}{L^2}\right) \tag{8.47}$$

All of the homogeneous boundary conditions have been employed at this point. We display a solution set for homogeneous conditions.

6. *Solution Set for Homogeneous Conditions.* From the separation substitution along with (8.46) and (8.47) we write a solution set

$$u_n(x, t) = \exp\left(-\frac{n^2 \pi^2 a^2 t}{L^2}\right) \sin \frac{n\pi x}{L}, \quad n \in \mathbf{N} \tag{8.48}$$

7. *Superposition.* The infinite linear combination of (8.48) is

$$u(x, t) = \sum_{n=1}^{\infty} A_n \exp\left(-\frac{n^2 \pi^2 a^2 t}{L^2}\right) \sin \frac{n\pi x}{L} \tag{8.49}$$

In order that we may compute the coefficients A_n of (8.49) we consider the following.

8. *Nonhomogeneous Boundary Condition.* From the last boundary condition

$$u(x, 0) = f(x) = \sum_{n=1}^{\infty} A_n \sin \frac{n\pi x}{L}, \quad 0 < x < L$$

The coefficients of the sine series are

$$A_n = \frac{2}{L} \int_0^L f(x) \sin \frac{n\pi x}{L} \, dx \qquad (8.50)$$

9. *Solution of the Original BVP*

$$u(x, t) = \frac{2}{L} \sum_{n=1}^{\infty} \left[\int_0^L f(\xi) \sin \frac{n\pi \xi}{L} \, d\xi \right] \exp\left(-\frac{n^2\pi^2 a^2 t}{L}\right) \sin \frac{n\pi x}{L}$$

or

$$u(x, t) = \frac{2}{L} \sum_{n=1}^{\infty} \int_0^L f(\xi) \exp\left(-\frac{n^2\pi^2 a^2 t}{L^2}\right) \sin \frac{n\pi x}{L} \sin \frac{n\pi \xi}{L} \, d\xi$$
$$(8.51)$$

In (8.51) we need to identify the variable of integration separate from the variables of the function. Other forms of the solution are possible. The series (8.49) with coefficients (8.50) comprise a simple form of the solution.

8.6. NUMERICAL SOLUTION OF THE HEAT EQUATION

We will use the one dimensional heat equation to illustrate the fundamental concepts of *finite difference* methods for obtaining numerical solutions to BVPs. These methods are based on finite difference approximations to derivatives such as those given by (1.48)–(1.50).

Consider the following nonhomogeneous BVP

$$u_t = a^2 u_{xx}, \quad 0 \leqslant x \leqslant L, \, t \geqslant 0$$
$$u(0, t) = \phi(t), \quad u(L, t) = \psi(t)$$
$$u(x, 0) = f(x)$$

Let us subdivide the x, t plane into a grid; that is, let

$$x_i = ih, \quad i = 0, 1, \ldots, m, \quad t_j = jk, \quad j = 0, 1, \ldots, n$$

where h and k are (small) positive numbers known as *step sizes*. Our objective will be to approximate the solution u at each of the grid points

(x_i, t_j). For brevity let us use the notation $u_{i,j}$ for the approximate value of $u(x_i, t_j)$. The derivatives in the PDE $u_t = a^2 u_{xx}$ will be approximated using the formulas (1.48) and (1.50) to give

$$u_t(x_i, t_j) \approx \frac{u_{i,j+1} - u_{i,j}}{k}$$

$$u_{xx}(x_i, t_j) \approx \frac{u_{i-1,j} - 2u_{i,j} + u_{i+1,j}}{h^2}$$

The PDE is approximated by the equation

$$\frac{u_{i,j+1} - u_{i,j}}{k} = a^2 \frac{u_{i-1,j} - 2u_{i,j} + u_{i+1,j}}{h^2} \tag{8.52}$$

Let $\lambda = a^2 k / h^2$. Solving (8.52) for $u_{i,j+1}$, we obtain

$$u_{i,j+1} = \lambda(u_{i-1,j} + u_{i+1,j}) + (1 - 2\lambda)u_{i,j} \tag{8.53}$$

The initial and boundary values can be written as

$$u_{i,0} = f(x_i), \quad i = 1, 2, \ldots, m - 1$$
$$u_{0,j} = \phi(t_j), \quad u_{m,j} = \psi(t_j), \quad j = 0, 1, \ldots, n$$

The approximate solution can then be stepped along in time by using (8.53) with (a) $j = 0$ to compute $u_{i,1}$, $i = 1, 2, \ldots, m - 1$, (b) $j = 1$ to compute $u_{i,2}$, $i = 1, 2, \ldots, m - 1$, and so forth. The method is stable only when $\lambda < 1/2$; so the step sizes must be selected with care. This technique is referred to as the *forward difference method*.

8.7. VERIFICATION AND UNIQUENESS OF THE SOLUTION FOR THE HEAT PROBLEM

We assume that (8.49) with coefficients (8.50) is the formal solution of the BVP (8.41). By using Bessel's inequality and a bit of the theory of infinite series, we can show that $A_n \to 0$ as $n \to \infty$. For all $n \in \mathbf{N}$, A_n is bounded or $|A_n| < A$ if A is some positive constant. For $t_0 > 0$,

$$\left| A_n \exp\left(-\frac{n^2 \pi^2 a^2 t}{L^2} \right) \sin \frac{n\pi x}{L} \right| < A \exp\left(\frac{n^2 \pi^2 a^2 t_0}{L^2} \right)$$

when $t \geq t_0$. Using the ratio test one finds that the series of constant terms $\exp(-n^2 \pi^2 a^2 t_0 / L^2)$ converges. According to the Weierstrass M-test, series (8.49) converges uniformly relative to x and t when $0 \leq x \leq L$, $t \geq t_0$. The series is made up of terms that are continuous functions. Series (8.49) converges to a continuous function $u(x, t)$ when $t \geq t_0 > 0$. In particular $u(x, t)$ is continuous at $x = 0$ and $x = L$. As a result if $x = 0$ and $x = L$ in (8.49), then

$$u(0, t) = u(L, t) = 0$$

when $t > 0$.

By termwise differentiation relative to t, we have

$$u_t = - \sum_{n=1}^{\infty} A_n \left(\frac{n^2 \pi^2 a^2}{L^2} \right) \exp \left(- \frac{n^2 \pi^2 a^2 t}{L^2} \right) \sin \frac{n \pi x}{L} \qquad (8.54)$$

By testing the series for u_t of (8.54) using a procedure similar to the one just employed, we find that the series converges uniformly in $0 \leq x \leq L$, $t \geq t_0 > 0$. In the same way one can differentiate (8.49) twice relative to x and obtain

$$u_{xx} = - \sum_{n=1}^{\infty} A_n \left(\frac{n^2 \pi^2}{L^2} \right) \exp \left(- \frac{n^2 \pi^2 a^2 t}{L^2} \right) \sin \frac{n \pi x}{L} \qquad (8.55)$$

This series is uniformly convergent also. If the series of (8.55) for u_{xx} is multiplied by a^2 one has the series for u_t. As a result,

$$u_t = a^2 u_{xx}$$

is satisfied by the solution (8.49).

We still need to show that $u(x, t)$ satisfies the nonhomogeneous condition

$$u(x, 0) = f(x), \quad 0 \leq x \leq L$$

The reader should see Churchill and Brown [15, pp. 268–271] for a discussion of *Abel's test for uniform convergence*. If f is continuous on $[0, L]$ and $f(0) = f(L) = 0$, and f is PWS on $(0, L)$, then

$$f(x) = \sum_{n=1}^{\infty} A_n \sin \frac{n \pi x}{L} \qquad (8.56)$$

By employing Abel's test, the series resulting from the product of the terms of the uniformly convergent series of (8.56) and the bounded monotone members of $\exp(-n^2 \pi^2 a^2 t / L^2)$ converges uniformly relative to t. Therefore,

$$u(x, t) = \sum_{n=1}^{\infty} A_n \exp \left(- \frac{n^2 \pi^2 a^2 t}{L^2} \right) \sin \frac{n \pi x}{L}$$

converges uniformly when $0 \leq x \leq L$, $t \geq 0$. The function $u(x, t)$ is continuous when $0 \leq x \leq L$, $t \geq 0$. Certainly u is continuous in t when $t \geq 0$. Therefore,

$$u(x, 0+) = u(x, 0)$$

Since $u(x, 0) = f(x)$, the initial nonhomogeneous condition is satisfied. The solution (8.49) is verified and a solution exists.

To show uniqueness, we assume that the BVP (8.41) has two solutions $u(x, t)$ and $v(x, t)$, and $w(x, t) = u(x, t) - v(x, t)$. Then $w(x, t)$ satisfies the BVP

$$w_t = a^2 w_{xx}, \quad 0 < x < L, \; t > 0$$
$$w(0, t) = w(L, t) = 0, \quad t \geq 0 \tag{8.57}$$
$$w(x, 0) = 0, \quad 0 \leq x \leq L$$

To show that $w(x, t) \equiv 0$ we follow an idea* used by Myint-U [35, pp. 147–148]. An integral function

$$H(t) = \frac{1}{2a^2} \int_0^L w^2(x, t) \, dx \tag{8.58}$$

is introduced. Differentiating (8.58) relative to t, one obtains

$$H'(t) = \frac{1}{a^2} \int_0^L w w_t \, dx = \int_0^L w w_{xx} \, dx$$

after employing the PDE of (8.57). After integrating by parts, one finds that

$$H'(t) = w(L, t) w_x(L, t) - w(0, t) w_x(0, t) - \int_0^L w_x^2(x, t) \, dx \tag{8.59}$$

From (8.57),

$$w(0, t) = w(L, t) = 0$$

so that (8.59) becomes

$$H'(t) = - \int_0^L w_x^2 \, dx \leq 0 \tag{8.60}$$

Since $w(x, 0) = 0$, we have $H(0) = 0$. From this result and (8.60) we conclude that $H(t)$ is a nonincreasing function. Therefore,

$$H(t) \leq 0 \tag{8.61}$$

and from (8.58)

---

*From [35, pp. 147–148], by permission of Elsevier/North-Holland.

$$H(t) \geqslant 0 \qquad (8.62)$$

To satisfy both (8.61) and (8.62)

$$H(t) = 0$$

Since $w(x, t)$ is continuous and $H(t) = 0$, we conclude from (8.58) that

$$w(x, t) \equiv 0$$

for $0 \leqslant x \leqslant L$, $t \geqslant 0$. Hence $u(x, t) = v(x, t)$ and the solution is unique.

For uniqueness of a more elaborate heat problem see Sagan [42, pp. 79–81].

Exercises 8.2

1. Both faces of a bar of length L are insulated. The lateral surface of the bar is also insulated. The initial temperature in the bar is $\cos(3\pi x/L)$. Find the lateral temperature at any point x and time t for the bar. The BVP is

$$u_t(x, t) = a^2 u_{xx}(x, t), \quad 0 < x < L, \, t > 0$$

$$u_x(0, t) = u_x(L, t) = 0, \quad t \geqslant 0$$

$$u(x, 0) = \cos \frac{3\pi x}{L}, \quad 0 < x < L$$

2. One face of a rod at $x = 0$ is kept at zero temperature and the face at $x = L$ is insulated. The initial temperature distribution is given by $f(x)$. Show that the related BVP is

$$u_t = a^2 u_{xx}, \quad 0 < x < L, \, t > 0$$

$$u(0, t) = u_x(L, t) = 0, \quad t \geqslant 0$$

$$u(x, 0) = f(x), \quad 0 < x < L$$

Find $u(x, t)$.

3. Consider the BVP

$$u_t = u_{xx}, \quad 0 \leqslant x \leqslant 4, \, t \geqslant 0$$

$$u(0, t) = 5t, \quad u(4, t) = 100 - 15t$$

$$u(x, 0) = 25x$$

(a) Find a function v satisfying the PDE and the boundary conditions but not the initial condition. Hint: Let $v(x, t) = (ax + b)t + g(x)$ and determine a, b, and g.

(b) Find a function w satisfying the PDE, homogeneous boundary conditions, and the initial condition $w(x, 0) = 25x - v(x, 0)$.

(c) Show that $u = v + w$ is a solution to the BVP.

(d) Approximate $u(3, 1)$ by adding three terms in the series for w to v.

(e) Use the forward difference method with $h = 1$ and $k = 0.25$ to approximate $u(3, 1)$.

4. Face $x = L$ of a rod is kept at zero temperature and face $x = 0$ is insulated. Initially the temperature distribution is $\cos 5\pi x/2L$. Write the related BVP and determine its solution.

5. Solve the BVP

$$u_t = 2u_{xx}, \quad 0 < x < 4, \ t > 0$$

$$u(0, t) = u(4, t) = 0, \quad t \geq 0$$

$$u(x, 0) = \begin{cases} x & \text{if } 0 < x < 2 \\ 4 - x & \text{if } 2 \leq x \leq 4 \end{cases}$$

6. Solve the BVP

$$u_t = u_{xx}, \quad 0 < x < 2, \ t > 0$$

$$u_x(0, t) = u(2, t) = 0, \quad t \geq 0$$

$$u(x, 0) = \begin{cases} 1 & \text{when } 0 < x < 1 \\ 2 - x & \text{when } 1 < x < 2 \end{cases}$$

8.8. GRAVITATIONAL POTENTIAL

Gravitational potential may be defined by the function

$$u(x, y, z) = \frac{c}{r}$$

where $c = GMm$. G is the gravitational constant, M is the mass of a particle at a fixed point (X, Y, Z), and m is the mass of a particle at (x, y, z). The distance r between the two points satisfies

$$r^2 = (x - X)^2 + (y - Y)^2 + (z - Z)^2$$

If A is a particle of mass M and B a particle of mass m, then A attracts B with a gravitational force that is the gradient of the function c/r.

To extend the idea, we let $u(x, y, z)$ be the *potential u* of a continuous body at a point (x, y, z) outside the body. Then the potential u is defined by

$$u(x, y, z) = k \int \int \int_V \frac{\rho \, dX \, dY \, dZ}{r}$$

where ρ is the density of a mass at (X, Y, Z) and k is a positive constant. In the remarks that follow, we assume that the resulting derived functions are continuous. Then

$$u_x = -k \int \int \int_V \frac{\rho(x - X)}{r^3} \, dX \, dY \, dZ$$

$$u_{xx} = -k \int \int \int_V \left[\frac{\rho}{r^3} - \frac{3\rho(x - X)^2}{r^5} \right] dX \, dY \, dZ \qquad (8.63)$$

$$u_{yy} = -k \int \int \int_V \left[\frac{\rho}{r^3} - \frac{3\rho(y - Y)^2}{r^5} \right] dX \, dY \, dZ \qquad (8.64)$$

$$u_{zz} = -k \int \int \int_V \left[\frac{\rho}{r^3} - \frac{3\rho(z - Z)^2}{r^5} \right] dX \, dY \, dZ \qquad (8.65)$$

Adding (8.63), (8.64), and (8.65), one finds that

$$\nabla^2 u = u_{xx} + u_{yy} + u_{zz} = 0 \qquad (8.66)$$

This is *Laplace's equation* in 3 space referenced to rectangular coordinates.

In much the same way as with gravitational potential, u of equation (8.66) may represent electric or magnetic potential functions at points not filled with electric charges or magnetic poles. Laplace's equation is associated with incompressible fluid flow problems. As we mentioned previously, the steady state heat problem involves $\nabla^2 u = 0$.

8.9. LAPLACE'S EQUATION

In this section we are concerned with BVPs associated with Laplace's equation (8.66). The same notation $\nabla^2 u$ is used to represent $u_{xx} + u_{yy}$ and equivalent forms relative to other coordinate systems.

The first type BVP, the *Dirichlet problem*, has the form

$$\nabla^2 u = 0, \quad u = f \text{ on } B$$

where B is the boundary of the domain D of the problem.

The second type BVP, the *Neumann problem*, has the form

$$\nabla^2 u = 0 , \quad \frac{\partial u}{\partial n} = g \text{ on } B$$

The $\partial u / \partial n$ is the outward normal derivative of u.

The thrid type BVP concerns

$$\nabla^2 u = 0 , \quad \frac{\partial u}{\partial n} + hu = p \text{ on } B$$

This is sometimes referred to as the *Robin* or *Churhcill problem*. It is a *mixed problem*. Some descriptions of the Robin problem, while still mixed, have

$$u = q_1$$

on part of the boundary and

$$\frac{\partial u}{\partial n} = q_2$$

on the remainder of the boundary of D.

The steady state heat problem with a fixed temperature distribution at all points on the boundary of the domain is an example of the boundary conditions of the Dirichlet problem. A steady state heat problem with the heat flux across the boundary given at all points has the Neumann boundary condition. No heat sources or sinks are assumed in these two examples.

A function u is *harmonic* in a domain D if it satisfies $\nabla^2 u = 0$ and the second order derivatives are continuous in D. Uniqueness of these special problems containing the Laplacian equation is considered by Young [54, pp. 253–256].

We investigate a steady state temperature distribution problem. The function u represents the temperature at any point (x, y) throughout a thin square plate (faces insulated) with its edges $x = 0$, $x = \pi$, $y = \pi$ all kept at zero temperatures. Side $y = 0$ is held at a temperature $f(x)$, $0 \leqslant x \leqslant \pi$. Figure 8.5 has a description of the boundary conditions. A statement of the BVP follows:

$$\nabla^2 u = u_{xx} + u_{yy} = 0 , \quad 0 < x < \pi , 0 < y < \pi$$

$$u(0, y) = u(\pi, y) = u(x, \pi) = 0 , \quad 0 \leqslant x \leqslant \pi, 0 \leqslant y \leqslant \pi$$

$$u(x, 0) = f(x) , \quad 0 \leqslant x \leqslant \pi$$

This is a Dirichlet problem in a square domain. We solve the problem using the Fourier method.

Figure 8.5. Temperature in a square plate.

1. *Separation of Variables.* Let $u(x, y) = X(x)Y(y)$

$$X''Y + XY'' = 0$$

Dividing by XY and recognizing that ratios must be constant, we have

$$\frac{X''}{X} = -\frac{Y''}{Y} = -\alpha^2 \tag{8.67}$$

2. *Related ODEs*

$$X'' + \alpha^2 X = 0 , \quad Y'' - \alpha^2 Y = 0$$

3. *Homogeneous Boundary Conditions*

$$u(0, y) = X(0)Y(y) = 0$$

If $Y(y) \neq 0$, then $X(0) = 0$.

$$u(\pi, y) = X(\pi)Y(y) = 0$$

If $Y(y) \neq 0$, then $X(\pi) = 0$.

$$u(x, \pi) = X(x)Y(\pi) = 0$$

If $X(x) \neq 0$, then $Y(\pi) = 0$.

4. *A Related SLP.* $X'' + \alpha^2 X = 0$, $X(0) = X(\pi) = 0$

$$X = C_1 \cos \alpha x + C_2 \sin \alpha x$$

is the general solution of the SLDE.

$$X(0) = C_1 + 0 = 0 , \quad X(\pi) = C_2 \sin \alpha x = 0$$

If $C_2 \neq 0$, then $\sin \alpha\pi = 0$. This means that $\alpha\pi = n\pi$ and $\alpha = n$. Then

$$\alpha^2 = n^2, \quad n \in \mathbf{N} \tag{8.68}$$

is a set of eigenvalues for the SLP. If $n = 0$, then $\alpha = 0$ and a trivial solution only results. We cannot add $n = 0$ in (8.68). Corresponding eigenfunctions are

$$X_n(x) = \sin nx, \quad n \in \mathbf{N} \tag{8.69}$$

5. *The New Y Equation.* Only a single constraint is available with this equation. Hence

$$Y'' - n^2 Y = 0, \quad Y(\pi) = 0$$

The general solution of the ODE is

$$Y = B_1 \cosh ny + B_2 \sinh ny$$

$$Y(\pi) = B_1 \cosh n\pi + B_2 \sinh n\pi = 0$$

$$B_1 = -B_2 \frac{\sinh n\pi}{\cosh n\pi}$$

$$Y_n(y) = \frac{\sinh n(y - \pi)}{\cosh n\pi}, \quad n \in \mathbf{N} \tag{8.70}$$

6. *Solution Set for Nonhomogeneous Conditions.* From (8.69) and (8.70), one can write

$$u_n(x, y) = \frac{\sinh n(y - \pi)}{\cosh n\pi} \sin nx, \quad n \in \mathbf{N}$$

7. *Superposition*

$$u(x, y) = \sum_{n=1}^{\infty} A_n \frac{\sinh n(y - \pi)}{\cosh n\pi} \sin nx$$

8. *Nonhomogeneous Boundary Condition*

$$u(x, 0) = f(x) = \sum_{n=1}^{\infty} A_n \frac{\sinh(-n\pi)}{\cosh n\pi} \sin nx$$

where

$$A_n \left(\frac{-\sinh n\pi}{\cosh n\pi} \right) = \frac{2}{\pi} \int_0^{\infty} f(x) \sin nx \, dx$$

$$A_n = -\frac{2 \cosh n\pi}{\pi \sinh n\pi} \int_0^{\pi} f(x) \sin nx \, dx$$

9. *Solution for the original BVP*

$$u(x, y) = \sum_{n=1}^{\infty} A_n^* \frac{\sinh n(\pi - y)}{\cosh n\pi} \sin nx$$

where

$$A_n^* = \frac{2 \cosh n\pi}{\pi \sinh n\pi} \int_0^{\pi} f(x) \sin nx \, dx$$

or

$$u(x, y) = \frac{2}{\pi} \sum_{n=1}^{\infty} \int_0^{\pi} f(\xi) \sin n\xi \frac{\sinh n(\pi - y)}{\sinh n\pi} \sin nx \, d\xi$$

if the coefficients A_n^* are inserted inside the summation.

If we assign a positive constant α^2 in (8.67), the SLP in X has a trivial solution only. This we wish to avoid, since u is trivial if X is zero.

8.10. NUMERICAL SOLUTION OF THE LAPLACE EQUATION

We will describe here a finite difference method, known as the *Liebmann* or *Gauss–Seidel* method for approximating solutions to the Laplace equation. The basic approximations are similar to those used for the one dimensional heat equation. Consider the BVP

$$u_{xx} + u_{yy} = 0, \quad 0 \le x \le L, \, 0 \le y \le L$$
$$u(x, 0) = f_1(x), \quad u(x, L) = f_2(x)$$
$$u(0, y) = g_1(y), \quad u(L, y) = g_2(y)$$

First we will subdivide the x, y plane by setting $x_i = ih$, $y_j = jh$, $i, j = 0, 1, \ldots, n$, where h is the step size. (For simplicity, we have chosen a BVP which is symmetric in x and y and hence we have used the same step size for both variables.) Approximating the derivatives in the Laplace equation by (1.50) we obtain

$$\frac{u_{i-1,j} - 2u_{i,j} + u_{i+1,j}}{h^2} + \frac{u_{i,j-1} - 2u_{i,j} + u_{i,j+1}}{h^2} = 0$$

Solving for $u_{i,j}$ we obtain

$$u_{i,j} = \frac{u_{i-1,j} + u_{i+1,j} + u_{i,j-1} + u_{i,j+1}}{4} \tag{8.71}$$

Now (8.71) represents a linear system of $(n-1)^2$ equations in $(n-1)^2$ unknowns which could be solved by standard techniques such as Gaussian elimination. However, because this system is sparce (that is, each row in the matrix of coefficients has only a few nonzero entries), it is normally more efficient to solve this system by iteration. The Gauss–Seidel method proceeds as follows.

1. Compute the boundary values $u_{i,0}$, $u_{i,n}$, $u_{0,j}$, $u_{n,j}$, $i, j = 0, 1, \ldots, n$.
2. Compute initial estimates for $u_{i,j}$, $i, j = 1, 2, \ldots, n-1$ at the interior grid points. For example, one could use linear interpolation between boundary values along each row of the grid.
3. Obtain a new value for $u_{1,1}$ using (8.71) with $i = 1$ and $j = 1$.
4. Obtain a new value for $u_{2,1}$ using (8.71) with $i = 2$, $j = 1$, and using the new value of $u_{1,1}$.
5. Systematically obtain new values for $u_{i,j}$ for all interior grid points using (8.71) and always using the most recent values for the u values on the right side of this equation.
6. Repeat steps 3–5 until the u values do not change significantly when they are recomputed.

Exercises 8.3

1. The function u is the temperature at (x, y) throughout a thin rectangular plate with sides $x = 0$, $x = 1$ and $y = 2$ all kept at zero temperatures. Side $y = 0$ is held at a temperature $f(x)$, $0 \leqslant x \leqslant 1$. Write the BVP and find the steady state temperature distribution throughout the rectangular plate. The plate has insulated faces.
2. A square plate with two units on an edge has three edges maintained at zero temperatures and the fourth edge at a temperature distribution $f(x)$. The BVP is

$$u_{xx} + u_{yy} = 0, \quad 0 < x < 2, \, 0 < y < 2$$

$$u(0, y) = u(2, y) = 0, \quad 0 < y < 2$$

$$u(x, 0) = 0, \, u(x, 2) = f(x), \quad 0 < x < 2$$

Solve the BVP.
3. Find the harmonic function which satisfies the BVP

$$\nabla^2 u = 0, \quad 0 < x < a, \, 0 < y < b$$

$$u(0, y) = u(a, y) = 0, \quad 0 < y < b$$

$$u(x, 0) = f(x), \, u(x, b) = g(x), \quad 0 < x < a$$

4. A square plate (Figure 8.6) with 2 units on each edge has zero temperature along the edge $x = 0$ and is insulated along $x = 2$. The edge $y = 0$ is kept at zero temperature, but at $y = 2$ the temperature is $f(x) = \sin 3\pi x/4$. Write a steady state BVP corresponding to this description and solve it.

5. Consider the BVP

$$u_{xx} + u_{yy} = 0, \quad 0 \leqslant x \leqslant 10, \ 0 \leqslant y \leqslant 10$$
$$u(x, 0) = 0 = u(0, y)$$
$$u(x, 10) = x^2, \quad u(10, y) = y^2$$

Find an approximate solution using the Gauss–Seidel method with $h = 2$.

6. Solve the Dirichlet problem

$$\nabla^2 u = 0, \quad 0 < x < \pi, \ 0 < y < 2\pi$$
$$u(0, y) = u(\pi, y) = 0, \quad 0 < y < 2\pi$$
$$u(x, 0) = 0, \ u(x, 2\pi) = 1, \quad 0 < x < \pi$$

7. Solve the Neumann problem

$$\nabla^2 u = 0, \quad 0 < x < \pi, \ 0 < y < \pi$$
$$u_x(0, y) = u_x(\pi, y) = 0, \quad 0 < y < \pi$$
$$u_y(x, 0) = \cos x, \quad 0 \leqslant x \leqslant \pi$$
$$u_y(x, \pi) = 0, \quad 0 \leqslant x \leqslant \pi$$

8. Solve the Neumann problem

$$\nabla^2 u = 0, \quad 0 < x < \pi, \ 0 < y < \pi$$
$$u_x(0, y) = 0, \quad 0 \leqslant y \leqslant \pi$$
$$u_x(\pi, y) = 2 \cos y, \quad 0 \leqslant y \leqslant \pi$$
$$u_y(x, 0) = 0, \quad u_y(x, \pi) = 0, \quad 0 \leqslant y \leqslant \pi$$

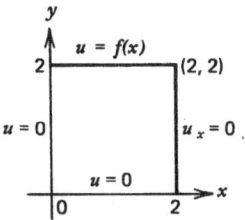

Figure 8.6. The insulated thin plate.

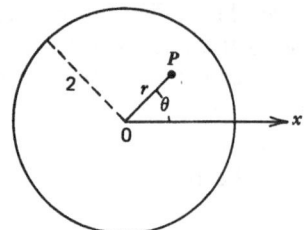

Figure 8.7. Polar coordinates for a circular disk.

8.11. TEMPERATURE IN A CIRCULAR DISK WITH INSULATED FACES

We assume that the radius of the disk is two units. The initial temperature, dependent only on the radius of the disk, is $f(r)$. The outer circumference is kept at zero temperature. As suggested in the title the plane faces are insulated. See Figure 8.7. It is our aim to find the temperature $u(r, t)$. First, it is wise for us to formulate the mathematical model or BVP.

$$u_t = a^2\left(u_{rr} + \frac{1}{r}u_r\right), \quad 0 < r < 2, \ t > 0$$

$$u(2, t) = 0, \quad t \geqslant 0 \tag{8.72}$$

$$u(r, 0) = f(r), \quad 0 < r < 2$$

$$|u(r, t)| < M, \quad 0 < r < 2, \ t \geqslant 0$$

In the heat equation of (8.72) we have another specialization of (5.67). In this problem u is dependent only on r and t. Therefore, the Laplacian $\nabla^2 u$ is dependent on r alone and is written

$$\nabla^2 u(r) = u_{rr} + \frac{1}{r}u_r$$

Our solution follows using the Fourier method.

1. *Separation of Variables.* Let $u(r, t) = R(r)T(t)$. The PDE becomes

$$RT' = a^2\left(R''T + \frac{1}{r}R'T\right)$$

$$\frac{T'}{a^2 T} = \frac{R''}{R} + \frac{1}{r}\frac{R'}{R} = -\alpha^2$$

2. *Related ODEs*

$$R'' + \frac{1}{r}R' + \alpha^2 R = 0$$

$$T' + \alpha^2 a^2 T = 0$$

3. *Homogeneous Boundary Condition*

$$u(2, t) = R(2)T(t) = 0$$

If $T(t) \neq 0$, then $R(2) = 0$.

There is insufficient information to write a SLP. We can consider the following.

4. *The R Equation*

$$rR'' + R' + \alpha^2 rR = 0 , \quad R(2) = 0$$

The differential equation may be written in the form

$$[rR']' + \alpha^2 rR = 0$$

This is the same type displayed in (5.48) with $n = 0$. Therefore,

$$R(r) = C_1 J_0(\alpha r) + C_2 Y_0(\alpha r)$$

Since Y_0 is unbounded at $r = 0$, we select $C_2 = 0$.

$$R(2) = 0 = C_1 J_0(2\alpha) = 0$$

If $C_1 \neq 0$, then $J_0(2\alpha) = 0$. Thus

$$R_k(r) = J_0(\alpha_k r) \tag{8.73}$$

where $2\alpha_k$ are the positive zeros of J_0.

5. *The T Equation.* The new T equation is

$$T' + \alpha_k^2 a^2 T = 0$$

It has a solution

$$T_k(t) = \exp\left(-\alpha_k^2 a^2 t\right) \tag{8.74}$$

6. *Solution Set for Homogeneous Conditions.* Using solutions (8.73) and (8.74) in the separation substitution, we have the solution set

$$u_k(r, t) = \exp\left(-\alpha_k^2 a^2 t\right) J_0(\alpha_k r) \tag{8.75}$$

7. *Superposition.* The infinite linear combination of (8.75) is the series

$$u(r, t) = \sum_{k=1}^{\infty} A_k \exp\left(-\alpha_k^2 a^2 t\right) J_0(\alpha_k r)$$

8. *Nonhomogeneous Boundary Condition*

$$u(r, 0) = f(r) = \sum_{k=1}^{\infty} A_k J_0(\alpha_k r)$$

is a Fourier-Bessel series with coefficients

$$A_k = \frac{2}{2^2 J_1^2(2\alpha_k)} \int_0^2 rf(r) J_0(\alpha_k r) \, dr$$

9. *Solution of Original BVP*

$$u(r, t) = \frac{1}{2} \sum_{k=1}^{\infty} \frac{1}{J_1^2(2\alpha_k)} \int_0^2 s J_0(\alpha_k s) f(s) \exp\left(-\alpha_k^2 a^2 t\right) J_0(\alpha_k r) \, ds$$

8.12. STEADY STATE TEMPERATURE IN A RIGHT SEMICIRCULAR CYLINDER

We assume that half the right circular cylinder has a radius a and a height b. It is bounded by the planes $z = 0$, $z = b$, and the face $y = 0$ which, in cylindrical coordinates, can be described by both $\theta = 0$ and $\theta = \pi$. We assume that the lower horizontal plane face is kept at temperature zero. The upper plane surface is kept at temperature $f(r, \theta)$. The plane vertical face remains at zero temperature. In this problem we wish to find the temperature distribution $u(r, \theta, z)$ (see Figure 8.8). The BVP follows:

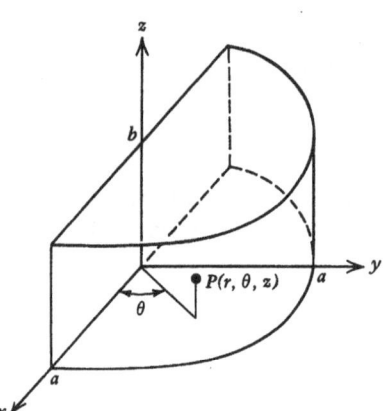

Figure 8.8. Half of a circular cylinder.

$$u_{rr} + \frac{1}{r} u_r + \frac{1}{r^2} u_{\theta\theta} + u_{zz} = 0, \quad 0 < r < a, \, 0 < \theta < \pi, \, 0 < z < b$$

$$u(r, 0, z) = u(r, \pi, z) = 0, \quad 0 < r < a, \, 0 < z < b$$

$$u(a, \theta, z) = 0, \quad 0 < \theta < \pi, \, 0 < z < b$$

$$u(r, \theta, 0) = 0, \quad 0 < r < a, \, 0 < \theta < \pi$$

$$u(r, \theta, b) = f(r, \theta), \quad 0 < r < a, \, 0 < \theta < \pi$$

$$|u(r, \theta, z)| < M$$

We begin the solution by

1. *Separation of Variables.* Let $u(r, \theta, z) = R(r)\Theta(\theta)Z(z)$.

$$R''\Theta Z + \frac{1}{r} R'\Theta Z + \frac{1}{r^2} R\Theta'' Z + R\Theta Z'' = 0$$

or

$$r^2 \frac{R''}{R} + r \frac{R'}{R} + r^2 \frac{Z''}{Z} = -\frac{\Theta''}{\Theta} = \alpha^2$$

If

$$\frac{Z''}{Z} = \beta^2$$

then

$$r^2 \frac{R''}{R} + r \frac{R'}{R} + r^2\beta^2 - \alpha^2 = 0$$

2. *Related ODEs*

$$rR'' + R' + \left(r\beta^2 - \frac{\alpha^2}{r} \right) R = 0$$

$$Z'' - \beta^2 Z = 0$$

$$\Theta'' + \alpha^2\Theta = 0$$

3. *Homogeneous Boundary Conditions*

$$u(r, 0, z) = R(r)\Theta(0)Z(z)$$

If $R(r) \neq 0$ and $Z(z) \neq 0$, then $\Theta(0) = 0$.

$$u(r, \pi, z) = R(r)\Theta(\pi)Z(z) = 0$$

If $R(r) \neq 0$ and $Z(z) \neq 0$, then $\Theta(\pi) = 0$.

$$u(a, \theta, z) = R(a)\Theta(\theta)Z(z) = 0$$

If $\Theta(\theta) \neq 0$ and $Z(z) \neq 0$, then $R(a) = 0$.

$$u(r, \theta, 0) = R(r)\Theta(\theta)Z(0) = 0$$

If $R(r) \neq 0$ and $\Theta(\theta) \neq 0$, then $Z(0) = 0$. We have enough information here to state a related SLP.

4. *Related SLP*

$$\Theta'' + \alpha^2\Theta = 0, \quad \Theta(0) = 0, \quad \Theta(\pi) = 0$$

The general solution is

$$\Theta(\theta) = C_1 \cos \alpha\theta + C_2 \sin \alpha\theta$$
$$\Theta(0) = C_1 + 0 = 0$$
$$\Theta(\pi) = C_2 \sin \alpha\pi = 0$$

If $C_2 \neq 0$, then $\sin \alpha\pi = 0$, $\alpha\pi = n\pi$, $\alpha = n$, and

$$\alpha_n^2 = n^2$$

is the set of eigenvalues for the SLP. Eigenfunctions are

$$\Theta_n(\theta) = \sin n\theta, \quad n \in \mathbf{N} \tag{8.76}$$

If $n = 0$, the SLP has only a trivial solution. Therefore the domain of n is adequate in (8.76).

5. *The Z Equation*

$$Z'' - \beta^2 Z = 0, \quad Z(0) = 0$$

The Z equation has a solution

$$Z = B_1 \cosh \beta z + B_2 \sinh \beta z$$
$$Z(0) = B_1 + 0 = 0$$

We fail to have a complete SLP, so the nature of β is undetermined at present and

$$Z(z) = \sinh \beta z$$

6. *The R Equation*

$$[rR']' + \left[r\beta^2 - \frac{n^2}{r}\right]R = 0, \quad R(a) = 0$$

This is a Bessel equation where λ is β^2 and $n = n$. A bounded solution may be expressed as

$$R(r) = J_n(\beta r)$$

However,

$$R(a) = J_n(\beta a) = 0$$

Therefore, $a\beta_{nk}$ are the zeros of J_n and

$$R_{nk}(r) = J_n(\beta_{nk}r), \quad k \in \mathbf{N} \tag{8.77}$$

Backing up a bit, we can write

$$Z_{nk}(z) = \sinh \beta_{nk}z \tag{8.78}$$

7. *Solution Set for Homogeneous Conditions.* From the single variable function solutions (8.76), (8.77), and (8.78) we write

$$u_{nk}(r, \theta, z) = \sin n\theta \sinh \beta_{nk}z J_n(\beta_{nk}r), \quad n, k \in \mathbf{N} \tag{8.79}$$

8. *Superposition.* A double sum is used in this case

$$u(r, \theta, z) = \sum_{n=1}^{\infty} \sum_{k=1}^{\infty} A_{nk} \sin n\theta \sinh \beta_{nk}z J_n(\beta_{nk}r)$$

9. *Nonhomogeneous Boundary Conditions*

$$u(r, \theta, b) = f(r, \theta) = \sum_{n=1}^{\infty} \sum_{k=1}^{\infty} A_{nk} \sin n\theta \sinh \beta_{nk}b \, J_n(\beta_{nk}r)$$

This may be rewritten

$$f(r, \theta) = \sum_{n=1}^{\infty} \sin n\theta \sum_{k=1}^{\infty} A_{nk} \sinh \beta_{nk}b \, J_n(\beta_{nk}r) \tag{8.80}$$

so that

$$\sum_{k=1}^{\infty} A_{nk} \sinh \beta_{nk}b \, J_n(\beta_{nk}r)$$

are the coefficients of the sine series in (8.80). Therefore,

$$\sum_{k=1}^{\infty} A_{nk} \sinh \beta_{nk}b \, J_n(\beta_{nk}r) = \frac{2}{\pi} \int_0^{\pi} f(r, \theta) \sin n\theta \, d\theta \quad (8.81)$$

However, (8.81) is a Fourier–Bessel series with $A_{nk} \sinh \beta_{nk}b$ as the coefficients in the series. Therefore,

$$A_{nk} \sinh \beta_{nk}b = \frac{2}{a^2 J_{n+1}^2(\beta_{nk}a)} \int_0^a rJ_n(\beta_{nk}r)$$

$$\times \left[\frac{2}{\pi} \int_0^{\pi} f(r, \theta) \sin n\theta \, d\theta \right] dr$$

and

$$A_{nk} = \frac{4}{a^2 \pi \sinh \beta_{nk}b \, J_{n+1}^2(\beta_{nk}a)} \int_0^a \int_0^{\pi} rf(r, \theta) J_n(\beta_{nk}r) \sin n\theta \, d\theta \, dr$$

10. *Solution of the Original BVP*

$$u(r, \theta, z) = \sum_{n=1}^{\infty} \sum_{k=1}^{\infty} A_{nk} \sin n\theta \sinh \beta_{nk}z \, J_n(\beta_{nk}r)$$

where

$$A_{nk} = \frac{4}{a^2 \pi \sinh \beta_{nk}b \, J_{n+1}^2(\beta_{nk}a)} \int_0^a \int_0^{\pi} rf(r, \theta) J_n(\beta_{nk}r) \sin n\theta \, d\theta \, dr$$

8.13. HARMONIC INTERIOR OF A RIGHT CIRCULAR CYLINDER

We assume that the cylinder is bounded by three surfaces $r = a$, $z = 0$, and $z = b$. If $u(r, z)$ is the harmonic function, it is assumed that $u = 0$ on $z = 0$ and u is $f(z)$ on the surface $r = a$, $0 < z < b$. We wish to find $u(r, z)$ for the BVP

$$u_{rr} + \frac{1}{r} u_r + u_{zz} = 0, \quad 0 < r < a, 0 < z < b$$

$$u(r, 0) = u(r, b) = 0, \quad 0 < r < a$$

$$u(a, z) = f(z), \quad 0 < z < b$$

$$|u(r, z)| < M$$

We begin the solution by

1. *Separation of Variables.* Let $u(r, z) = R(r)Z(z)$

$$R''Z + \frac{1}{r} R'Z + RZ'' = 0$$

and

$$\frac{R''}{R} + \frac{1}{r}\frac{R'}{R} = -\frac{Z''}{Z} = \alpha^2$$

if Z is to be bounded.

2. *Related ODEs*

$$Z'' + \alpha^2 Z = 0$$
$$rR'' + R' - \alpha^2 rR = 0$$

3. *Homogeneous Boundary Conditions*

$$u(r, 0) = R(r)Z(0) = 0$$

If $R(r) \neq 0$, then $Z(0) = 0$.

$$u(r, b) = R(r)Z(b) = 0$$

If $R(r) \neq 0$, $Z(b) = 0$.

4. *A Related SLP*

$$Z'' + \alpha^2 Z = 0, \quad Z(0) = Z(b) = 0$$

The SLP has a general solution

$$Z = C_1 \cos \alpha z + C_2 \sin \alpha z$$
$$Z(0) = C_1 + 0 = 0$$
$$Z(b) = C_2 \sin \alpha b = 0$$

If $C_2 \neq 0$, $\sin \alpha b = 0$, $ab = n\pi$, $\alpha = n\pi/b$, so that

$$\alpha_n^2 = \frac{n^2\pi^2}{b^2}$$

for the eigenvalues. The eigenfunctions are

$$Z_n(z) = \sin \frac{n\pi z}{b}, \quad n \in \mathbb{N} \tag{8.82}$$

If $n = 0$, $Z = 0$ for the problem. Therefore, n is adequately described in (8.82).

5. *The Related R Equation*

$$[rR']' - \alpha^2 rR = 0 \tag{8.83}$$

This equation is not quite the same as (5.14) where we considered the solution of the modified Bessel equation. If we let $x = \alpha r$ in (5.14) we obtain

$$r \frac{d^2y}{dr^2} + \frac{dy}{dr} - \left(\alpha^2 r + \frac{n^2}{r}\right) y = 0$$

If y is replaced by R and $n = 0$, we have (8.83). We must not confuse this n with n in (8.82). A general solution of (8.83) is

$$R = C_1 I_0(\alpha r) + C_2 K_0(\alpha r)$$

However, K_0 is unbounded at $r = 0$ and C_2 needs to be zero. The parameter α has already been determined as $n\pi/b$. Therefore,

$$R_n(r) = I_0\left(\frac{n\pi r}{b}\right)$$

6. *Solution Set for Homogeneous Conditions*

$$u_n(r, z) = I_0\left(\frac{n\pi r}{b}\right) \sin \frac{n\pi z}{b}, \quad n \in \mathbf{N}$$

7. *Superposition.* The linear combination is written as a series

$$u(r, z) = \sum_{n=1}^{\infty} A_n I_0\left(\frac{n\pi r}{b}\right) \sin \frac{n\pi z}{b}$$

8. *Nonhomogeneous Boundary Condition*

$$u(a, z) = f(z) = \sum_{n=1}^{\infty} A_n I_0\left(\frac{n\pi a}{b}\right) \sin\left(\frac{n\pi z}{b}\right), \quad 0 < z < b$$

This is a sine series with coefficients

$$A_n I_0\left(\frac{n\pi a}{b}\right) = \frac{2}{b} \int_0^b f(z) \sin \frac{n\pi z}{b} \, dz$$

and

$$A_n = \frac{2}{b I_0\left(\dfrac{n\pi a}{b}\right)} \int_0^b f(z) \sin \frac{n\pi z}{b} \, dz$$

9. Solution for the Original BVP

$$u(r, z) = \frac{2}{b} \sum_{n=1}^{\infty} \frac{I_0(n\pi r/b)}{I_0(n\pi a/b)} \int_0^b f(\xi) \sin\frac{n\pi\xi}{b} \sin\frac{n\pi z}{b} \, d\xi$$

Exercises 8.4

1. In a cylindrical region, $r<1$, $0<z<2$, solve the steady state temperature problem

$$u_{rr} + \frac{1}{r} u_r + u_{zz} = 0, \quad 0<r<1, \, 0<z<2$$

$$u(1, z) = 0, \quad 0<z<2$$

$$u(r, 2) = 0, \quad 0<r<1$$

$$u(r, 0) = f(r), \quad 0<r<1$$

$$|u(r, z)| < M$$

2. Determine the steady state solution for the temperature distribution $u(r, z)$ in a cylinder of radius 1 and height h given that

$$u_{rr} + \frac{1}{r} u_r + u_{zz} = 0, \quad 0<r<1, \, 0<z<h$$

$$u(1, z) = 0, \quad 0<z<h$$

$$u(r, h) = 0, \quad 0<r<1$$

$$u(r, 0) = T_0, \quad 0<r<1$$

$$|u(r, z)| < M$$

3. Find a harmonic function $u(r, z)$ for the inside of a cylinder bounded by $r = a$, $z = 0$ and $z = h$ if $u = 0$ on the surface $r = a$ and $z = 0$, and $u = f(r)$ on the plane surface $z = h$.

4. Determine the steady state temperature in the cylindrical region so that

$$u_{rr} + \frac{1}{r} u_r + u_{zz} = 0, \quad 0<r<1, \, 0<z<2$$

$$u(r, 2) = 0, \quad 0<r<1$$

$$u_r(1, z) = ku(1, z), \quad 0<z<2, \, k>0$$

$$u(r, 0) = f(r), \quad 0<r<1$$

$$|u(r, z)| < M$$

5. A solid is bounded by long concentric cylinders. The inner cylinder has a radius p and the outer cylinder has a radius q. Diffusivity is a^2. Inner and outer surfaces are kept at zero temperatures and the initial temperature

is dependent on r alone, given by $f(r)$. Find the temperature $u(r, t)$. The BVP follows:

$$u_t = a^2\left(u_{rr} + \frac{1}{r} u_r\right), \quad p < r < q, \ t > 0$$

$$u(p, t) = u(q, t) = 0, \quad t \geq 0$$

$$u(r, 0) = f(r), \quad p < r < q$$

$$|u(r, t)| < M$$

8.14. STEADY STATE TEMPERATURE DISTRIBUTION IN A SPHERE

We assume that the temperature distribution on the surface of a sphere, having radius a, is preserved so that $u(a, \phi) = f(\phi)$. We wish to determine the steady state temperature $u(\rho, \phi)$ for the sphere.

In this problem, the Laplacian is not dependent on θ and the heat equation is without the time variable t. Therefore, the steady state equation is

$$u_{\rho\rho} + \frac{2}{\rho} u_\rho + \frac{1}{\rho^2} u_{\phi\phi} + \frac{\cot \phi}{\rho^2} u_\phi = 0 \tag{8.84}$$

To simplify the equation in the BVP, we multiply (8.84) by ρ^2 and write the problem

$$\rho^2 u_{\rho\rho} + 2\rho u_\rho + u_{\phi\phi} + \frac{\cos \phi}{\sin \phi} u_\rho = 0, \quad 0 < \rho < a, \ 0 < \phi < \pi$$

$$u(a, \phi) = f(\phi), \quad 0 \leq \phi \leq \pi$$

$$|u(\rho, \phi)| \leq M$$

On the conical surfaces, Figure 8.9, the steady state temperature is dependent only on ρ and ϕ. We proceed with the Fourier method.

1. *Separation of Variables*. Let

$$u(\rho, \phi) = R(\rho)\Phi(\phi)$$

Then

$$\rho^2 R''\Phi + 2\rho R'\Phi + R\Phi'' + \frac{\cos \phi}{\sin \phi} R\Phi' = 0$$

and

$$\frac{\rho^2 R''}{R} + \frac{2\rho R'}{R} = -\left(\frac{\Phi''}{\Phi} + \frac{\cos \phi}{\sin \phi} \frac{\Phi'}{\Phi}\right) = \lambda$$

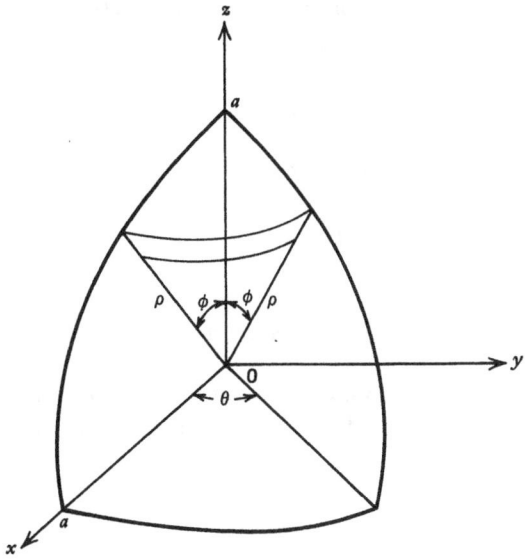

Figure 8.9. Conical surface bounded by a spherical surface.

2. *Related ODEs.* If $\lambda = n(n+1)$, then

$$\rho^2 R'' + 2\rho R' - n(n+1)R = 0$$

$$\sin \phi \, \Phi'' + \cos \phi \, \Phi' + n(n+1) \sin \phi \, \Phi = 0$$

There are no homogeneous boundary conditions in this problem.

3. *The R Equation.* This is an Euler equation. If we let $\rho = e^t$ or $t = \ln \rho$, then

$$\frac{dR}{d\rho} = \frac{1}{\rho} \frac{dR}{dt}$$

and

$$\frac{d^2 R}{d\rho^2} = \frac{1}{\rho^2} \left[\frac{d^2 R}{dt^2} - \frac{dR}{dt} \right]$$

The transformed equation is

$$\frac{d^2 R}{dt^2} + \frac{dR}{dt} - n(n+1)R = 0$$

and the characteristic equation

$$m^2 + m - n(n + 1) = 0$$

or

$$m = n, \quad m = -(n + 1)$$

The solution as a function of t is

$$R(t) = A_1 e^{nt} + A_2 e^{-(n+1)t}$$

and

$$R(\rho) = A_1 \rho^n + A_2 \rho^{-(n+1)} \tag{8.85}$$

We have no formal boundary condition except that u must be bounded. To ensure this condition we assign $A_2 = 0$ to accommodate $\rho = 0$. Therefore,

$$R_n(\rho) = \rho^n \tag{8.86}$$

4. *The Φ Equation.* The second equation of the related ODEs is exactly (6.22) with Φ replacing y. It has a solution (6.23),

$$\Phi_n(\phi) = P_n(\cos \phi) \tag{8.87}$$

5. *Solution Set of the Homogeneous Differential Equation.* Using the separation substitution and the two ODE solutions (8.86) and (8.87), we have

$$u_n(\rho, \phi) = \rho^n P_n(\cos \phi), \quad n \in \mathbf{N_0}$$

6. *Superposition.* The infinite linear combination is the series

$$u(\rho, \phi) = \sum_{n=0}^{\infty} C_n \rho^n P_n(\cos \phi)$$

7. *The Nonhomogeneous Boundary Condition*

$$u(a, \phi) = f(\phi) = \sum_{n=0}^{\infty} C_n a^n P_n(\cos \phi)$$

According to the result of No. 6 of Exercises 6.3,

$$C_n a^n = \frac{2n + 1}{2} \int_0^{\pi} f(\phi) \sin \phi P_n(\cos \phi) \, d\phi$$

or

$$C_n = \frac{2n + 1}{2a^n} \int_0^{\pi} f(\phi) \sin \phi P_n(\cos \phi) \, d\phi$$

8. *Solution for the Original BVP*

$$u(\rho, \phi) = \frac{1}{2} \sum_{n=0}^{\infty} \left[\frac{2n+1}{a^n} \int_0^{\pi} f(\xi) \sin \xi P_n(\cos \xi) \, d\xi \right] \rho^n P_n(\cos \phi)$$

8.15. POTENTIAL FOR A SPHERE

Let us assume that the sphere has a radius a. We wish to find the potential $v(\rho, \phi)$ if (a) $\rho < a$ and (b) $\rho > a$ with the condition $\lim_{\rho \to a} v(\rho, \phi) = f(\phi)$. For part (a), the BVP is

$$\nabla^2 v(\rho, \phi) = 0, \quad 0 < \rho < a, \, 0 < \phi < \pi$$

$$\lim_{\rho \to a^-} v(\rho, \phi) = f(\phi), \quad 0 \leq \phi \leq \pi$$

$$|v(\rho, \phi)| < M$$

If $\lim_{\rho \to a^-} v(\rho, \phi) = v(a, \phi)$, the problem is the same as the problem of Section 8.14 for the steady state temperature distribution. The discussion follows the form of the preceding problem and the solution may be stated

$$v(\rho, \phi) = \frac{1}{2} \sum_{n=0}^{\infty} (2n+1) \left(\frac{\rho}{a} \right)^n P_n(\cos \phi) \int_0^{\pi} f(\xi) \sin \xi P_n(\cos \xi) \, dt$$

For part (b) the BVP has a change in the boundary condition and a change in the domain. The BVP follows:

$$\nabla^2 v(\rho, \phi) = 0, \quad \rho > a, \, 0 < \phi < \pi$$

$$\lim_{\rho \to a^+} v(\rho, \phi) = f(\phi), \quad 0 \leq \phi \leq \pi$$

$$|v(\rho, \phi)| < M$$

If $\lim_{\rho \to a^+} v(\rho, \phi) = v(a, \phi)$, then the solution discussion is the same up to the selection of the arbitrary constant in (8.85). Here, to satisfy the condition of boundedness, we select $A_1 = 0$. The R solution becomes

$$R_n(\rho) = \rho^{-(n+1)} \tag{8.88}$$

A solution for the Φ equation is still

$$\Phi_n(\phi) = P_n(\cos \phi) \tag{8.89}$$

5. *Solution Set for the Homogeneous Differential Equation*

$$v_n(\rho, \phi) = \rho^{-(n+1)} P_n(\cos \phi), \quad n \in \mathbf{N}_0$$

6. *Superposition*

$$v(\rho, \phi) = \sum_{n=0}^{\infty} C_n^* \rho^{-(n+1)} P_n(\cos \phi)$$

7. *Nonhomogeneous Boundary Condition*

$$v(a, \phi) = \sum_{n=0}^{\infty} C_n^* a^{-(n+1)} P_n(\cos \phi)$$

where

$$C_n^* = \frac{a^{n+1}(2n+1)}{2} \int_0^{\pi} f(\phi) \sin \phi P_n(\cos \phi) \, d\phi$$

8. *Solution for the Original BVP Part (b)*

$$v(\rho, \phi) = \frac{1}{2} \sum_{n=0}^{\infty} \left(\frac{a}{\rho}\right)^{n+1} (2n+1) P_n(\cos \phi) \int_0^{\pi} f(\xi) \sin \xi P_n(\cos \xi) \, d\xi$$

This is the potential outside the sphere.

Exercises 8.5

1. For a sphere of radius a, the upper half of the surface has a temperature $u(a, \phi) = 30°C$. The temperature on the lower half is kept at $0°C$. Determine the steady state temperature $u(\rho, \phi)$.

2. The spherical surface of a hemisphere is kept at a temperature T_0, but the base is kept at a temperature zero. Solve for the steady state temperature $u(\rho, \phi)$ in the hemisphere. Show the first two terms of the series solution.

3. For the BVP

$$\nabla^2 u(\rho, \phi) = 0, \quad 0 < \rho < 1, \, 0 < \phi < \pi$$

$$u(1, \phi) = \begin{cases} T_0 & \text{when } 0 < \phi < \pi/2 \\ 0 & \text{when } \pi/2 < \phi < \pi \end{cases}$$

$$|u(\rho, \phi)| < M$$

show that

$$u(\rho, \phi) = \frac{T_0}{2} + \frac{T_0}{2} \sum_{n=0}^{\infty} [P_{2n}(0) - P_{2n+2}(0)] \rho^{2n+1} P_{2n+1}(\cos \phi)$$

Give a spherical interpretation for the stated BVP.

4. Solve the BVP for $v(\rho, \phi)$ in

$$\nabla^2 v(\rho, \phi) = 0, \quad 0 < \rho < 1, \, 0 < \phi < \pi$$

$$v(1, \phi) = 0, \quad 0 < \phi < \pi$$

$$v \rightarrow -V_0 \rho \cos \phi \quad \text{as } \rho \rightarrow \infty$$

In a uniform field this is a potential problem of a grounded conducting sphere.

5. If $u(\rho, \phi)$ represents the steady state temperature between concentric spheres $2 \leq \rho \leq 3$, where $u(2, \phi) = f(\cos \phi)$ and $u(3, \phi) = 0$, $0 < \phi < \pi$, show that

$$u(\rho, \phi) = \sum_{n=0}^{\infty} C_n \frac{3^{2n+1} - \rho^{2n+1}}{3^{2n+1} - 2^{2n+1}} \left(\frac{2}{\rho}\right)^{n+1} P_n(\cos \phi)$$

where

$$C_n = \frac{2n+1}{2} \int_0^{\pi} f(\cos \phi) \sin \phi P_n(\cos \phi) \, d\phi$$

6. Solve the BVP for $u(\rho, \phi)$

$$\nabla^2 u(\rho, \phi) = 0, \quad 0 < \rho < 1, \, 0 < \phi < \pi$$

$$u(1, \phi) = 1 - 2 \cos^2 \phi$$

$$|u(\rho, \phi)| < M$$

Give a physical interpretation for the BVP.

9

ADDITIONAL APPLICATIONS

In this chapter we consider BVPs with solutions expressed as multiple Fourier series, Fourier integrals, and special functions. The problems with multiple Fourier series solutions generally involve multidimensional geometry while those with Fourier integral solutions involve BVPs with unbounded domains. We will also consider BVPs having nonhomogeneous PDEs or boundary conditions. Transformations are suggested which change certain problems to fit the framework of the Fourier method. In addition, several problems are considered for which transform methods (especially the Laplace transform) are useful.

9.1. MECHANICAL AND ELECTRICAL OSCILLATIONS

The equation

$$m \frac{d^2x}{dt^2} + b \frac{dx}{dt} + Rx = f(t) \tag{9.1}$$

describes the displacement x of a vibrating mass m on a vertical spring. Although we shall omit the details, Hooke's law and Newton's law of motion are used in formulation (9.1). The term $m(d^2x/dt^2)$ is the result of the *time rate of change in momentum $d(mv)/dt$*, (m constant) from Newton's law. For small speeds the magnitude of the *damping force* is assumed to be approximately proportional to the instantaneous speed, $b(dx/dt)$ (b a constant of proportionality). The spring constant k comes from the proportionality of force to stretch in Hooke's law. The function $f(t)$ is an external or driving force for the system.

The equation

$$L \frac{d^2Q}{dt^2} + R \frac{dQ}{dt} + \frac{1}{C} Q = E(t) \tag{9.2}$$

describes the charge Q in a simple series circuit, the current $I = dQ/dt$, and the electromotive force (emf) $E(t)$ in the circuit. Flow of the current is determined by Kirchhoff's law, which may be stated: *voltage supplied is equal to the sum of the voltage drops around a circuit*. From elementary laws of electricity, the voltage drop across: (a) the resistance is IR; (b) the capacitor is Q/C; (c) the inductance is $L(dI/dt)$. Thus, the result

$$L \frac{dI}{dt} + RI + \frac{1}{C} Q = E(t)$$

which can be written in the form of (9.2).

An analog of mechanical and electrical systems having ODEs (9.1) and (9.2) follows:

Position x	compared to	Charge Q
Mass m	compared to	Inductance L
Damping Constant b	compared to	Resistance R
Velocity $v = \dfrac{dx}{dt}$	compared to	Current $I = \dfrac{dQ}{dt}$
Spring Constant k	compared to	Inverse Capacitance $1/C$
External Force $F(t)$	compared to	Electromotive Force $E(t)$

If the coefficients in (9.1) and (9.2) are constants we have described two systems that have the same basic type of IVP; namely,

$$\alpha y''(t) + \beta y'(t) + \gamma y(t) = g(t) , \quad y(0) = y_0, \ y'(0) = v_0$$

For mechanical problems, we will be using the centimeter, gram, second (CGS) system; or the foot, pound, second (FPS) system. A dyne is a force of $1 \, g \, cm/s^2$. For a body acted upon by its weight W, the corresponding acceleration is that due to gravity g. The force W is given by Newton's law: $W = mg$. We assume that $g = 980 \, cm/s^2$ in the CGS system and $g = 32 \, ft/s^2$ in the FPS system. If units are not specified, any system can be used if consistency is maintained.

We will offer only one set of units in our electrical problems. Voltage or potential also may be referred to as electromotive force (emf). The units of our system are:

Voltage, Potential, emf	E	Volt
Resistance	R	Ohm
Inductance	L	Henry
Capacitance	C	Farad
Current	I	Ampere
Charge	Q	Coulomb

Since we are using capital letters (such as Q) to describe certain quantities in electrical circuits, we suggest using corresponding lowercase letters to represent Laplace transforms. Thus $\mathcal{L}\{Q(t)\} = q(s)$.

Exercises 9.1

1. A force of 50 dyn stretches a spring 2 cm (see Figure 9.1). A mass of 1 g is suspended on the spring. As soon as equilibrium is reached, the mass is raised 5 cm above the equilibrium position and then released. We assume no damping takes place. This may be called a *free vibrating motion*. Show that the IVP for this situation is

$$\frac{d^2x}{dt^2} + 25x = 0$$

$$x(0) = -5, \quad x'(0) = 0$$

Assume the position axis has its positive direction pointed downward. Use the Laplace transform method to solve the IVP. Then check your solution.

2. A force of 50 dyn stretches a spring 2 cm. A mass of 1 g is suspended on the spring. As soon as equilibrium is achieved, there is no motion at time $t = 0$. After 2 s the mass is struck by an impulsive force of 25 dyn in the upward direction. Assume no damping in the system. Form an IVP which describes the motion and then solve the problem. Show that the ODE of the IVP is

$$\frac{d^2x}{dt^2} + 25x = -25\delta(t - 2)$$

if the position axis is directed downward.

Figure 9.1. Spring (a) without weight; (b) with weight attached in equilibrium position; (c) weight displaced from equilibrium position.

(a)   (b)   (c)

Figure 9.2. *R-L-C* circuit.

3. A 32 lb weight hangs from a vertical spring having a spring constant equal to 0.5 lb/ft. The weight is acted upon by a damping force (in lb) which is numerically equal to 6 times the instantaneous velocity (in ft/s). The weight is pulled 6 in below the equilibrium position and released. Write the IVP for the motion and solve the problem. Assume that the position axis is directed so that positive is downward. The Laplace transform method is suggested for obtaining the solution.

4. An 8 lb weight stretches a vertical spring 6 in. No noticeable damping is present. An impressed or driving force of cos 8*t* acts upon the spring. The 8 lb weight is started in motion from an equilibrium position with an upward velocity of 6 ft/s. Write an IVP that describes this motion and solve it. This problem exhibits a phenomenon known as *resonance*. What are some of the features of resonance?

5. A series circuit has an emf of 100 V, a resistor of 10 Ω and a capacitor of 10^{-3} F (see Figure 9.2). The switch in the circuit is closed at $t = 0$, and there is no charge at this instant. Determine the IVP and solve for the charge and the current for $t > 0$. Solve the IVP using Laplace transforms and then check your solution.

6. A series circuit has an emf of 100 V, a resistor of 100 Ω, and a capacitor of 10^{-3} F. The switch is closed at $t = 0$, and there is no charge at this instant. After 10 s the switch is opened so there is no further emf in the system. Determine the IVP and solve for the charge and current for $t > 0$. Compare the solution with the one obtained for the preceding exercise.

7. A generator having an emf given by 360 cos 40*t* connected in series with a 0.2 H inductor and a 2×10^{-3} F capacitor. Write an IVP describing this condition if both I and Q are zero when $t = 0$. Find Q and I when $t > 0$.

8. A 10 Ω resistor, a 2 H inductor, and a 0.04 F capacitor are in series with an emf of 200 V. When $t = 0$, $Q = I = 0$. Write the IVP and use Laplace transforms to solve for $Q(t)$ and $I(t)$ when $t > 0$.

9.2. THE VIBRATING MEMBRANE

A vibrating membrane, such as a rectangular drumhead, has displacements that satisfy the two dimensional wave equation. Instead of the one dimensional geometry displayed in the string problem, we are concerned now with the two dimensional geometry of the membrane in a plane. Before preparing a model for this problem, we describe a few assumptions concerning the material and behavior of the membrane.

1. The membrane is homogeneous. The density ρ is constant.
2. The membrane is composed of a perfectly flexible material which offers no resistance to deformation perpendicular to the xy plane. Motion of each element is perpendicular to the xy plane.
3. The membrane is stretched and fixed along a boundary in the xy plane.
4. Tension per unit length T due to stretching is the same in every direction and is constant during motion. Weight of the membrane is negligible.
5. Deflection $u(x, y, t)$ of the membrane while in motion is relatively small in comparison to the size of the membrane. The angles of inclination are small.

In Figure 9.3, an element of the membrane $ABCD$ is projected into a small rectangle with edges Δx and Δy parallel to the x and y axes. Deflections and angles of inclination are small enough that sides of the element are approximated by Δx and Δy. According to 4, forces acting on the edges are approximately $T\,\Delta x$ and $T\,\Delta y$, and are tangent to the membrane. Horizontal components involve cosines of very small angles of inclination. Since these forces are directed in opposite directions they add to zero approximately. The sum of the horizontal forces in the x direction is

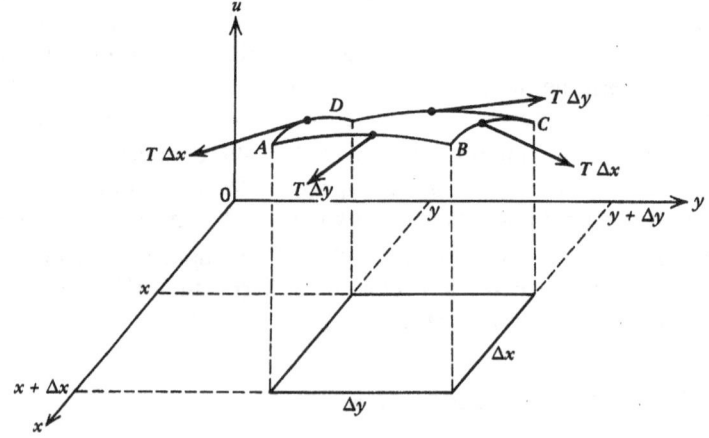

Figure 9.3. An element and projection of a stretched membrane.

$$T \, \Delta y (\cos \beta - \cos \alpha) = 0 \tag{9.3}$$

and in the y direction the sum is

$$T \, \Delta x (\cos \delta - \cos \gamma) = 0 \tag{9.4}$$

See Figure 9.4 for cross sections in the xu and yu planes. If the horizontal component of $T \, \Delta y$ is T_{hx}, then from (9.3)

$$T_{hx} = T \, \Delta y \cos \beta = T \, \Delta y \cos \alpha \tag{9.5}$$

and if T_{hy} is the horizontal component of $T \, \Delta x$, from (9.4) we have

$$T_{hy} = T \, \Delta x \cos \delta = T \, \Delta x \cos \gamma \tag{9.6}$$

From (9.5) and (9.6)

$$T \, \Delta y = \frac{T_{hx}}{\cos \beta} = \frac{T_{hx}}{\cos \alpha} \tag{9.7}$$

and

$$T \, \Delta x = \frac{T_{hy}}{\cos \delta} = \frac{T_{hy}}{\cos \gamma} \tag{9.8}$$

Adding the forces in the vertical direction and using Newton's second law of motion, one obtains

$$T \, \Delta y (\sin \beta - \sin \alpha) + T \, \Delta x (\sin \delta - \sin \gamma) = \rho \, \Delta x \, \Delta y u_{tt} \tag{9.9}$$

If $T \, \Delta y$ and $T \, \Delta x$ in (9.9) are replaced by (9.7) and (9.8), then

$$T_{hx}[\tan \beta - \tan \alpha] + T_{hy}[\tan \delta - \tan \gamma] = \rho \, \Delta x \, \Delta y u_{tt} \tag{9.10}$$

Recognizing that

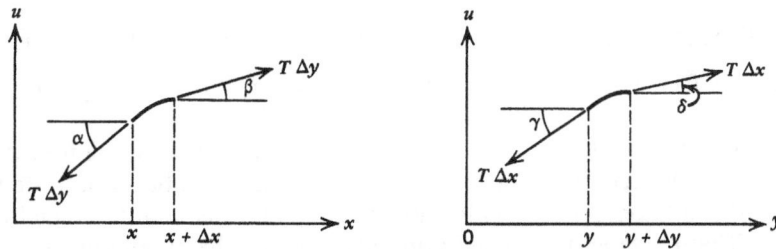

Figure 9.4. Cross sections of a membrane showing angles of inclination.

$$\tan \beta = u_x(x + \Delta x, y, t) \quad \text{and} \quad \tan \alpha = u_x(x, y, t)$$
$$\tan \delta = u_y(x, y + \Delta y, t) \quad \text{and} \quad \tan \gamma = u_y(x, y, t)$$

(9.10) becomes

$$T_{hx}[u_x(x + \Delta x, y, t) - u_x(x, y, t)]$$
$$+ T_{hy}[u_y(x, y + \Delta y, t) - u_y(x, y, t)] = \rho \, \Delta x \, \Delta y u_{tt} \qquad (9.11)$$

If the cosines of the inclinations are all approximately 1, then (9.11) is

$$T \, \Delta y[u_x(x + \Delta x, y, t) - u_x(x, y, t)]$$
$$+ T \, \Delta x[u_y(x, y + \Delta y, t) - u_y(x, y, t)] = \rho \, \Delta x \, \Delta y u_{tt} \qquad (9.12)$$

Division of (9.12) by $\rho \, \Delta x \, \Delta y$ permits the form

$$\frac{T}{\rho} \left[\frac{u_x(x + \Delta x, y, t) - u_x(x, y, t)}{\Delta x} \right]$$
$$+ \frac{T}{\rho} \left[\frac{u_y(x, y + \Delta y, t) - u_y(x, y, t)}{\Delta y} \right] = u_{tt}(x, y, t) \qquad (9.13)$$

As $\Delta x \to 0$ and $\Delta y \to 0$ in (9.13), then

$$u_{tt}(x, y, t) = a^2 \nabla^2 u(x, y, t) \qquad (9.14)$$

where

$$\nabla^2 u = u_{xx} + u_{yy} \quad \text{and} \quad a^2 = \frac{T}{\rho}$$

Equation (9.14) is the *wave equation in two dimensions*.

As an example of a vibrating membrane with appropriate constraints we display the following BVP:

$$u_{tt} = c^2(u_{xx} + u_{yy}), \quad 0 < x < a, \; 0 < y < b, \; t > 0$$
$$u(0, y, t) = u(a, y, t) = u(x, 0, t) = u(x, b. t) = 0, \quad t \geq 0 \qquad (9.15)$$
$$u(x, y, 0) = f(x, y), \quad u_t(x, y, 0) = g(x, y), \quad 0 < x < a, \; 0 < y < b$$

The first item in (9.15) is the wave equation. The second set of four items indicates no deflection along the edges where the membrane is fastened. The third condition gives an indication of the initial shape of the membrane surface. The fourth and last condition indicates that the membrane is at rest initially. The Fourier method for the solution follows:

1. *Separation of Variables.* u is a function of three variables in this problem. Therefore, we let $u(x, y, t) = X(x)Y(y)T(t)$. The PDE of (9.15) becomes

$$XYT'' = c^2[X''YT + XY''T]$$

Division by c^2XYT permits the result

$$\frac{T''}{c^2T} = \frac{X''}{X} + \frac{Y''}{Y} \tag{9.16}$$

If X''/X is assigned $-\alpha^2$ and Y''/Y is assigned $-\beta^2$, then

$$\frac{T''}{c^2T} = -(\alpha^2 + \beta^2)$$

and the three related ODEs follow.

2. *Related ODEs*

$$X'' + \alpha^2 X = 0$$
$$Y'' + \beta^2 Y = 0$$
$$T'' + (\alpha^2 + \beta^2)c^2T = 0$$

result from (9.16) and the constant assignments.

3. *Homogeneous Boundary Conditions*

$$u(0, y, t) = X(0)Y(y)T(t) = 0$$

If $Y(y) \neq 0$ and $T(t) \neq 0$, then

$$X(0) = 0$$
$$u(a, y, t) = X(a)Y(y)T(t) = 0$$

If $Y(t) \neq 0$ and $T(t) \neq 0$, then

$$X(a) = 0$$
$$u(x, 0, t) = X(x)Y(0)T(t) = 0$$

If $X(x) \neq 0$ and $T(t) \neq 0$, then

$$Y(0) = 0$$
$$u(x, b, t) = X(x)Y(b)T(t) = 0$$

If $X(x) \neq 0$, $T(t) \neq 0$, then

$$Y(b) = 0$$

4. *A SLP in X*

$$X'' + \alpha^2 X = 0, \quad X(0) = X(a) = 0$$

$$X = C_1 \cos \alpha x + C_2 \sin \alpha x$$

$$X(0) = C_1 + 0 = 0$$

$$X(a) = C_2 \sin a\alpha = 0$$

C_1 is zero. If C_2 is also zero, then we obtain a trivial solution for the BVP. If $C_2 \neq 0$, then $\sin a\alpha = 0$ and $a\alpha = n\pi$.

$$\alpha_n^2 = \frac{n^2\pi^2}{a^2}, \quad n \in \mathbf{N} \tag{9.17}$$

are eigenvalues. If $\alpha = n = 0$, only the trivial solution follows. The set of eigenvalues in (9.17) is adequate. The corresponding set of eigenfunctions is

$$X_n(x) = \sin \frac{n\pi x}{a}, \quad n \in \mathbf{N}$$

5. *A SLP in Y*

$$Y'' + \beta^2 Y = 0, \quad Y(0) = Y(b) = 0$$

This BVP has two complete SLPs associated with its solution. The Y problem is exactly the same as the X problem with b replacing a, Y replacing X, and β replacing α. Therefore, the eigenvalues are

$$\beta_n^2 = \frac{m^2\pi^2}{b^2}, \quad m \in \mathbf{N}$$

and the eigenfunctions are

$$Y_m(y) = \sin \frac{m\pi y}{b}, \quad m \in \mathbf{N}$$

6. *The New T Equation*

$$T'' + \left(\frac{n^2\pi^2}{a^2} + \frac{m^2\pi^2}{b^2} \right) c^2 T = 0$$

No homogeneous conditions accompany the new T equation. The solution may be expressed

$$T_{mn}(t) = A \cos h_{mn}t + B \sin h_{mn}t$$

where

$$h_{mn} = c\pi \left(\frac{n^2}{a^2} + \frac{m^2}{b^2} \right)^{1/2}$$

7. *Solution Set for Homogeneous Conditions*

$$u_{mn}(x, y, t) = [A \cos h_{mn}t + B \sin h_{mn}t] \sin \frac{n\pi x}{a} \sin \frac{m\pi y}{b}$$

8. *Superposition.* In this problem we use a double series representation. The coefficients A_{mn} and B_{mn} absorb A and B in the previous statement.

$$u(x, y, t) = \sum_{m=1}^{\infty} \sum_{n=1}^{\infty} [A_{mn} \cos h_{mn}t + B_{mn} \sin h_{mn}t] \sin \frac{n\pi x}{a} \sin \frac{m\pi y}{b}$$

9. *Nonhomogeneous Boundary Conditions.* The first condition in this category is

$$u(x, y, 0) = f(x, y) = \sum_{m=1}^{\infty} \sum_{n=1}^{\infty} A_{mn} \sin \frac{n\pi x}{a} \sin \frac{m\pi y}{b}$$

This is the double Fourier sine series of Section 3.10. The coefficients are

$$A_{mn} = \frac{4}{ab} \int_0^a \int_0^b f(x, y) \sin \frac{n\pi x}{a} \sin \frac{m\pi y}{b} \, dy \, dx, \quad m, n \in \mathbf{N}$$

It is necessary to compute the partial derivative for the last boundary condition. Thus

$$u_t(x, y, t) = \sum_{m=1}^{\infty} \sum_{n=1}^{\infty} h_{mn}[-A_{mn} \sin h_{mn}t + B_{mn} \cos h_{mn}t]$$
$$\times \sin \frac{n\pi x}{a} \sin \frac{m\pi y}{b}$$

The final boundary condition is

$$u_t(x, y, 0) = g(x, y) = \sum_{m=1}^{\infty} \sum_{n=1}^{\infty} h_{mn}B_{mn} \sin \frac{n\pi x}{a} \sin \frac{m\pi y}{b}$$

The coefficients

$$h_{mn}B_{mn} = \frac{4}{ab} \int_0^a \int_0^b g(x, y) \sin \frac{n\pi x}{a} \sin \frac{m\pi y}{b} \, dy \, dx$$

are from the double sine series of Section 3.10.

$$B_{mn} = \frac{4}{abh_{mn}} \int_0^a \int_0^b g(x, y) \sin \frac{n\pi x}{a} \sin \frac{m\pi y}{b} \, dy \, dx \,, \quad m, n \in \mathbf{N}$$

10. *Solution for the Original BVP*

$$u(x, y, t) = \sum_{m=1}^{\infty} \sum_{n=1}^{\infty} [A_{mn} \cos h_{mn}t + B_{mn} \sin h_{mn}t] \sin \frac{n\pi x}{a} \sin \frac{m\pi y}{b}$$

where

$$A_{mn} = \frac{4}{ab} \int_0^a \int_0^b f(x, y) \sin \frac{n\pi x}{a} \sin \frac{m\pi y}{b} \, dy \, dx \,, \quad m, n \in \mathbf{N}$$

$$B_{mn} = \frac{4}{abh_{mn}} \int_0^a \int_0^b g(x, y) \sin \frac{n\pi x}{a} \sin \frac{m\pi y}{b} \, dy \, dx \,, \quad m, n \in \mathbf{N}$$

and

$$h_{mn} = c\pi \left(\frac{n^2}{a^2} + \frac{m^2}{b^2} \right)^{1/2}$$

9.3. VIBRATIONS OF A CIRCULAR MEMBRANE DEPENDENT ON DISTANCE FROM CENTER

The displacement of the membrane, represented by $u(r, t)$ is independent of the vectorial angle θ. We assume that initially the displacement is $f(r)$ and the velocity is $g(r)$. The membrane is attached along the circumference of the circle $r = b$ in the plane of the membrane. The BVP follows:

$$u_{tt} = a^2 \left[u_{rr} + \frac{1}{r} u_r \right], \quad 0 < r < b, \, t > 0$$

$$u(b, t) = 0 \,, \quad t \geq 0$$

$$u(r, 0) = f(r) \,, \quad 0 < r < b \tag{9.18}$$

$$u_t(r, 0) = g(r) \,, \quad 0 < r < b$$

$$|u(r, t)| < M \,, \quad 0 < r < b, \, t \geq 0$$

The solution follows.

1. *Separation of Variables.* Let $u(r, t) = R(r)T(t)$.

$$RT'' = a^2\left[R''T + \frac{1}{r}R'T\right]$$

$$\frac{T''}{aT} = \frac{R''}{R} + \frac{1}{r}R' = -\alpha^2$$

2. *Related ODEs*

$$rR'' + R' + \alpha^2 rR = 0$$

$$T'' + \alpha^2 a^2 T = 0$$

3. *Homogeneous Boundary Condition*

$$u(b, t) = R(b)T(t) = 0$$

If $T(t) \neq 0$, then $R(b) = 0$.

4. *The R Equation*

$$[rR']' + \alpha^2 rR = 0, \quad R(b) = 0$$

The solution of this ODE is

$$R(r) = C_1 J_0(\alpha r) + C_2 Y_0(\alpha r)$$

but C_2 must be assigned zero, since Y_0 is unbounded at $r = 0$. If the boundary condition is used, then

$$R(b) = C_1 J_0(\alpha b) = 0$$

If $C_1 \neq 0$, then $J_0(\alpha b) = 0$ and

$$R_k(r) = J_0(\alpha_k r) \tag{9.19}$$

where $\alpha_k b$ are the zeros of J_0.

5. *The T Equation.* The T equation

$$T'' + \alpha_k^2 a^2 T = 0$$

has solutions

$$T_k(t) = B_1 \cos \alpha_k at + B_2 \sin \alpha_k at \tag{9.20}$$

6. *Solution Set for Homogeneous Conditions*. According to the separation substitution and (9.19) and (9.20) we have

$$u_k(r, t) = [B_1 \cos \alpha_k at + B_2 \sin \alpha_k at] J_0(\alpha_k r) \qquad (9.21)$$

7. *Superposition*. We write the linear combination of (9.21) as the series

$$u(r, t) = \sum_{k=1}^{\infty} [K_k \cos \alpha_k at + M_k \sin \alpha_k t] J_0(\alpha_k r)$$

B_1 and B_2 of (9.21) are absorbed into K_k and M_k.

8. *Nonhomogeneous Boundary Conditions*. One of these boundary conditions requires the derivative

$$u_t(r, t) = \sum_{k=1}^{\infty} \alpha_k a[-K_k \sin \alpha_k at + M_k \cos \alpha_k at] J_0(\alpha_k r)$$

At time $t = 0$, the two boundary conditions become

$$u(r, 0) = f(r) = \sum_{k=1}^{\infty} K_k J_0(\alpha_k r) \qquad (9.22)$$

and

$$u_t(r, 0) = g(r) = \sum_{k=1}^{\infty} \alpha_k a M_k J_0(\alpha_k r) \qquad (9.23)$$

From (9.22) and (9.23) we write

$$K_k = \frac{2}{b^2 J_1^2(\alpha_k b)} \int_0^b rf(r) J_0(\alpha_k r) \, dr$$

and

$$M_k = \frac{2}{\alpha_k a b^2 J_1^2(\alpha_k b)} \int_0^b rg(r) J_0(\alpha_k r) \, dr$$

9. *Solution of Original BVP*

$$u(r, t) = \sum_{k=1}^{\infty} [K_k \cos \alpha_k at + M_k \sin \alpha_k at] J_0(\alpha_k r)$$

where

$$K_k = \frac{2}{b^2 J_1^2(\alpha_k b)} \int_0^b rf(r) J_0(\alpha_k r) \, dr$$

and

$$M_k = \frac{2}{\alpha_k ab^2 J_1^2(\alpha_k b)} \int_0^b rg(r) J_0(\alpha_k r) \, dr$$

9.4. THE VIBRATING STRING WITH AN EXTERNAL FORCE

In this example, we assume that a string is stretched between two fixed points as in (8.8). An external force is applied to the string in such a way that the force is dependent on the position x along the string. This addition to the problem makes the PDE nonhomogeneous. Superposition is part of the Fourier method. To be certain that it may be employed, we transform the problem to a new one which has a homogeneous PDE. The following problem illustrates the procedure. As with past problems we assume that units are compatible with the BVP.

$$y_{tt}(x, t) = y_{xx}(x, t) + \gamma \sin \frac{\pi x}{L}, \quad 0 < x < L, \ t > 0, \ \gamma \text{ constant}$$

$$y(0, t) = y(L, t) = 0, \quad t \geq 0 \tag{9.24}$$

$$y_t(x, 0) = 0, \quad y(x, 0) = 0, \quad 0 \leq x \leq L$$

Our first effort is to select an appropriate transformation which will change the BVP to one similar to (8.8).

1. *Transformation.* Our selection is

$$y(x, t) = v(x, t) + \psi(x) \tag{9.25}$$

where $\psi(x)$ is at this time an undetermined function of x alone. Substituting (9.25) into the PDE of (9.24), we obtain the new equation

$$v_{tt} = v_{xx} + \psi''(x) + \gamma \sin \frac{\pi x}{L}$$

If we let $\psi''(x) + \gamma \sin(\pi x/L) = 0$, the new v equation becomes

$$v_{tt} = v_{xx} \tag{9.26}$$

which is homogeneous. If it can be arranged, we prefer that

$$v(0, t) = v(L, t) = 0$$

According to the transformation

$$y(0, t) = v(0, t) + \psi(0) = 0$$

If $\psi(0) = 0$, then

$$v(0, t) = 0 \tag{9.27}$$

Similarly,

$$y(L, t) = v(L, t) + \psi(L) = 0$$

If $\psi(L) = 0$, then

$$v(L, t) = 0 \tag{9.28}$$

In the transformation process we have found a related BVP.

2. *Related BVP in* $\psi(x)$

$$\psi''(x) + \gamma \sin \frac{\pi x}{L} = 0, \quad \psi(0) = \psi(L) = 0 \tag{9.29}$$

Solving the ODE of (9.29), we have

$$\psi(x) = \frac{\gamma L^2}{\pi^2} \sin \frac{\pi x}{L} + K_1 x + K_2$$

If $\psi(0) = 0$, then $K_2 = 0$; if $\psi(L) = 0$, then $K_1 = 0$. The solution of (9.29),

$$\psi(x) = \frac{\gamma L^2}{\pi^2} \sin \frac{\pi x}{L} \tag{9.30}$$

transforms our original PDE and the first two boundary conditions properly. The remaining two conditions imply that

$$y_t(x, 0) = v_t(x, 0) = 0 \tag{9.31}$$

and

$$y(x, 0) = v(x, 0) + \psi(x) = 0$$

Therefore,

$$v(x, 0) = -\psi(x) = -\frac{\gamma L^2}{\pi^2} \sin \frac{\pi x}{L} \tag{9.32}$$

Using the information from (9.26), (9.27), (9.28), (9.31), and (9.32), we state the new BVP.

3. *Related BVP in* $v(x, t)$

$$v_{tt} = v_{xx}, \quad 0 < x < L, \ t > 0$$
$$v(0, t) = v(L, t) = 0, \quad t \geq 0$$
$$v_t(x, 0) = 0, \quad 0 < x < L \tag{9.33}$$
$$v(x, 0) = -\psi(x) = -\frac{\gamma L^2}{\pi^2} \sin \frac{\pi x}{L}, \quad 0 < x < L$$

To solve (9.33), we observe that the problem is the same as (8.8) with $a^2 = 1$ and $f(x) = -\psi(x) = (-\gamma L^2/\pi^2) \sin (\pi x/L)$. Therefore, after superposition

$$v(x, t) = \sum_{n=1}^{\infty} C_n \sin \frac{n\pi x}{L} \cos \frac{n\pi t}{L}$$

$$v(x, 0) = -\frac{\gamma L^2}{\pi^2} \sin \frac{\pi x}{L} = \sum_{n=1}^{\infty} C_n \sin \frac{n\pi x}{L}$$

where

$$C_n = 0 \quad \text{if } n \neq 1$$

and

$$C_1 = -\frac{\gamma L^2}{\pi^2}$$

Therefore, the solution of the v problem is

$$v(x, t) = -\frac{\gamma L^2}{\pi^2} \sin \frac{\pi x}{L} \cos \frac{\pi t}{L} \tag{9.34}$$

4. *Solution of the Original BVP.* Substituting (9.30) and (9.34) in the transformation (9.25), we obtain

$$y(x, t) = -\frac{\gamma L^2}{\pi^2} \sin \frac{\pi x}{L} \cos \frac{\pi t}{L} + \frac{\gamma L^2}{\pi^2} \sin \frac{\pi x}{L}$$

or

$$y(x, t) = \frac{\gamma L^2}{\pi^2} \sin \frac{\pi x}{L} \left(1 - \cos \frac{\pi t}{L}\right)$$

for the solution of the BVP (9.24).

Exercises 9.2

1. Solve the BVP for the vibrating membrane if

$$u_{tt} = a^2(u_{xx} + u_{yy}), \quad 0 < x < \pi, \, 0 < y < \pi, \, t > 0$$

$$u(0, y, t) = u(\pi, y, t) = u(x, 0, t) = u(x, \pi, t) = 0, \quad t \geqslant 0$$

$$u_t(x, y, 0) = 0, \quad u(x, y, 0) = 0.01xy, \quad 0 < x < \pi, \, 0 < y < \pi$$

2. A thin elastic circular membrane vibrates transversely so that the following BVP models its behavior. Find $u(r, t)$.

$$u_{tt} = a^2\left[u_{rr} + \frac{1}{r}u_r\right], \quad 0 < r < 2, \, t > 0$$

$$u_t(r, 0) = 0, \quad 0 < r < 2$$

$$u(2, t) = 0, \quad t \geqslant 0$$

$$u(r, 0) = f(r), \quad 0 < r < 2$$

$$|u(r, t)| < M$$

3. A membrane is stretched over a circular frame and attached along the circumference of the frame. The radius of the frame is c. The membrane is struck in such a manner that its initial displacement is $f(r, \theta)$. It is released from rest. Determine the displacement $u(r, \theta, t)$. The BVP follows:

$$u_{tt} = a^2\left[u_{rr} + \frac{1}{r}u_r + \frac{1}{r^2}u_{\theta\theta}\right], \quad 0 < r < c, \, 0 < \theta < 2\pi, \, t > 0$$

$$u(c, \theta, t) = 0, \quad 0 < \theta < 2\pi, \, t \geqslant 0$$

$$u_t(r, \theta, 0) = 0, \quad 0 < r < c, \, 0 < \theta < 2\pi$$

$$u(r, \theta, 0) = f(r, \theta), \quad 0 < r < c, \, 0 < \theta < 2\pi$$

$$|u(r, \theta, t)| < M$$

4. Write the BVP for the motion of a vibrating membrane in Figure 9.5. Assume that the membrane is fixed along the quarter of the circle $r = 2$ and along the line segments $\theta = 0$ and $\theta = \pi/2$. It is released from rest at $t = 0$ from the given position $f(r, \theta)$. Find the displacement $u(r, \theta, t)$.

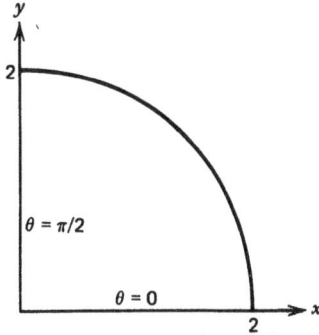

Figure 9.5. Quarter circle vibrating membrane.

5. A string vibrates in a substance that resists motion. If the resistive force is proportional to the velocity, the following BVP results:

$$y_{xx} = y_{tt} + ky_t, \quad 0 < x < 2, \; t > 0$$

$$y(0, t) = y(2, t) = 0, \quad t \geqslant 0$$

$$y_t(x, 0) = 0, \; y(x, 0) = 0.1, \quad 0 < x < 2$$

Find the displacement $y(x, t)$.

6. A flexible wire is stretched between $(0, 0)$ and $(\pi, 0)$ along an x axis in a horizontal position and is initially at rest. The force of gravity is taken into account and the y axis is directed upward. The equation of motion is

$$y_{tt} = a^2 y_{xx} - g, \quad 0 < x < \pi, \; t > 0, \; g = \text{gravitational constant}$$

7. Determine a solution for the BVP

$$y_{tt} = a^2 y_{xx} - hy, \quad 0 < x < c, \; t > 0$$

$$y(0, t) = y(c, t) = 0, \quad t \geqslant 0$$

$$y_t(x, 0) = 0, \; y(x, 0) = ke^{-x}, \quad 0 < x < c$$

8. Solve the BVP

$$y_{tt} = y_{xx} - y, \quad 0 < x < 2, \; t > 0$$

$$y(0, t) = y(2, t) = 0, \quad t \geqslant 0$$

$$y_t(x, 0) = 0, \; y(x, 0) = 0.1 \begin{cases} x & \text{if } 0 < x < 1 \\ 2 - x & \text{if } 1 < x < 2 \end{cases}$$

9. Given the BVP

$$y_{tt} = y_{xx} - y_t , \quad 0 < x < \pi, \ t > 0$$

$$y_x(0, t) = y(\pi, t) = 0 , \quad t \geqslant 0$$

$$y_t(x, 0) = k \cos \frac{3x}{2} , \quad y(x, 0) = 0 , \quad 0 < x < \pi$$

Solve for $y(x, t)$.

10. Under appropriate conditions the transverse vibrations of a beam are given by the equation

$$y_{tt} + a^2 y_{xxxx} = 0$$

It is assumed that vibrations are small and perpendicular to the x axis. If E is the modulus of elasticity, I is the moment of inertia of a cross section about the x axis, A is the cross sectional area, and ρ is the mass per unit length, then

$$a^2 = \frac{EI}{A\rho}$$

The transverse deflection at any point x on the length of the beam and at any time t is represented by $y(x, t)$. For a definite set of conditions let us assume that a uniform beam of length L, fixed at each end, begins to vibrate with initial deflection $f(x)$ and initial velocity zero. The BVP follows:

$$y_{tt} + a^2 y_{xxxx} = 0 , \quad 0 < x < L, \ t > 0$$

$$y(0, t) = y(L, t) = y_{xx}(0, t) = y_{xx}(L, t) = 0 , \quad t \geqslant 0$$

$$y_t(x, 0) = 0, \ y(x, 0) = f(x) , \quad 0 < x < L$$

As is our custom a bounded solution is requested.

11. Solve the BVP using the finite Fourier transform method

$$u_{tt} = u_{xx} , \quad 0 < x < \pi, \ t > 0$$

$$u(0, t) = u(\pi, t) = 0 , \quad t \geqslant 0$$

$$u_t(x, 0) = 0, \ u(x, 0) = f(x) , \quad 0 < x < \pi$$

Finite sine transforms are recommended.

9.5. NONHOMOGENEOUS END TEMPERATURES IN A ROD

We have discussed a heat conduction problem (8.41) with homogeneous boundary conditions at the ends of the rod. In the problem that follows all constraints are nonhomogeneous, even though the PDE is homogeneous.

$$u_t(x, t) = a^2 u_{xx}(x, t), \quad 0 < x < L, \, t > 0$$
$$u(0, t) = T_0, \, u(L, t) = T_1, \quad t \geq 0 \tag{9.35}$$
$$u(x, 0) = f(x), \quad 0 < x < L$$

As in Section 9.4 we use the following.

1. *Transformation.* Let

$$u(x, t) = v(x, t) + \psi(x) \tag{9.36}$$

Then

$$v_t(x, t) = a^2 [v_{xx}(x, t) + \psi''(x)]$$

is the transformed equation. If we select $\psi''(x)$ as zero, then

$$v_t(x, t) = a^2 v_{xx}(x, t) \tag{9.37}$$

is the new homogeneous PDE.

The first boundary condition can be expressed as

$$u(0, t) = v(0, t) + \psi(0) = T_0$$

If we elect to have

$$v(0, t) = 0 \tag{9.38}$$

then

$$\psi(0) = T_0$$

Likewise,

$$u(L, t) = v(L, t) + \psi(L) = T_1$$

If we choose

$$v(L, t) = 0 \tag{9.39}$$

then

$$\psi(L) = T_1$$

From the last boundary condition

$$u(x, 0) = v(x, 0) + \psi(x) = f(x)$$

we obtain

$$v(x, 0) = f(x) - \psi(x) \tag{9.40}$$

As a result of the transformation, we have a related BVP.

2. *Related BVP in* $\psi(x)$

$$\psi''(x) = 0, \quad \psi(0) = T_0, \quad \psi(L) = T_1 \tag{9.41}$$

From the ODE of (9.41), we obtain the general solution

$$\psi(x) = K_1 x + K_2$$

If $\psi(0) = T_0 = 0 + K_2$ and

$$\psi(L) = T_1 = K_1 L + K_2$$

then

$$K_2 = T_0, \quad K_1 = \frac{T_1 - T_0}{L}$$

As a result,

$$\psi(x) = \frac{T_1 - T_0}{L} x + T_0 \tag{9.42}$$

From (9.37), (9.38), (9.39), and (9.40), we have the following.

3. *Related BVP in* $v(x, t)$

$$v_t(x, t) = a^2 v_{xx}(x, t), \quad 0 < x < L, \, t > 0$$
$$v(0, t) = v(L, t) = 0, \quad t \geq 0 \tag{9.43}$$
$$v(x, 0) = f(x) - \psi(x) = f(x) - \frac{T_1 - T_0}{L} x - T_0, \quad 0 < x < L$$

BVP (9.43) is the same as (8.41) except that $f(x)$ is replaced by $f(x) - (T_1 - T_0)x/L - T_0$ in (9.43). The solution of the BVP in v is

$$v(x, t) = \sum_{n=1}^{\infty} C_n \exp\left(-\frac{n^2\pi^2a^2t}{L^2}\right) \sin \frac{n\pi x}{L} \tag{9.44}$$

$$v(x, 0) = f(x) + \frac{T_0 - T_1}{L}x - T_0 = \sum_{n=1}^{\infty} C_n \sin \frac{n\pi x}{L}$$

where

$$C_n = \frac{2}{L}\int_0^L \left[f(x) + \frac{T_0 - T_1}{L}x - T_0\right] \sin \frac{n\pi x}{L} dx$$

or

$$C_n = \frac{2}{L}\int_0^L f(x) \sin \frac{n\pi x}{L} dx + \frac{2(T_0 - T_1)}{L^2}\int_0^L x \sin \frac{n\pi x}{L} dx$$

$$- \frac{2T_0}{L}\int_0^L \sin \frac{n\pi x}{L} dx$$

Evaluation of the last two integrals permits the form

$$C_n = \frac{2}{L}\int_0^L f(x) \sin \frac{n\pi x}{L} dx + \frac{2}{n\pi}[T_1(-1)^n - T_0], \quad n \in \mathbb{N}$$

4. *Solution of the Original BVP.* By substituting (9.42) and (9.44) in (9.36), we have the solution

$$u(x, t) = \frac{T_1 - T_0}{L}x + T_0 + \sum_{n=1}^{\infty} C_n \exp\left(-\frac{n^2\pi^2a^2t}{L^2}\right) \sin \frac{n\pi x}{L}$$

for (9.35), where

$$C_n = \frac{2}{n\pi}[T_1(-1)^n - T_0] + \frac{2}{L}\int_0^L f(x) \sin \frac{n\pi x}{L} dx$$

9.6. A ROD WITH INSULATED ENDS

The lateral surfaces and the two ends of a rod of length π are insulated. The initial temperature distribution is $f(x)$. If $a^2 = 1$, the BVP accompanying the description follows:

$$u_t = u_{xx}, \quad 0 < x < \pi, \, t > 0$$

$$u_x(0, t) = u_x(\pi, t) = 0, \quad t \geq 0 \tag{9.45}$$

$$u(x, 0) = f(x), \quad 0 < x < \pi$$

This is a problem resembling No. 1 of Exercises 8.2. The initial temperature distribution is f and $L = \pi$ in our current BVP. With this problem we demonstrate the use of the finite Fourier transformation for solving a BVP. Let $U(n, t)$ be the finite cosine transform of $u(x, t)$. Transforming the heat equation, we have (see No. 1 of Exercises 7.6)

$$U_t(n, t) = (-1)^n u_x(\pi, t) - u_x(0, t) - n^2 U(n, t) \qquad (9.46)$$

According to the first two boundary conditions of (9.45), we can write (9.46)

$$U_t(n, t) + n^2 U(n, t) = 0$$

The solution of this equation is

$$U(n, t) = K e^{-n^2 t}, \quad U(n, 0) = K$$

Therefore

$$U(n, t) = U(n, 0) e^{-n^2 t}$$

According to the definition and the third boundary condition of (9.45), we obtain

$$U(n, 0) = \int_0^\pi u(x, 0) \cos nx \, dx$$

$$= \int_0^\pi f(x) \cos nx \, dx$$

Therefore,

$$U(n, t) = \left[\int_0^\pi f(\xi) \cos n\xi \, d\xi \right] e^{-n^2 t}$$

The inverse of $U(n, t)$ is

$$u(x, t) = \frac{U(0, 0)}{\pi} + \frac{2}{\pi} \sum_{n=1}^\infty U(n, t) \cos nx$$

or

$$u(x, t) = \frac{1}{\pi} \int_0^\pi f(\xi) \, d\xi + \frac{2}{\pi} \sum_{n=1}^\infty \left[\int_0^\pi f(\xi) \cos n\xi \, d\xi \right] e^{-n^2 t} \cos nx$$

In this method the boundary conditions are incorporated in the transformation process. Once the transform of the solution is determined, its inverse is the solution of the BVP. The method has particular advantages when boundary conditions match exactly parts of the transform of the derivatives. In the problem above, the transformed equation (9.46) was noticeably simplified by the two end conditions of (9.45). Frequently, problems involving a nonhomogeneous PDE, as in Section 9.4, can be solved with the transformation method. For further information on finite Fourier transforms see Churchill [14, Chapter 11].

Exercises 9.3

1. The ends of a rod are at temperatures 10°C and 50°C. The rod is four units long and has an initial temperature distribution of 30°C. The diffusivity constant a^2 is 4. Verify that the BVP is

$$u_t = 4u_{xx}, \quad 0 < x < 4, \; t > 0$$

$$u(0, t) = 10, \; u(4, t) = 50, \quad t \geq 0$$

$$u(x, 0) = 30, \quad 0 < x < 4$$

   Find the temperature $u(x, t)$.

2. Solve the BVP

$$u_t = u_{xx}, \quad 0 < x < \pi, \; t > 0$$

$$u(0, t) = 0, \; u(\pi, t) = T_0, \quad t \geq 0$$

$$u(x, 0) = 0, \quad 0 < x < \pi$$

3. Find a solution for the BVP

$$u_t = u_{xx}, \quad 0 < x < 1, \; t > 0$$

$$u_x(0, t) = 0, \; u(1, t) = 3, \quad t \geq 0$$

$$u(x, 0) = x, \quad 0 < x < 1$$

4. If each cross section of a slender wire has a uniform temperature, the linear law of surface heat transfer between the wire and its neighboring environment is applicable. We assume that the neighboring environment has a temperature zero. Temperature of the wire is $u(x, t)$. The wire is placed along the x axis and the heat conduction equation is

$$u_t(x, t) = a^2 u_{xx}(x, t) - hu(x, t), \quad 0 < x < L, t > 0, h > 0 \text{ constant}$$

   We assume that the ends are insulated, so that

$$u_x(0, t) = u_x(L, t) = 0 , \quad t \geqslant 0$$

The initial temperature is $f(x)$. Therefore

$$u(x, 0) = f(x) , \quad 0 < x < L$$

(a) Solve for $u(x, t)$.
(b) Solve the BVP

$$v_t(x, t) = a^2 v_{xx}(x, t) , \quad 0 < x < L, t > 0$$

with boundary conditions as in (a),

$$v_x(0, t) = v_x(L, t) = 0, v(x, 0) = f(x)$$

Notice that

$$u(x, t) = e^{-ht} v(x, t)$$

5. Solve the BVP

$$u_t = a^2 u_{xx} + k e^{-x} , \quad 0 < x < 1, t > 0$$
$$u(0, t) = u(1, t) = 0 , \quad t \geqslant 0$$
$$u(x, 0) = - \frac{k}{a^2} e^{-x}$$

Give a physical explanation of the problem.
6. Given the BVP

$$u_t = u_{xx} + k \sin x , \quad 0 < x < \pi, t > 0$$
$$u(0, t) = u(\pi, t) = 0 , \quad t \geqslant 0$$
$$u(x, 0) = \sin x , \quad 0 \leqslant x \leqslant \pi$$

Determine $u(x, t)$.
7. Show that the BVP

$$u_{xx} + u_{yy} = 0, \quad 0 < x < 1, y > 0$$
$$u(x, 0) = 1, \quad 0 < x < 1$$
$$u_x(0, y) = 0, \quad u_x(1, y) + u(1, y) = 0, \quad y > 0$$
$$|u(x, y)| < M$$

has a solution

$$u(x, y) = 2 \sum_{n=1}^{\infty} \frac{\sin \alpha_n}{\alpha_n(1 + \sin^2 \alpha_n)} e^{-\alpha_n y} \cos \alpha_n x$$

where α_n are the positive roots of $\alpha \tan \alpha = 1$. In this problem the norm, $\|\cos \alpha_n x\|$, depends on α_n. See (2.44a). Show a few α_n graphically.

8. The temperature in a sphere of a certain substance is given by $u(\rho, t)$. In spherical coordinates the surface of the sphere has equation $\rho = 1$. Initially the substance is at a uniform temperature π throughout the sphere, and the surface of the sphere is kept at zero temperature. The function u satisfies the PDE

$$u_t = \frac{a^2}{\rho} (\rho u_{\rho\rho} + 2u_\rho), \quad 0 < \rho < 1, \, t > 0$$

and boundary conditions

$$u(1, t) = 0, \quad u(\rho, 0) = \pi$$

Transform the problem so that u is replaced by w where

$$w(\rho, t) = \rho u(\rho, t)$$

If u is continuous at $\rho = 0$, then $w(0, t) = 0$. Solve.

9. Using the finite transform method solve the BVP

$$u_t = u_{xx} + h(x, t), \quad 0 < x < \pi, \, t > 0$$
$$u(0, t) = u(\pi, t) = 0, \quad t \geqslant 0$$
$$u(x, 0) = f(x), \quad 0 < x < \pi$$

This is similar to (9.45) with heat being generated in the rod at a rate $h(x, t)$ per unit time.

9.7. A SEMI-INFINITE BAR

Until now we have been concerned with BVPs where lengths of rods, areas of membranes, and so on, were all finite. Here we assume that a bar fails to have a finite length. This concept may be impossible to produce physically, but if the bar is very long a semi-infinite length may be a good modeling approximation. The interval in this case, $0 < x < \infty$ represents all $x > 0$. Assume that a bar has its surface insulated for its entire length. A temperature zero is imposed on the end $x = 0$. This temperature is held constant for all time t. The initial temperature distribution is given by $f(x)$. The BVP follows:

$$u_t = a^2 u_{xx}, \quad 0 < x < \infty, \ t > 0$$
$$u(0, t) = 0, \quad t \geqslant 0$$
$$u(x, 0) = f(x), \quad 0 < x < \infty \qquad (9.47)$$
$$|u(x, t)| < M, \quad 0 < x < \infty, \ t > 0$$

1. *Separation of Variables.* Let $u(x, t) = X(x)T(t)$

$$XT' = a^2 X''T, \quad \frac{T'}{a^2 T} = \frac{X''}{X} = -\alpha^2$$

2. *Related ODEs*

$$X'' + \alpha^2 X = 0, \quad T' + \alpha^2 a^2 T = 0$$

3. *Homogeneous Constraint*

$$u(0, t) = X(0)T(t) = 0$$

If $T(t) \not\equiv 0$, then

$$X(0) = 0$$

There is only one homogeneous constraint. We fail to have a complete SLP as a result.

4. *Related X Equation and Constraint*

$$X'' + \alpha^2 X = 0, \quad X(0) = 0$$
$$X = C_1 \cos \alpha x + C_2 \sin \alpha x$$
$$X(0) = C_1 + 0 = 0$$
$$X_\alpha(x) = \sin \alpha x, \quad \alpha > 0 \qquad (9.48)$$

5. *Related T Equation*

$$T' + \alpha^2 a^2 T = 0, \quad T_\alpha(t) = \exp(-\alpha^2 a^2 t) \qquad (9.49)$$

No boundary condition accompanies the T equation.

6. *Solution Set for Homogeneous Conditions.* Using (9.48) and (9.49) in the separation formula, we have

$$u_\alpha(x, t) = \exp(-\alpha^2 a^2 t) \sin \alpha x, \quad \alpha > 0 \qquad (9.50)$$

where solutions depend on the parameter α. The solutions (9.50) satisfy the homogeneous problem for all real α. Negative values of α need not be included, since they offer no new independent solutions. Because α is not restricted to a set of natural numbers, superposition resulting in a series is inappropriate. In this case *superposition* is accomplished by *integration relative to the parameter α*.

7. *Superposition by Integration.* The infinite linear combination in this form is

$$u(x, t) = \int_0^\infty B(\alpha) \exp\left(-\alpha^2 a^2 t\right) \sin \alpha x \, d\alpha \qquad (9.51)$$

8. *Nonhomogeneous Constraint*

$$u(x, 0) = f(x) = \int_0^\infty B(\alpha) \sin \alpha x \, d\alpha \qquad (9.52)$$

We observe that (9.52) is a Fourier sine integral. The coefficients (4.21a) are

$$B(\alpha) = \frac{2}{\pi} \int_0^\infty f(\xi) \sin \alpha \xi \, d\xi \qquad (9.53)$$

9. *Solution of the Original BVP.* From (9.51) and (9.53)

$$u(x, t) = \int_0^\infty B(\alpha) \exp\left(-\alpha^2 a^2 t\right) \sin \alpha x \, d\alpha$$

where

$$B(\alpha) = \frac{2}{\pi} \int_0^\infty f(\xi) \sin \alpha \xi \, d\xi$$

is a formal solution of the BVP (9.47).

9.8. AN INFINITE BAR

In this problem we consider the heat conduction in the middle of a very long bar. For our model we have only the heat equation and the initial distribution of temperatures given. We assume that the bar is insulated laterally. The two items of the BVP follow:

$$u_t = a^2 u_{xx}, \quad -\infty < x < \infty, \ t > 0$$
$$u(x, 0) = f(x), \quad -\infty < x < \infty \qquad (9.54)$$

To solve the problem we begin with

1. *Separation of Variables and Related ODEs.* Let $u(x, t) = X(x) T(t)$. The initial work is exactly that of the preceding section through items 1 and 2. There are no homogeneous boundary conditions.

2. *Related X Equation.* Note that

$$X_\alpha(x) = C_1 \cos \alpha x + C_2 \sin \alpha x \qquad (9.55)$$

is the solution.

3. *Related T Equation.* The solution is given by

$$T_\alpha(t) = \exp\left(-\alpha^2 a^2 t\right) \qquad (9.56)$$

4. *Solution Set for Homogeneous Conditions.* The PDE is the only homogeneous part of the BVP. Therefore, (9.55) and (9.56) in the separation formula permit us to write the solution set

$$u_\alpha(x, t) = \exp\left(-\alpha^2 a^2 t\right)[C_1 \cos \alpha x + C_2 \sin \alpha x], \quad \alpha \text{ real}$$

5. *Superposition by Integration*

$$u(x, t) = \int_0^\infty \exp(-\alpha^2 a^2 t)[A(\alpha) \cos \alpha x + B(\alpha) \sin \alpha x]\, d\alpha$$

where C_1 and C_2 are absorbed in $A(\alpha)$ and $B(\alpha)$.

6. *Nonhomogeneous Boundary Condition*

$$u(x, 0) = f(x) = \int_0^\infty [A(\alpha) \cos \alpha x + B(\alpha) \sin \alpha x]\, d\alpha \qquad (9.57)$$

The result (9.57) is the Fourier integral of (4.14). The coefficients (4.15) are

$$A(\alpha) = \frac{1}{\pi} \int_{-\infty}^\infty f(\xi) \cos \alpha\xi\, d\xi$$

$$B(\alpha) = \frac{1}{\pi} \int_{-\infty}^\infty f(\xi) \sin \alpha\xi\, d\xi \qquad (9.58)$$

7. *Solution of the Original BVP*

$$u(x, t) = \int_0^\infty \exp\left(-\alpha^2 a^2 t\right)[A(\alpha) \cos \alpha x + B(\alpha) \sin \alpha x]\, d\alpha \qquad (9.59)$$

where

$$A(\alpha) = \frac{1}{\pi} \int_{-\infty}^{\infty} f(\xi) \cos \alpha\xi \, d\xi$$

$$B(\alpha) = \frac{1}{\pi} \int_{-\infty}^{\infty} f(\xi) \sin \alpha\xi \, d\xi$$

is a formal solution for the BVP (9.54).

We show another form of the solution at this time. If $A(\alpha)$ and $B(\alpha)$ are inserted in their integral forms in (9.59) we have

$$u(x, t) = \frac{1}{\pi} \int_{0}^{\infty} \exp(-\alpha^2 a^2 t) \int_{-\infty}^{\infty} f(\xi)[\cos \alpha\xi \cos \alpha x + \sin \alpha\xi \sin \alpha x] \, d\xi \, d\alpha$$

$$(9.60)$$

Using a trigonometric identity in (9.60), we obtain

$$u(x, t) = \frac{1}{\pi} \int_{0}^{\infty} \int_{-\infty}^{\infty} f(\xi) \exp(-\alpha^2 a^2 t) \cos \alpha(\xi - x) \, d\xi \, d\alpha$$

If the order of integration is changed, then

$$u(x, t) = \frac{1}{\pi} \int_{-\infty}^{\infty} f(\xi) \left[\int_{0}^{\infty} \exp(-\alpha^2 a^2 t) \cos \alpha(\xi - x) \, d\alpha \right] d\xi \quad (9.61)$$

The inside integral of (9.61) has the form*

$$w(s) = \int_{0}^{\infty} e^{-\alpha^2 b} \cos \alpha s \, d\alpha, \quad b > 0 \tag{9.62}$$

See Churchill and Brown [15, p. 203, Problem 19] for suggestions of (9.62).

$$w'(s) = \int_{0}^{\infty} \frac{\partial}{\partial s} (e^{-\alpha^2 b} \cos \alpha s) \, d\alpha$$

$$= -\int_{0}^{\infty} \alpha e^{-\alpha^2 b} \sin \alpha s \, d\alpha \tag{9.63}$$

In (9.63) we integrate by parts. As a result

$$w'(s) = \left[\frac{1}{2b} e^{-\alpha^2 b} \sin \alpha s \right]_{0}^{\infty} - \frac{s}{2b} \int_{0}^{\infty} e^{-\alpha^2 b} \cos \alpha s \, d\alpha$$

or

*From Churchill and Brown [15], by permission McGraw-Hill Book Company.

$$w'(s) = -\frac{s}{2b}\,w(s) \tag{9.64}$$

The solution of (9.64) may be expressed as

$$\ln w(s) = -\frac{s^2}{4b} + K \tag{9.65}$$

or

$$w(s) = C \exp\left(-\frac{s^2}{4b}\right) \tag{9.66}$$

We see that

$$w(0) = C$$

in (9.66); but in (9.62)

$$w(0) = \int_0^\infty e^{-\alpha^2 b}\,d\alpha$$

Likewise,

$$w(0) = \int_0^\infty e^{-\beta^2 b}\,d\beta$$

Therefore,

$$[w(0)]^2 = \int_0^\infty \int_0^\infty e^{-b(\alpha^2+\beta^2)}\,d\alpha\,d\beta$$

Using polar coordinates, $r^2 = \alpha^2 + \beta^2$, we have formally

$$[w(0)]^2 = \int_0^{\pi/2}\int_0^\infty e^{-br^2} r\,dr\,d\theta = \int_0^{\pi/2}\frac{1}{2b}\,d\theta = \frac{\pi}{4b}$$

Thus

$$w(0) = \frac{1}{2}\sqrt{\frac{\pi}{b}} \tag{9.67}$$

Using (9.66) and (9.67), we find that

$$w(s) = \frac{1}{2}\sqrt{\frac{\pi}{b}}\,\exp\left(-\frac{s^2}{4b}\right) \tag{9.68}$$

and employing (9.62) and (9.68), we have

$$\int_0^\infty e^{-\alpha^2 b} \cos \alpha s \, d\alpha = \frac{1}{2} \sqrt{\frac{\pi}{b}} \exp\left(-\frac{s^2}{4b}\right), \quad b > 0 \qquad (9.69)$$

The solution (9.61) may be written

$$u(x, t) = \frac{1}{2a\sqrt{\pi t}} \int_{-\infty}^\infty f(\xi) \exp\left[-\frac{(\xi - x)^2}{4a^2 t}\right] d\xi \qquad (9.70)$$

if one uses the result (9.69). Introducing the new variable $\gamma = (\xi - x)/2a\sqrt{t}$, (9.60) becomes

$$u(x, t) = \frac{1}{\sqrt{\pi}} \int_{-\infty}^\infty f(x + 2a\sqrt{t}\gamma) \exp(-\gamma^2) \, d\gamma \qquad (9.71)$$

Under appropriate conditions it can be shown that (9.70) or (9.71) satisfy the BVP (9.54).

As a special case of (9.54), we let f be a constant temperature T_0 over the interval $-1 < x < 1$, and zero elsewhere. Then

$$f(x) = \begin{cases} T_0 & \text{when } -1 < x < 1 \\ 0 & \text{when } x < -1 \text{ and } x > 1 \end{cases}$$

and (9.70) becomes

$$u(x, t) = \frac{T_0}{2a\sqrt{\pi t}} \int_{-1}^1 \exp\left[-\frac{(\xi - x)^2}{4a^2 t}\right] d\xi$$

If the substitution $\gamma = (\xi - x)/2a\sqrt{t}$ is made, then

$$u(x, t) = \frac{T_0}{\sqrt{\pi}} \int_{-(1+x)/2a\sqrt{t}}^{(1-x)/2a\sqrt{t}} \exp(-\gamma^2) \, d\gamma \ . \qquad (9.72)$$

In Chapter 7, the *error function*, erf(x), is defined by the integral

$$\text{erf}(x) = \frac{2}{\sqrt{\pi}} \int_0^x \exp(-\gamma^2) \, d\gamma$$

Therefore, (9.72) may be written

$$u(x, t) = \frac{T_0}{2} \left[\frac{2}{\sqrt{\pi}} \int_{-(1+x)/2a\sqrt{t}}^0 \exp(-\gamma^2) \, d\gamma \right.$$
$$\left. + \frac{2}{\sqrt{\pi}} \int_0^{(1-x)/2a\sqrt{t}} \exp(-\gamma^2) \, d\gamma \right]$$

We replace γ in the first integral by $-\beta$. Then

$$u(x, t) = \frac{T_0}{2} \left[\frac{2}{\sqrt{\pi}} \int_0^{(1+x)/2a\sqrt{t}} \exp(-\beta^2) \, d\beta \right.$$

$$\left. + \frac{2}{\sqrt{\pi}} \int_0^{(1-x)/2a\sqrt{t}} \exp(-\gamma^2) \, d\gamma \right]$$

In terms of error functions, we write

$$u(x, t) = \frac{T_0}{2} \left[\operatorname{erf}\left(\frac{1+x}{2a\sqrt{t}}\right) + \operatorname{erf}\left(\frac{1-x}{2a\sqrt{t}}\right) \right], \quad t > 0$$

The error function is recorded in tables and displayed graphically. See Abramowitz and Stegun [1, pp. 297–316] for graphs, tables, and other information concerning this function.

Exercises 9.4

1. By direct verification, show that

$$u(x, t) = \operatorname{erf}\left(\frac{x}{2a\sqrt{t}}\right)$$

is a solution of the BVP

$$u_t = a^2 u_{xx}, \quad 0 < x < \infty, \ t > 0$$
$$u(0, t) = 0, \quad t > 0$$
$$u(x, 0) = 1, \quad 0 < x < \infty$$

2. Test the error function to determine whether it is even or odd, or neither even nor odd.

3. (a) Find α and β so that

$$\int_a^b \exp(-\gamma^2) \, d\gamma = \frac{\sqrt{\pi}}{2} [\operatorname{erf}(\alpha) - \operatorname{erf}(\beta)]$$

(b) Determine ξ so that

$$\int_{-b}^b \exp(-\gamma^2) \, d\gamma = \sqrt{\pi} \operatorname{erf}(\xi)$$

4. The face $x = 0$ of a semi-infinite bar is held at a temperature zero. The function $f(x)$ represents the initial temperature distribution. Show that the temperature at any point x and time t is given by

$$u(x, t) = \frac{1}{\sqrt{\pi}} \left[\int_{-x/s}^\infty f(s\gamma + x) e^{-\gamma^2} \, d\gamma - \int_{x/s}^\infty f(s\gamma - x) e^{-\gamma^2} \, d\gamma \right]$$

where $s = 2a\sqrt{t}, \ t > 0$.

5. (a) Solve the BVP

$$u_t = a^2 u_{xx}, \quad 0 < x < \infty, \ t > 0$$
$$u(0, t) = 0, \quad t > 0$$
$$u(x, 0) = f(x)$$

Use Fourier sine transforms to solve the problem.
(b) If $f(x) = 1$, $x > 0$, in the solution for (a) show that

$$u(x, t) = \text{erf}\left(\frac{x}{2a\sqrt{t}}\right)$$

6. Show that

$$\int_{-\infty}^{\infty} \exp(-\alpha^2 t + i\alpha x) \, d\alpha = \sqrt{\frac{\pi}{t}} \exp\left(-\frac{x^2}{4t}\right)$$

As in (9.62), we now let

$$w(s) = \int_{-\infty}^{\infty} e^{-\alpha^2 b} e^{i\alpha s} \, d\alpha$$

By differentiating and integration by parts, we have

$$w'(s) = -\frac{s}{2b} w(s)$$

$$w(s) = C \exp\left(-\frac{s^2}{4b}\right)$$

$$w(0) = C = \int_{-\infty}^{\infty} e^{-\alpha^2 b} \, d\alpha$$

$$[w(0)]^2 = \frac{\pi}{b}$$

Then

$$w(s) = \sqrt{\frac{\pi}{b}} \exp\left(-\frac{s^2}{4b}\right)$$

The result requested follows immediately.
7. (a) The function $f(x)$ represents the temperature distribution initially in an infinite bar. Its sides are insulated. Using a Fourier exponential transformation find the temperature $u(x, t)$. The BVP follows:

$$u_t = a^2 u_{xx}, \quad -\infty < x < \infty, \ t > 0$$

$$u(x, 0) = f(x), \quad -\infty < x < \infty$$

Show that

$$u(x, t) = \frac{1}{2a\sqrt{\pi t}} \int_{-\infty}^{\infty} f(\xi) \exp\left[-\frac{(\xi - x)^2}{4a^2 t}\right] d\xi$$

(b) Show that if

$$f(x) = \begin{cases} 0 & \text{if } -\infty < x < 0 \\ 2 & \text{if } 0 < x < \infty \end{cases}$$

in (a), then

$$u(x, t) = \text{erf}\left(\frac{1}{2a\sqrt{t}}\right) + 1$$

In (a) we find that the transformed equation is

$$U_t(\alpha, t) = -\alpha^2 a^2 U(\alpha, t)$$

if $U(\alpha, t)$ is the transform of $u(x, t)$.

$$U(\alpha, t) = C(\alpha)e^{-\alpha^2 a^2 t}$$

$$U(\alpha, 0) = \int_{-\infty}^{\infty} f(x) \exp(-i\alpha x) \, dx = F_e(\alpha)$$

$$U(\alpha, t) = F_e(\alpha)e^{-\alpha^2 a^2 t}$$

If we let

$$H_e(\alpha) = e^{-\alpha^2 a^2 t}$$

then according to Exercise 6

$$h(x) = \frac{1}{2\pi} \int_{-\infty}^{\infty} e^{-\alpha^2 a^2 t} e^{i\alpha x} \, d\alpha$$

$$= \frac{1}{2a\sqrt{\pi t}} \exp\left(-\frac{x^2}{4a^2 t}\right)$$

$$U(\alpha, t) = F_e(\alpha)H_e(\alpha)$$

The inverse is the convolution integral

$$u(x, t) = \int_{-\infty}^{\infty} f(\xi)h(\xi - x) \, d\xi$$

and the desired result is apparent.

For $f(x)$ in part (b) the solution may be written

$$u(x, t) = \frac{1}{a\sqrt{\pi t}} \int_0^{\infty} \exp\left[-\frac{(\xi - x)^2}{4a^2 t}\right] d\xi$$

A process similar to the procedure following (9.71) is beneficial.

8. A semi-infinitely long rod has a unit constant source of heat applied at $x = 0$. Initially throughout its length the rod has a temperature zero. Far away from $x = 0$ on the rod the temperature has a limit zero, and the temperature change along the rod also has a limit zero. Show that the BVP describing these conditions is

$$u_t = a^2 u_{xx}, \quad t > 0, \, x > 0, \, a > 0$$

$$u(x, 0) = 0, \quad x > 0, \quad u(0, t) = 1, \quad t > 0$$

$$\lim_{x \to \infty} u(x, t) = \lim_{x \to \infty} u_x(x, t) = 0, \quad t > 0$$

Using the Laplace transform method, solve the BVP and then check your solution. Show that

$$u(x, t) = 1 - \operatorname{erf}\left(\frac{x}{2a\sqrt{t}}\right) = 1 - \frac{2}{\sqrt{\pi}} \int_0^{x/(2a\sqrt{t})} e^{-\alpha^2} d\alpha$$

9.9. DISCRETE FOURIER TRANSFORM SOLUTIONS

The discrete Fourier transform (DFT), $F_d(\omega)$, can be used to approximate the Fourier transform, $F(\omega)$, of a given function f. Consequently, the DFT can be used to obtain approximate solutions to BVPs which are solvable by Fourier transform methods. To see this let $[A, B]$ be an interval such that the integral of $|f|$ over the complement of $[A, B]$ is negligible. Let n be a positive integer, $\Delta x = (B - A)/n$, and $x_j = A + j \Delta x$, $j = 0, 1, \ldots, n - 1$. Then for Δx sufficiently small

$$F(\omega) = \int_{-\infty}^{\infty} f(x)e^{-i\omega x} dx \doteq \int_A^B f(x)e^{-i\omega x} \, dx$$

$$\doteq \sum_{j=0}^{n-1} f(x_j)e^{-i\omega x_j} \Delta x = (B - A)F_d(\omega)$$

Conversely, suppose that $F(\omega)$ is known and we wish to determine $f(x)$. If the integral of $|F|$ over the complement of $[0, 2\pi]$ is negligible and $\Delta\omega = 2\pi/n$ is sufficiently small,

$$f(x) = \frac{1}{2\pi} \int_{-\infty}^{\infty} F(\omega)e^{i\omega x}\, d\omega \doteq \frac{1}{2\pi} \int_{0}^{2\pi} F(\omega)e^{i\omega x}\, d\omega$$

$$\doteq \frac{1}{2\pi} \sum_{j=0}^{n-1} f(\omega_j)e^{i\omega_j x}\, \Delta\omega = \frac{1}{n} \sum_{j=0}^{n-1} F(\omega_j)e^{i\omega_j x} = \frac{1}{n} J^{-1}(F(\omega))$$

where $\omega_j = 2\pi j/n$ and J^{-1} represents the inverse DFT.

For practical computation one must take into account the concept of aliasing; that is, the fact that two different frequencies may be indistinguishable to the DFT because of its discrete nature. Specifically, note that

$$\omega_{n-j} = \frac{2\pi(n-j)}{n} = 2\pi - \omega_j$$

Then Theorem 7.4 implies that

$$F_d(\omega_{n-j}) = F_d(-\omega_j) = \overline{F_d(\omega_j)}$$

where $\overline{F_d}$ is the complex conjugate of F_d. Thus, $F_d(\omega_{n-j})$ will not be a good approximation to $F(\omega_{n-j})$ in cases where $F(\omega_{n-j})$ is significantly different from $\overline{F(\omega_j)}$. To handle this problem Brigham [9, p. 135] recommends folding the transform about $n/2$ before inverting; that is, set

$$F(\omega_{n-j}) = \overline{F(\omega_j)}, \quad j = 0, 1, \ldots, \frac{n}{2} - 1$$

before computing the inverse DFT.

Exercises 9.5

1. Consider the BVP used in Problem 7 of Exercises 9.4:

$$u_t = a^2 u_{xx}, \quad -\infty < x < \infty, \ t > 0$$

$$u(x, 0) = f(x), \quad -\infty < x < \infty$$

Let

$$U(\omega, t) = \int_{-\infty}^{\infty} e^{-i\omega x} u(x, t)\, dx$$

be the Fourier transform of the solution with respect to x, and let F represent the Fourier transform of f. One can show (see Problem 7) that

$$U(\omega, t) = F(\omega)e^{-a^2\omega^2 t}$$

Write a computer program to compute an approximate solution to this BVP by means of the DFT. The steps needed are:

(i) Compute the DFT of f at the Fourier frequencies $\omega_j = 2\pi j/n$.
(ii) Multiply $F_d(\omega_j)$ by $\exp(-a^2\omega_j^2 t)$ to obtain an approximation for

$$U(\omega_j, t), \quad j = 0, 1, \ldots, n/2.$$

(iii) Fold the approximation to $U(\omega_j, t)$ about $n/2$.
(iv) Approximate the solution by computing the inverse DFT of the folded transform.

Run your program for the following data: $n = 20$, $a = 1$, $A = 0$, $B = 20$, $t = 0, 1, 4, 9, 16, 25$, using

(a) $f(x) = e^{-(x-10)^2/2}$,
(b) $f(x) = 1$, $4 \le x \le 16$, 0 otherwise.

A separate computation is required for each value of t. The results for (a) can be compared to the analytic solution which is given by

$$u(x, t) = \frac{1}{\sqrt{\lambda}} \exp\left[-\frac{(x-10)^2}{2\lambda}\right]$$

where $\lambda = 1 + 2a^2 t$. An analytic solution for part (b) can be computed using the method outlined in Problem 7.

9.10. A SEMI-INFINITE STRING

We assume that one end of the string is fixed and stretched along the positive x axis. For convenience we attach the string at the origin. We assume that the string is at rest initially and has an initial position $f(x)$. Some of these conditions may be a bit difficult to realize in practice, but they form our assumptions. We let $y(x, t)$ represent displacements perpendicular to the x axis. A statement of the BVP follows:

$$y_{tt} = a^2 y_{xx}, \quad 0 < x < \infty, \ t > 0$$

$$y(0, t) = 0, \quad t \ge 0 \tag{9.73}$$

$$y_t(x, 0) = 0, \ y(x, 0) = f(x), \quad 0 < x < \infty$$

We proceed with

1. *Separation of Variables.* Let $y(x, t) = X(x)T(t)$

$$\frac{T''}{a^2 T} = \frac{X''}{X} = -\alpha^2$$

2. *Related ODEs*

$$X'' + \alpha^2 X = 0, \quad T'' + \alpha^2 a^2 T = 0$$

3. *Homogeneous Boundary Conditions*

$$y(0, t) = X(0)T(t) = 0$$

If $T(t) \not\equiv 0$, then

$$X(0) = 0$$
$$y_t(x, 0) = X(x)T'(0) = 0$$

If $X(x) \not\equiv 0$, then

$$T'(0) = 0$$

4. *Related X Equation and Constraint*

$$X'' + \alpha^2 X = 0, \quad X(0) = 0$$

$$X = C_1 \cos \alpha x + C_2 \sin \alpha x$$

$$X(0) = C_1 + 0 = 0$$

$$X_\alpha(x) = \sin \alpha x, \quad \alpha > 0 \qquad (9.74)$$

5. *Related T Equation and Constraint*

$$T'' + \alpha^2 a^2 T = 0, \quad T'(0) = 0$$

$$T = K_1 \cos \alpha a t + K_2 \sin \alpha a t$$

$$T' = \alpha a [-K_1 \sin \alpha a t + K_2 \cos \alpha a t]$$

$$T'(0) = 0 + \alpha a K_2 = 0$$

If $\alpha \neq 0$, then

$$K_2 = 0, \quad T_\alpha(t) = \cos \alpha a t \qquad (9.75)$$

6. *Solution Set for Homogeneous Conditions.* Employing (9.74) and (9.75) in the separation formula, one obtains the solution set

$$y_\alpha(x, t) = \sin \alpha x \cos \alpha a t$$

7. *Superposition by Integration*

$$y(x, t) = \int_0^\infty B(\alpha) \sin \alpha x \cos \alpha a t \, d\alpha$$

8. *Nonhomogeneous Boundary Condition*

$$y(x, 0) = f(x) = \int_0^\infty B(\alpha) \sin \alpha x \, d\alpha \tag{9.76}$$

Result (9.76) is a Fourier sine integral with

$$B(\alpha) = \frac{2}{\pi} \int_0^\infty f(\xi) \sin \alpha \xi \, d\xi$$

9. *Solution for the Original BVP*

$$y(x, t) = \int_0^\infty B(\alpha) \sin \alpha x \cos \alpha a t \, d\alpha \tag{9.77}$$

where

$$B(\alpha) = \frac{2}{\pi} \int_0^\infty f(\xi) \sin \alpha \xi \, d\xi \tag{9.78}$$

Integral (9.77) with coefficients (9.78) comprise the solution for (9.73). If one inserts the integral for $B(\alpha)$ in (9.77), then

$$y(x, t) = \frac{2}{\pi} \int_0^\infty \int_0^\infty f(\xi) \sin \alpha x \cos \alpha a t \sin \alpha \xi \, d\xi \, d\alpha$$

Using

$$\sin \alpha(x + at) + \sin \alpha(x - at) = 2 \sin \alpha x \cos \alpha a t$$

$$y(x, t) = \frac{1}{\pi} \int_0^\infty \int_0^\infty f(\xi)[\sin \alpha(x + at) + \sin \alpha(x - at)] \sin \alpha \xi \, d\xi \, d\alpha$$

or

$$y(x, t) = \frac{1}{2} \int_0^\infty \left[\frac{2}{\pi} \int_0^\infty f(\xi) \sin \alpha \xi \, d\xi \right] \sin \alpha(x + at) \, d\alpha$$

$$+ \frac{1}{2} \int_0^\infty \left[\frac{2}{\pi} \int_0^\infty f(\xi) \sin \alpha \xi \, d\xi \right] \sin \alpha(x - at) \, d\alpha \tag{9.79}$$

Result (9.79) is the same as

$$y(x, t) = \frac{1}{2} \int_0^\infty B(\alpha) \sin \alpha(x + at) \, d\alpha + \frac{1}{2} \int_0^\infty B(\alpha) \sin \alpha(x - at) \, d\alpha \tag{9.80}$$

In (9.76), if x is replaced by $x + at$, then

$$f(x + at) = \int_0^\infty B(\alpha) \sin \alpha(x + at) \, d\alpha \tag{9.81}$$

and if x is replaced by $x - at$, then

$$f(x - at) = \int_0^\infty B(\alpha) \sin \alpha(x - at) \, d\alpha \tag{9.82}$$

Observing the integrals in (9.80) and those of (9.81) and (9.82), we can write the solution

$$y(x, t) = \tfrac{1}{2}[f(x + at) + f(x - at)], \quad x > 0, \ t > 0 \tag{9.83}$$

If f' is smooth it is easy to verify that (9.83) is a solution of the BVP (9.73) if the extension of f is odd.

9.11. A SEMI-INFINITE STRING WITH INITIAL VELOCITY

One end of the string is fixed at the origin. Initially the string is positioned along the x axis and has a velocity $f(x)$. If $y(x, t)$ represents displacements perpendicular to the x axis and $a^2 = 1$, then the BVP is

$$y_{tt} = y_{xx}, \quad 0 < x < \infty, \ t > 0$$
$$y(0, t) = 0, \quad t \geq 0 \tag{9.84}$$
$$y(x, 0) = 0, \ y_t(x, 0) = f(x), \quad 0 < x < \infty$$

We include this problem to demonstrate the use of the Fourier sine transformation for solving a BVP. Let $Y(\alpha, t)$ be the Fourier sine transform of $y(x, t)$. Transforming the wave equation of (9.84), we obtain (see No. 7(a) of Exercises 7.7)

$$Y_{tt}(\alpha, t) = \alpha y(0, t) - \alpha^2 Y(\alpha, t)$$

Since $y(0, t) = 0$, then

$$Y_{tt}(\alpha, t) + \alpha^2 Y(\alpha, t) = 0$$

and

$$Y(\alpha, t) = A(\alpha) \cos \alpha t + B(\alpha) \sin \alpha t$$

$$Y(\alpha, 0) = A(\alpha)$$

From the definition of the transform and the given boundary condition

$$Y(\alpha, 0) = \int_0^\infty y(x, 0) \sin \alpha x \, dx = 0$$

Therefore,

$$A(\alpha) = 0$$

and

$$Y(\alpha, t) = B(\alpha) \sin \alpha t$$

Differentiating, we have

$$Y_t(\alpha, t) = \alpha B(\alpha) \cos \alpha t \tag{9.85}$$

$$Y_t(\alpha, 0) = \int_0^\infty y_t(x, 0) \sin \alpha x \, dx$$

$$= \int_0^\infty f(x) \sin \alpha x \, dx$$

From (9.85),

$$Y_t(\alpha, 0) = \alpha B(\alpha)$$

and

$$B(\alpha) = \frac{1}{\alpha} \int_0^\infty f(x) \sin \alpha x \, dx$$

Thus

$$Y(\alpha, t) = \left[\frac{1}{\alpha} \int_0^\infty f(\xi) \sin \alpha \xi \, d\xi \right] \sin \alpha t$$

and the inverse is

$$y(x, t) = \frac{2}{\pi} \int_0^\infty \left\{ \left[\frac{1}{\alpha} \int_0^\infty f(\xi) \sin \alpha \xi \, d\xi \right] \sin \alpha t \right\} \sin \alpha x \, d\alpha$$

The solution may be displayed

$$y(x, t) = \frac{2}{\pi} \int_0^\infty \int_0^\infty \frac{1}{\alpha} f(\xi) \sin \alpha x \sin \alpha t \sin \alpha \xi \, d\xi \, d\alpha$$

For more information on Fourier transform methods see Churchill [14, Chapters 12 and 13].

Exercises 9.6

1. (a) Solve the BVP

$$y_{tt} = a^2 y_{xx}, \quad -\infty < x < \infty, \ t > 0$$

$$y(x, 0) = f(x), \quad y_t(x, 0) = 0, \quad -\infty < x < \infty$$

   (b) Show that the solution in (a) is equivalent to

$$y(x, t) = \tfrac{1}{2}[f(x + at) + f(x - at)]$$

2. A string fixed at $x = 0$ lies along the entire positive x axis satisfying the equation $y = 0.01xe^{-x}$ initially. It has an initial velocity zero. Assuming no gravitational forces act, find the displacement $y(x, t)$. The BVP follows.

$$y_{tt} = a^2 y_{xx}, \quad 0 < x < \infty, \ t > 0$$

$$y(0, t) = 0, \quad t \geq 0$$

$$y_t(x, 0) = 0, \ y(x, 0) = 0.01xe^{-x}, \quad 0 < x < \infty$$

3. (a) Find the harmonic function $v(x, y)$ for the half-plane $y > 0$ if the function v has the value $f(x)$ for all points on the x axis. We write the BVP as follows:

$$v_{xx} + v_{yy} = 0, \quad 0 < y < \infty, \ -\infty < x < \infty$$

$$v(x, 0) = f(x), \quad -\infty < x < \infty$$

   (b) Show that the solution in (a) may be written

$$v(x, y) = \frac{1}{\pi} \int_0^\infty \int_{-\infty}^\infty e^{-\alpha y} f(\xi) \cos \alpha(\xi - x) \, d\xi \, d\alpha$$

4. Write the solution of Exercise 3 in the form

$$v(x, y) = \frac{1}{\pi} \int_{-\infty}^\infty \frac{yf(\xi)}{y^2 + (\xi - x)^2} \, d\xi$$

5. (a) A region is bounded so that $x > 0$ and $y > 0$. The edge $y = 0$ is maintained at zero potential, while the edge $x = 0$ is kept at potential $f(y)$. Show that the potential at any point (x, y) is

$$v(x, y) = \frac{1}{\pi} \int_0^\infty x f(\xi) \left[\frac{1}{(\xi - y)^2 + x^2} - \frac{1}{(\xi + y)^2 + x^2} \right] d\xi$$

(b) If $f(y) = 1$, demonstrate that

$$v(x, y) = \frac{2}{\pi} \arctan \frac{y}{x}$$

6. An infinite strip is bounded by $y = 0$ and $y = 1$. Along $y = 0$, the potential is kept at zero and along $y = 1$, the potential is maintained at $f(x)$. Find the potential $v(x, y)$ between the two lines. The accompanying BVP is

$$v_{xx} + v_{yy} = 0, \quad 0 < y < 1, \ -\infty < x < \infty$$
$$v(x, 0) = 0, \ v(x, 1) = f(x), \quad -\infty < x < \infty$$

Show that

$$v(x, y) = \frac{1}{\pi} \int_0^\infty \int_{-\infty}^\infty \frac{f(\xi) \sinh \alpha y \cos \alpha(x - \xi)}{\sinh \alpha} \, d\xi \, d\alpha$$

7. (a) Determine a solution for the BVP

$$u_{xx} + u_{yy} = 0, \quad 0 < y < \infty, \ -\infty < x < \infty$$
$$u(x, 0) = \begin{cases} -T_0 & \text{if } -1 < x < 0 \\ T_0 & \text{if } 0 < x < 1, \ T_0 > 0 \\ 0 & \text{if } -\infty < x < -1 \text{ and } 1 < x < \infty \end{cases}$$

Give a physical interpretation of the problem.

(b) Solve (a) if

$$u(x, 0) = \begin{cases} 0 & \text{if } -\infty < x < -1 \\ T_0 & \text{if } -1 < x < 1 \\ 0 & \text{if } 1 < x < \infty \end{cases}$$

8. (a) Determine the potential $v(x, y)$ in the semi-infinite strip $y > 0$, $0 < x < 1$ satisfying the conditions

$$v_y(x, 0) = 0, \quad v_x(0, y) = 0, \quad v(1, y) = f(y)$$

when

$$f(y) = \begin{cases} 1 & \text{if } 0 < y < 1 \\ 0 & \text{if } 1 < y < \infty \end{cases}$$

(b) Find $v(x, y)$ if $f(y) = e^{-y}$, $y > 0$.

9. Solve the BVP

$$v_{xx} + v_{yy} = 0, \quad 0 < y < 1, \; -\infty < x < \infty$$
$$v(x, 0) = 0, \quad v(x, 1) = f(x), \quad -\infty < x < \infty$$

10. As another vibrating string problem, suppose a long elastic string is initially stretched between a fixed point, $x = 0$, and a point so far away on the x axis that we may consider, $x \to \infty$. Transverse displacements, $y(x, t)$ satisfy the wave equation. At time $t = 0$, assume that the displacement is zero, $y = 0$, and the string is at rest, $y_t = 0$, for $x \geq 0$. The end of the string at $x = 0$ is allowed to move up and down along the y axis in a way prescribed by a continuous, twice differentiable function $y = f(t)$ when $x = 0$ for $t \geq 0$. Our problem is to describe the ensuing displacement $y(x, t)$ for $x > 0$, $t > 0$. The motion in the xy plane is illustrated in Figure 9.6. The description suggests a BVP:

$$y_{tt}(x, t) = a^2 y_{xx}(x, t), \quad x > 0, \, t > 0, \, a > 0$$
$$y(x, 0) = y_t(x, 0) = 0, \quad x \geq 0$$
$$y(0, t) = f(t), \quad \lim_{x \to \infty} y(x, t) = 0, \quad t > 0$$

Using Laplace transforms, solve the BVP and then check the solution. Show that

$$y(x, t) = f\left(t - \frac{x}{a}\right) U\left(t - \frac{x}{a}\right)$$

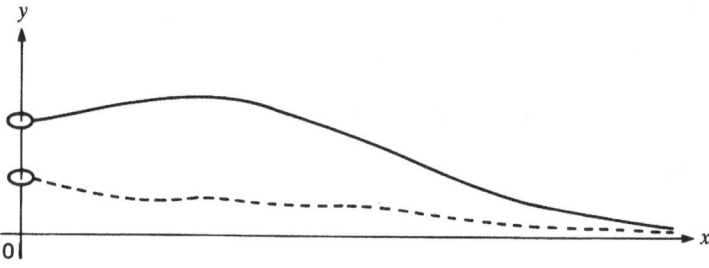

Figure 9.6. Motion of the elastic infinite string.

REFERENCES

1. Abramowitz, M., and I. Stegun (Eds.), *Handbook of Mathematical Functions with Formulas, Graphs and Mathematical Tables*, National Bureau of Standards, AMS 55, U.S. Government Printing Office, Washington, DC, 1964.

2. Birkhoff, G., and G. Rota, *Ordinary Differential Equations*, 3rd ed., Wiley, New York, 1978.

3. Bloomfield, P., *Fourier Analysis of Time Series*, Wiley, New York, 1976.

4. Bôcher, M., *Introduction to the Theory of Fourier's Series*, Annals of Mathematics, Second Series, Vol. 7, No. 2, Publications Office, Harvard University, Cambridge, MA, 1906.

5. Bowman, F., *Introduction to Bessel Functions*, Dover, New York, 1958.

6. Boyce, W., and R. DiPrima, *Elementary Differential Equations and Boundary Value Problems*, 4th ed., Wiley, New York, 1986.

7. Bracewell, R., *The Fourier Transform and Its Applications*, 2nd ed., McGraw-Hill, New York, 1986.

8. Brand, L., *Differential and Difference Equations*, Wiley, New York, 1966.

9. Brigham, E., *The Fast Fourier Transform*, Prentice Hall, Englewood Cliffs, NJ, 1974.

10. Budak, B., A. Samarskii, and A. Tikhohnov, *A Collection of Problems on Mathematical Physics*, Pergamon, New York, 1964.

11. Byerly, W., *An Elementary Treatise on Fourier Series*, Dover, New York, 1959.

12. Cajori, F., *A History of Mathematics*, 2nd ed., Macmillan, New York, 1919.

13. Carlslaw, H., *Introduction to the Theory of Fourier Series and Integrals*, 3rd ed., Dover, New York, 1930.

14. Churchill, R., *Operational Mathematics*, 3rd ed., McGraw-Hill, New York, 1972.

15. Churchill, R., and J. Brown, *Fourier Series and Boundary Value Problems*, 4th ed., McGraw-Hill, New York, 1987.

16. Conte, S., and C. de Boor, *Elementary Numerical Analysis: An Algorithmic Approach*, 3rd ed., McGraw-Hill, 1980.

17. Danese, A., *Advanced Calculus*, Vols. 1 and 2, Allyn and Bacon, Boston, 1965.

18. Davis, H., *Fourier Series and Orthogonal Functions*, Allyn and Bacon, Boston, 1963.

19. Erdelyi, A. (Ed.), *Tables of Integral Transforms*, Vols. I and II, McGraw-Hill, New York, 1954.

20. Fourier, J., *The Analytical Theory of Heat*, Dover, New York, 1955.

21. Fulks, W., *Advanced Calculus*, 3rd ed., Wiley, New York, 1978.

22. Haberman, R., *Elementary Applied Partial Differential Equations*, 2nd ed., Prentice Hall, Englewood Cliffs, NJ, 1987.

23. Hadamard, J., *Lectures on Cauchy's Problem*, Dover, New York, 1952.

24. Hardy, G., and W. Rogosinski, *Fourier Series*, 3rd ed., Cambridge University Press, London, 1956.

25. Hildebrand, F., *Advanced Calculus for Applications*, 2nd ed., Prentice Hall, Englewood Cliffs, NJ, 1976.

26. Ince, E., *Ordinary Differential Equations*, Dover, New York, 1956.

27. Jackson, D., *Fourier Series and Orthogonal Polynomials*, Carus Mathematical Monographs, No. 6, Mathematical Association of America, Washington, DC, 1941.

28. Jahnke, E., and F. Emde, *Tables of Functions with Formulae and Curves*, 4th ed., Dover, New York, 1945.

29. Kovach, L., *Boundary Value Problems*, Addison-Wesley, Reading, MA, 1983.

30. Kreyszig, E., *Advanced Engineering Mathematics*, 6th ed., Wiley, New York, 1988.

31. Lanczos, C., *Discourse on Fourier Series*, Hafner, New York, 1966.

32. Langer, R., *Fourier's Series*, Slaught Memorial Papers, No. 1, *American Mathematical Monthly*, Vol. 54, No. 7, Part 2, 1947.

33. MacRobert, T., *Spherical Harmonics*, 3rd ed., Pergamon, London, 1967.

34. Miles, J., *Integral Transforms in Applied Mathematics*, Cambridge University Press, London, 1971.

35. Myint-U, T., *Partial Differential Equations of Mathematical Physics*, 2nd ed., North Holland, New York, 1980.

36. Myint-U, T., and L. Debnath, *Partial Differential Equations for Scientists and Engineers*, 3rd ed., North Holland, New York, 1987.

37. Petrovskii, I., *Partial Differential Equations*, Saunders, Philadelphia, 1967.

38. Pickering, M., *An Introduction to Fast Fourier Transform Methods for Partial Differential Equations*, Wiley, New York, 1986.

39. Pipes, L., and L. Harvill, *Applied Mathematics for Engineers and Physicists*, 3rd ed., McGraw-Hill, New York, 1970.

40. Powers, D., *Boundary Value Problems*, 2nd ed., Academic, New York, 1979.

41. Rainville, E., *The Laplace Transform*, Macmillan, New York, 1963.

42. Sagan, H., *Boundary and Eigenvalue Problems in Mathematical Physics*, Wiley, New York, 1961.

43. Sneddon, I., *Fourier Transforms*, McGraw-Hill, New York, 1951.

44. Sommerfeld, A., *Partial Differential Equations in Physics*, Academic, New York, 1953.

45. Spiegel, M., *Fourier Analysis*, Schaum's Outline Series, McGraw-Hill, New York, 1974.

46. Szidarovszky, F., and S. Yakowitz, *Principles and Procedures of Numerical Analysis*, Plenum, New York, 1978.

47. Tolstov, G., *Fourier Series*, Prentice Hall, Englewood Cliffs, NJ, 1962.

48. Tranter, C., *Integral Transforms in Mathematical Physics*, 3rd ed., Chapman and Hall, London, 1966.

49. Vichnevetsky, R., and J. Bowles, *Fourier Analysis of Numerical Approximations of Hyperbolic Equations*, SIAM, Philadelphia, 1982.

50. Watson, G., *A Treatise on the Theory of Bessel Functions*, 2nd ed., Cambridge University Press, London, 1966.

51. Weaver, H., *Applications of Discrete and Continuous Fourier Analysis*, Wiley, New York, 1983.

52. Weaver, H., *Theory of Discrete and Continuous Fourier Analysis*, Wiley, New York, 1989.

53. Whittaker, E., and G. Watson, *A Course in Modern Analysis*, 4th ed., Cambridge University Press, London, 1952.

54. Young, E., *Partial Differential Equations*, Allyn and Bacon, Boston, 1972.

55. Zachmanoglou, E., and D. Thoe, *Introduction to Partial Differential Equations with Applications*, Williams and Wilkins, Baltimore, MD, 1976.

56. Zygmund, A., *Trigonometric Series*, Dover, New York, 1955.

ANSWERS TO EXERCISES

CHAPTER 1

Exercises 1.1

3. Yes
5. (b) $W(y_1, y_2) = 1/[(x+1)^2(x+2)^2]$. Yes.
   (c) $C_1 = 1$ or 0.
   (d) $C_2 = 1$ or 0.
   (e) No. Theorem 1.2 is not violated.
6. (a) Yes.
   (b) Theorem 1.2 fails to tell us. Equation is not satisfied.
7. (a) Theorem 1.2 fails to apply. Equation is not satisfied.
   (b) No, unless $C_1 + C_2 = 1$. (a) Nonlinear, homogeneous;
      (b) Linear, not homogeneous.
8. (a) Yes.
   (b) $W(e^x, e^{2x}, xe^{2x}) \neq 0$. Set is linearly independent.
   (d) Yes, but not a general solution.
   (e) Yes, and $C_1 e^x + C_2 e^{2x} + C_3 xe^{2x}$ is a general solution.
9. (b) $W(2\cos^2 x - 1, 1 - 2\sin^2 x) = 0$. Set is linearly dependent.
   (c) $C_1 y_1 + C_2 y_2$ is a solution, but not a general solution.
10. $C_1 y_1 + C_2 y_2$ is not a solution unless $C_1 + C_2 = 1$. Theorem 1.2 is not violated.

Exercises 1.2

1. $y = C_1 e^{-2x} + C_2 e^{-3x}$.
2. $y = (C_1 + C_2 x)e^{2x}$.
3. $y = e^{-x}(C_1 \cos x + C_2 \sin x)$.
4. $y = C_1 e^{2x} + e^{-x}(C_2 \cos x + C_3 \sin x)$.

5. $y = \dfrac{e^x - e^{-x}}{e^\pi + e^{-\pi}}$.

6. $y = C_1 e^x + C_2 e^{-x} + C_3 \cos x + C_4 \sin x$.

7. $y = C_1 + C_2 e^{3x} + C_3 e^{2x}$.

8. $y = C_1 x^3 + C_2 x$.

9. $y = x^2 \ln x$.

10. $y = x[C_1 \cos(2 \ln x) + C_2 \sin(2 \ln x)]$.

11. $y = \pi \sin(\ln x)$.

Exercises 1.3

1. (a) If $a = 0$, $y = C \displaystyle\sum_{k=0}^{\infty} \dfrac{(-x)^k}{k!} + x - 1 = Ce^{-x} + x - 1$.

(b) If $a = 0$, $y = 3 \displaystyle\sum_{k=0}^{\infty} \dfrac{(x^2/2)^k}{k!} = 3e^{x^2/2}$.

(c) $y = C_0 \left[1 - \dfrac{(x-1)^2}{2!} + \dfrac{(x-1)^3}{3!} - \dfrac{(x-1)^4}{4!} + \dfrac{2(x-1)^5}{5!} \right.$

$\left. - \dfrac{7(x-1)^6}{6!} + \cdots \right]$

$+ C_1 \left[(x-1) - \dfrac{(x-1)^3}{3!} + \dfrac{2(x-1)^4}{4!} - \dfrac{5(x-1)^5}{5!} \right.$

$\left. + \dfrac{18(x-1)^6}{6!} + \cdots \right]$.

(d) If $a = 0$, $y = x - \dfrac{x^3}{3!} - \dfrac{x^5}{4!} - \dfrac{95x^7}{7!} + \cdots$.

(e) If $a = 0$, $y = C - 1 + \dfrac{x^2}{2} + \displaystyle\sum_{k=0}^{\infty} \dfrac{x^k}{k!} = C - 1 + \dfrac{x^2}{2} + e^x$.

(f) If $a = 0$, $y = C_0 \left[1 + \dfrac{x^3}{3!} + \dfrac{4x^6}{6!} + \dfrac{28x^9}{9!} + \cdots \right]$

$+ C_1 \left[x + \dfrac{2x^4}{4!} + \dfrac{10x^7}{7!} + \dfrac{8x^{10}}{9!} + \cdots \right]$.

If $a = 1$, $y = C_0 \left[1 + \dfrac{(x-1)^2}{2!} + \dfrac{(x-1)^3}{3!} + \dfrac{(x-1)^4}{4!} + \dfrac{4(x-1)^5}{5!} \right.$

$\left. + \dfrac{5(x-1)^6}{6!} \cdots \right] + C_1 \left[(x-1) + \dfrac{(x-1)^3}{3!} \right.$

$\left. + \dfrac{2(x-1)^4}{4!} + \dfrac{(x-1)^5}{5!} + \dfrac{6(x-1)^6}{6!} + \cdots \right]$.

(g) $y = 1 + \dfrac{x^2}{2!} - \dfrac{x^4}{4!} + \dfrac{3x^6}{6!} - \dfrac{5 \cdot 3x^8}{8!} + \dfrac{7 \cdot 5 \cdot 3x^{10}}{10!} - \cdots$.

2. (a) $y = 2 + \displaystyle\sum_{k=1}^{\infty} \dfrac{2x^k}{k!} = 2e^x$.

(b) $y = 2 + 3(x - 1) - (x - 1)^2 - \dfrac{5(x - 1)^3}{3!} - \dfrac{4(x - 1)^4}{4!}$

$\qquad + \dfrac{11(x - 1)^5}{5!} + \cdots$.

(c) $y = 1 + \dfrac{(x - 2)^2}{8} - \dfrac{(x - 2)^3}{24} + \dfrac{7(x - 2)^4}{16 \cdot 4!} - \dfrac{(x - 2)^5}{5!} + \cdots$.

(d) $y = A\left[1 - x^2 + \dfrac{x^4}{3} - \dfrac{x^6}{15} + \cdots\right]$

$\qquad + B\left[x - \dfrac{x^3}{2} + \dfrac{x^5}{8} - \cdots\right]$.

3. (a) $x = 0$ is a regular singular point.
   The indicial equation is $(2r - 1)^2 = 0$.
   The indicial roots are $\frac{1}{2}, \frac{1}{2}$.

(b) $x = 0$ is a regular singular point.
   The indicial equation is $r(r - 2) = 0$.
   The indicial roots are $0, 2$.

(c) $x = 0$ is a regular singular point.
   The indicial equation is $2r^2 - r - 1 = 0$.
   The indicial roots are $-\frac{1}{2}, 1$.

(d) $x = 0$ is a regular singular point.
   The indicial equation is $r(2r - 1) = 0$.
   The indicial roots are $0, \frac{1}{2}$.
   $y = K_1 \cos \sqrt{x} + K_2 \sin \sqrt{x}$.

(e) $x = 0$ is a regular singular point.
   The indicial equation is $r(2r - 1) = 0$.
   The indicial roots are $0, \frac{1}{2}$.

   $y_1 = C_1 x^0\left[1 - \dfrac{x^2}{3!} + \dfrac{x^4}{7 \cdot 4!} - \dfrac{5x^6}{11 \cdot 7!} + \cdots\right]$.

   $y_2 = C_2 x^{1/2}\left[1 - \dfrac{x^2}{5 \cdot 2} + \dfrac{x^4}{9 \cdot 8 \cdot 5} - \dfrac{x^6}{13 \cdot 9 \cdot 8 \cdot 6 \cdot 5} + \cdots\right]$.

   $y = y_1 + y_2$.

(f) $x = 0$ is a regular singular point.
   The indicial equation is $r^2 - \frac{1}{9} = 0$.
   The indicial roots are $\frac{1}{3}, -\frac{1}{3}$.

$$y_1 = C_1 x^{1/3} \left[1 - \frac{3x^2}{2^4} + \frac{3^2 x^4}{7 \cdot 2^7} - \frac{3^2 x^6}{7 \cdot 5 \cdot 2^{10}} + \cdots \right].$$

$$y_2 = C_2 x^{-1/3} \left[1 - \frac{3x^2}{2^3} + \frac{3^2 x^4}{5 \cdot 2^6} - \frac{3^2 x^6}{5 \cdot 2^{11}} + \cdots \right].$$

$$y = y_1 + y_2.$$

(g) $x = 0$ is a regular singular point.
The indicial equation is $r^2 = 0$.
The indicial roots are $0, 0$.
$y = (C_1 + C_2 \ln x) e^x$.

(h) $x = 0$ is a regular singular point.
The indicial equation is $r^2 - 4 = 0$.
The indicial roots are $2, -2$.
$y = C_1 x^{-2} + C_2 x^2$.

Exercises 1.4

1. $y(1.4) \doteq 26.02662$; $y(1.8) \doteq 92.62422$.
2. $y_1(0.1) \doteq 1.51890$; $y_2(0.1) \doteq 2.29760$.
3. (a) (i) $f'(3) \doteq 0.32790$; (ii) $f'(3) \doteq 0.33346$.
   (b) $f''(3) \doteq -0.11117$.
   (c) (i) $f'(3) \doteq 0.33277$; (ii) $f'(3) \doteq 0.33333$.
   $f''(3) \doteq -0.11206$ single precision).
   $f''(3) \doteq -0.11111$ (double precision).

Exercises 1.5

1. $u = \sin x + \sin y - 1$.

2. $u = \dfrac{x^3(y^2 - 1) + 2y^3 + 6 \cos x - 2}{6}$

3. $u = \dfrac{x[\pi \sin y - (2 + y^2)] - \pi[\cos x - (y^2 + 1)]}{\pi}$.

4. (a) Hyperbolic if $xy < 0$; elliptic if $xy > 0$; parabolic if $x = 0$ or $y = 0$.
   (b) Parabolic.
   (c) Hyperbolic; $u = f(y - 3x) + g(y + x)$.
   (d) Parabolic; $u = f(y + x) + xg(y + x)$ or $u = f(y + x) + yg(y + x)$.
   (e) Elliptic; $u = f(y + iax) + g(y - iax)$.
   (f) Elliptic; $u = f(y + (1 + i)x) + g(y + (1 - i)x)$.

5. $u = \left(\dfrac{1}{2c} \right) \displaystyle\int_{x-ct}^{x+ct} \phi(\alpha) \, d\alpha$.

6. (a) $u(r, s) = f(s) + g(r)$; $u(x, y) = f(y - 3x) + g(y + x)$.

7. $u = f(x - y) + e^{-y}g(x - y)$.

8. (a) $u = e^{\alpha(x+2y)} + e^{2y}e^{\alpha(x-2y)}$.

(b) $u = f(x + 2y) + e^{2y}g(x - 2y)$.

9. $u = \dfrac{y^4}{12} + H(x) + x^{-3}G(y)$.

10. (a) $u = f(y + 3x) + g(y - x) + e^x$.

(b) $u = f(y + 2x) + g(y - x) + \dfrac{\sin y}{2}$.

Exercises 1.6

1. (a) $u = Ce^{\lambda x + y/\lambda}$.

(b) $u = (C_1 \cos \alpha t + C_2 \sin \alpha t)(C_3 \cos \alpha x + C_4 \sin \alpha x)$.

(c) $u = e^{-y}(C_1 \cos \alpha x + C_2 \sin \alpha x)(C_3 \cos \sqrt{\alpha^2 - 1}\,y + C_4 \sin \sqrt{\alpha^2 - 1}\,y)$.

(d) $u = e^{-(x+y)}(C_1 \cos \sqrt{\alpha^2 - 1}\,x + C_2 \sin \sqrt{\alpha^2 - 1}\,x)$
$\times (C_3 \cos \sqrt{2 - \alpha^2}\,y + C_4 \sin \sqrt{2 - \alpha^2}\,y)$.

(e) $u = (tx)^{1/2}[C_1(tx)^{\sqrt{1 - 4\alpha^2}/2} + C_2(t/x)^{\sqrt{1 - 4\alpha^2}/2}$
$+ C_3(x/t)^{\sqrt{1 - 4\alpha^2}/2} + C_4(tx)^{-\sqrt{1 - 4\alpha^2}/2}]$.

(f) Not separable.

(g) $u = [C_1 \cos \alpha x + C_2 \sin \alpha x][C_3 \cos \alpha(\ln y) + C_4 \sin \alpha(\ln y)]$.

(h) Not separable in the usual sense either $X' = 0$ or $Y' = 0$.

(i) Not separable.

(j) $u = e^{-y/2}[C_1 \cos \alpha x + C_2 \sin \alpha x]\left[C_3 \cos \dfrac{\sqrt{4\alpha^2 - 1}\,y}{2} \sin \dfrac{\sqrt{4\alpha^2 - 1}\,y}{2}\right]$.

(k) $u = e^{-\alpha^2 t}(C_1 \cos \alpha x + C_2 \sin \alpha x)$.

2. (a) $u = C \sin \alpha x \sin \alpha t$.

(b) $u = Ce^{-y} \cos \alpha x \sin \sqrt{\alpha^2 - 1}\,y$.

(c) $u = Ce^{-\alpha^2 t} \cos \alpha x$.

3. (b) $u = e^{-\lambda(y + x/2)}\left[C_1 \cos \dfrac{\lambda\sqrt{3}x}{2} + C_2 \sin \dfrac{\lambda\sqrt{3}x}{2}\right]$.

CHAPTER 2

Exercises 2.1

1. $\frac{5}{2}\mathbf{e}_1 - \frac{1}{2}\mathbf{e}_3$.

2. (a) $\alpha = 5$, $\beta = -1$.

(b) $\mathbf{V}_{ijk} = \langle -6, 14, 8 \rangle$.

3. (b) $\mathbf{K}_1 = \langle 1, 2, 1 \rangle$, $\mathbf{K}_2 = \langle \frac{5}{6}, -\frac{8}{6}, \frac{11}{6} \rangle$, $\mathbf{K}_3 = \langle \frac{8}{7}, -\frac{8}{35}, -\frac{24}{35} \rangle$.

4. (b) Set is orthonormal.

5. (a) $\alpha = -3$.

   (b) $\left\{ \dfrac{1}{\sqrt{2}}, \sqrt{\tfrac{3}{2}}x, \sqrt{\tfrac{5}{8}}(1 - 3x^2) \right\}$.

6. (b) $\left\{ \dfrac{1}{\sqrt{2L}}, \left(\dfrac{1}{\sqrt{L}}\right) \cos \dfrac{n\pi x}{L}, \dfrac{1}{\sqrt{L}} \sin \dfrac{m\pi x}{L} \right\}$, $n, m \in \mathbf{N}$.

7. (a) Not orthonormal.

   (b) $\left\{ \dfrac{1}{\sqrt{2}}, \cos \dfrac{n\pi x}{2} \right\}$, $n \in \mathbf{N}$.

10. $f_1 = 1$, $f_2 = x$, $f_3 = x^2 - \tfrac{1}{3}$, $f_4 = x^3 - \dfrac{3x}{5}$.

Exercises 2.2

2. (b) $\sqrt{2\pi}$.

4. (a) $L_0(x) = 1$, $L_1(x) = 1 - x$, $L_2(x) = \dfrac{x^2}{2} - 2x + 1$.

   (c) $H_0(x) = 1$, $H_1(x) = 2x$, $H_2(x) = 4x^2 - 2$.

Exercises 2.3

1. (a) $\lambda_n = n^2\pi^2$, $y_n(x) = \sin n\pi x$, $n \in \mathbf{N}$.

   (b) $\lambda_n = \left(\dfrac{(2n-1)\pi}{4}\right)^2$, $y_n(x) = \sin \dfrac{(2n-1)\pi x}{4}$, $n \in \mathbf{N}$.

   (c) $\lambda_n = 4n^2$, $y_n(x) = \cos 2nx$, $n \in \mathbf{N}$.

   (d) λ_n positive roots of $\sqrt{\lambda_n} + \tan \sqrt{\lambda_n} = 0$, $y_n(x) = \sin \sqrt{\lambda_n}x$

2. (a) $\lambda_n = n^2$, $\{y_n(x)\} = \{1, \cos nx, \sin nx\}$, $n \in \mathbf{N}_0$.

   (b) $\lambda_n = 4n^2\pi^2$, $\{y_n(x)\} = \{1, \cos 2n\pi x, \sin 2n\pi x\}$, $n \in \mathbf{N}_0$.

   (c) $\lambda_n = n^2\pi^2$, $\{y_n(x)\} = \{1, \cos n\pi x, \sin n\pi x)$, $n \in \mathbf{N}_0$.

3. (a) $\lambda_n = n^2\pi^2 + 4$, $y_n(x) = x^{-2} \sin (n\pi \ln x)$.

   (b) $\lambda_n = \left(\dfrac{n\pi}{\ln 4}\right)^2 + \dfrac{1}{4}$, $y_n(x) = (3+x)^{-1/2} \sin \dfrac{n\pi \ln (3+x)}{\ln 4}$, $n \in \mathbf{N}$.

   (c) $\lambda_n = \left(\dfrac{n\pi}{\ln 2}\right)^2$, $y_n(x) = \sin \dfrac{n\pi \ln x}{\ln 2}$, $n \in \mathbf{N}$.

4. (c) Yes.

   (d) Yes.

Exercises 2.4

1. Yes, $-\infty < x < \infty$.
3. (a) Not uniformly convergent.
   (b) Not uniformly convergent.
4. Series uniformly convergent for all $x \geq a > 0$.
5. $A_1 = \dfrac{4}{\pi}$, $A_2 = 0$, $A_3 = \dfrac{4}{3\pi}$.
6. $\alpha_0 = \pi$, $\alpha_1 = \dfrac{-4}{\pi}$, $\beta_1 = 0$, $\alpha_2 = 0$, $\beta_2 = 0$.
7. (a) Yes, $x - \dfrac{x^2}{2} = \dfrac{16}{\pi^3} \displaystyle\sum_{n=1}^{\infty} \dfrac{1}{(2n-1)^3} \sin \dfrac{(2n-1)\pi x}{2}$
8. $\dfrac{2}{\pi} \displaystyle\sum_{n=1}^{\infty} \dfrac{(-1)^{n-1} \sin n\pi x}{n}$.
9. $\dfrac{4}{\pi} \displaystyle\sum_{n=1}^{\infty} \dfrac{1}{2n-1} \sin \dfrac{(2n-1)x}{2}$.
10. 0.

CHAPTER 3

Exercises 3.1

1. Function is PWC and PWS but not continuous at $x = 0$. Not smooth.
2. (a) Continuous at $x = 0$.
   (b) $f'_+(0) = 1$, $f'_-(0) = -1$.
   (c) No.
3. -2.
4. (a) $f'_+(0) = 1$, $f'(0+) = 1$.
   (b) $f'_-(0) = -1$, $f'(0-) = -1$.
   (c) $f'(0)$ fails to exist.
5. (a) $f'(x) = -x \cos \left(\dfrac{1}{x}\right) + 3x^2 \sin \left(\dfrac{1}{x}\right)$ if $x \neq 0$.
   (b) $f'(x)$ is not differentiable at $x = 0$.

Exercises 3.2

1. $\dfrac{8}{\pi} \displaystyle\sum_{n=1}^{\infty} \dfrac{1}{2n-1} \sin \dfrac{(2n-1)\pi x}{2}$; convergence at $x = 0$ is 0.
2. $1 + \dfrac{4}{\pi} \displaystyle\sum_{n=1}^{\infty} \left(\dfrac{1}{n} \sin \dfrac{n\pi}{2}\right) \cos \dfrac{n\pi x}{2}$; convergence at $x = -1$ is $\frac{1}{2}$.

3. $\dfrac{2}{\pi} \sum\limits_{n=1}^{\infty} \dfrac{(-1)^{n+1}}{n} \sin n\pi x$; convergence at $x = 1$ is 0.

4. $\dfrac{a_0}{2} + \sum\limits_{n=1}^{\infty} a_n \cos \dfrac{n\pi x}{2}$, where $a_0 = \tfrac{3}{2}$; $a_n = \dfrac{2}{n\pi} \sin \dfrac{n\pi}{2} - \dfrac{4}{n^2\pi^2}\left(\cos \dfrac{n\pi}{2} - 1\right)$;

convergence at $x = 1$ is $\tfrac{1}{2}$, and at $x = 2$ is 0.

5. $\sinh 1 + 2\sinh 1 \sum\limits_{n=1}^{\infty} \dfrac{(-1)^n}{1+n^2\pi^2} [\cos n\pi x + n\pi \sin n\pi x]$;   convergence   at

$\qquad$ $x = -1$ is $\dfrac{e + e^{-1}}{2}$, and at $x = 1$ is $\dfrac{e + e^{-1}}{2}$.

9. (a) $\dfrac{1}{2} + \dfrac{2}{\pi} \sum\limits_{n=1}^{\infty} \dfrac{(-1)^{n-1}}{2n-1} \cos(2n-1)x$.

10. (a) $\dfrac{2}{\pi} + \dfrac{4}{\pi} \sum\limits_{n=1}^{\infty} \dfrac{\cos 2nx}{1-4n^2}$.

11. (a) $\cos x$.

$\qquad$ **(b)** $\dfrac{8}{\pi} \sum\limits_{n=1}^{\infty} \dfrac{n}{4n^2-1} \sin 2nx$.

12. $\dfrac{1}{\pi} + \dfrac{\cos x}{2} + \dfrac{2}{\pi} \sum\limits_{n=1}^{\infty} \dfrac{(-1)^{n+1}}{4n^2-1} \cos 2nx$.

13. $2 \sum\limits_{n=1}^{\infty} \dfrac{(-1)^{n-1}}{n} \sin nx$.

14. $\dfrac{\pi}{2} - \dfrac{4}{\pi} \sum\limits_{n=1}^{\infty} \dfrac{1}{(2n-1)^2} \cos(2n-1)x$.

15. $\dfrac{2}{\pi} \sum\limits_{n=1}^{\infty} \left[\dfrac{2}{n^3}((-1)^n - 1)) - \dfrac{\pi^2}{n}(-1)^n\right] \sin nx$.

16. $\dfrac{L^2}{3} + \dfrac{4L^2}{\pi^2} \sum\limits_{n=1}^{\infty} \dfrac{(-1)^n}{n^2} \cos \dfrac{n\pi x}{L}$.

17. $\sum\limits_{n=-\infty}^{\infty} C_n e^{inx}$, where $C_n = \dfrac{1-(-1)^n}{in\pi}$, for $n \neq 0$, $C_0 = 0$.

19. $\sum\limits_{n=-\infty}^{\infty} C_n e^{inx}$, where $C_n = \dfrac{1}{\pi}\left(\dfrac{3+in}{9+n^2}\right)(-1)^n \sinh 3\pi$.

22. $1 - e^{-1} + 2(1 - e^{-1}) \sum\limits_{n=1}^{\infty} \dfrac{\cos 2n\pi x + 2n\pi \sin 2n\pi x}{1 + 4n^2\pi^2}$.

23. $\sinh 1 \sum\limits_{n=-\infty}^{\infty} \dfrac{(-1)^n}{1+n^2\pi^2} \exp(n\pi ix)$.

25. (a) None.

(b) $S_0(x) = \dfrac{\pi}{2}$; $S_1(x) = \dfrac{\pi}{2} - \dfrac{4}{\pi}\cos x$;

$$S_3(x) = \frac{\pi}{2} - \frac{4}{\pi}\cos x - \frac{4}{9\pi}\cos 3x.$$

Exercises 3.3

1. (b) Yes.

(c) Yes; isolated points $\dfrac{\pi}{2} \pm 2n\pi$, $n \in \mathbf{N}_0$.

2. $\dfrac{4}{\pi}\displaystyle\sum_{n=1}^{\infty}\dfrac{\sin(2n-1)x}{2n-1}$; for $x>0$, converges to 1, except points 0, π, 2π, etc.

3. $2\displaystyle\sum_{n=1}^{\infty}(-1)^{n-1}\cos nx$; series fails to converge.

4. $\dfrac{x^2}{4} = \dfrac{\pi^2}{12} - \displaystyle\sum_{n=1}^{\infty}\dfrac{(-1)^{n-1}}{n^2}\cos nx$; integration valid.

5. (a) $\dfrac{4}{\pi}\displaystyle\sum_{n=1}^{\infty}\dfrac{\sin(2n-1)x}{2n-1}$

(b) $\dfrac{\pi}{2} - \dfrac{4}{\pi}\displaystyle\sum_{n=1}^{\infty}\dfrac{\cos(2n-1)x}{(2n-1)^2}$; converges to x.

(c) No.

11. $\dfrac{16}{\pi^2}\displaystyle\sum_{m=1}^{\infty}\sum_{n=1}^{\infty}\dfrac{1}{(2m-1)(2n-1)}\sin\dfrac{(2m-1)\pi x}{a}\sin\dfrac{(2n-1)\pi y}{b}$.

12. $\dfrac{2\pi^2}{3}\displaystyle\sum_{n=1}^{\infty}\dfrac{(-1)^{m+1}}{m}\sin mx + 8\displaystyle\sum_{m=1}^{\infty}\sum_{n=1}^{\infty}\dfrac{(-1)^{m+n+1}}{mn^2}\sin mx\cos ny$.

13. $\dfrac{\pi^4}{9} + \dfrac{4\pi^2}{3}\displaystyle\sum_{m=1}^{\infty}\dfrac{(-1)^m}{m^2}\cos mx + \dfrac{4\pi^2}{3}\displaystyle\sum_{n=1}^{\infty}\dfrac{(-1)^n}{n^2}\cos ny$

$\qquad + 16\displaystyle\sum_{m=1}^{\infty}\sum_{n=1}^{\infty}\dfrac{(-1)^{m+n}}{m^2n^2}\cos mx\cos ny$.

14. $\dfrac{\sin 2}{\pi}\displaystyle\sum_{m=1}^{\infty}\dfrac{(-1)^{m+1}}{m}\sin m\pi x + \dfrac{8\sin 2}{\pi}\displaystyle\sum_{m=1}^{\infty}\sum_{n=1}^{\infty}\left[\dfrac{(-1)^{m+n+1}}{m(4-n^2\pi^2)}\sin m\pi x\right.$

$\qquad \left. \times\cos\dfrac{n\pi y}{2}\right]$.

CHAPTER 4

Exercises 4.1

3. (a) $\dfrac{1}{x}$.

Exercises 4.2

1. $\dfrac{2}{\pi} \displaystyle\int_0^\infty \dfrac{\sin 2\alpha \cos \alpha x}{\alpha}\, d\alpha.$

2. $\dfrac{2}{\pi} \displaystyle\int_0^\infty \dfrac{\sin \alpha\pi \sin \alpha x}{1 - \alpha^2}\, d\alpha.$

4. (c) $f(0) = \frac{1}{2},\ f(\pi) = -\frac{1}{2}.$

5. (d) Twice the expansion of (c).

   (e) $\frac{1}{2}$.

6. (b) 1.

CHAPTER 5

Exercises 5.1

12. (b) $y = C_1 J_{1/2} + C_2 J_{-1/2}.$

13. (e) $y = C_1 J_{3/2} + C_2 J_{-3/2}.$

Exercises 5.3

1. $1 \approx \displaystyle\sum_{k=1}^{\infty} \dfrac{J_0(\lambda_k x)}{\lambda_k J_1(2\lambda_k)}.$

2. $f(x) \approx \dfrac{1}{2} \displaystyle\sum_{k=1}^{\infty} \dfrac{J_1(2\lambda_k) J_0(\lambda_k x)}{\lambda_k J_1^2(4\lambda_k)}.$

4. $f(x) \sim A_0 + \displaystyle\sum_{k=1}^{\infty} A_k J_0(\lambda_0 x)$ where $A_0 = 1$ and $A_k = 0$ when $k \in \mathbf{N}.$

CHAPTER 6

Exercises 6.1

2. (b) $\frac{1}{16}[231x^6 - 315x^4 + 105x^2 - 5].$

5. (a) $y = C P_2(x).$

   (b) $y = -4 P_2(x).$

Exercises 6.2

1. (a) $\dfrac{3x^2 - 1}{2}$.

   (b) $\dfrac{5x^3 - 3x}{2}$.

   (c) $\dfrac{35x^4 - 30x^2 + 3}{8}$.

7. $x^2 = \frac{1}{3}P_0(x) + \frac{2}{3}P_2(x)$.

8. $x^3 = \frac{3}{5}P_1(x) + \frac{2}{5}P_3(x)$.

10. (a) $\frac{2}{5}$.

    (b) 0.

Exercises 6.3

1. (a) 0.

   (b) 0.

Exercises 6.4

1. $1 \sim \sum\limits_{n=0}^{\infty} C_n P_n(x)$ where $C_n = 0$ if $n \in \mathbf{N}$, $C_0 = 1$. Expansion: $1 = P_0(x)$.

2. $|x| \sim \frac{1}{2}P_0(x) + \frac{5}{8}P_2(x) - \frac{3}{16}P_4(x) + \cdots$.

3. $x^3 = \frac{3}{5}P_1(x) + \frac{2}{5}P_3(x)$.

4. $f(x) \sim \frac{3}{4}P_0(x) - \frac{1}{4}P_1(x) + \frac{5}{16}P_2(x) + \cdots$.

5. $f(x) \sim \frac{3}{2}P_0(x) - \frac{3}{4}P_1(x) + \frac{7}{16}P_3(x) + \cdots$.

8. $x \sim \frac{1}{2}P_0(x) + \frac{5}{8}P_2(x) - \frac{3}{16}P_4(x) + \cdots$; $-x$ is represented on $-1 < x < 0$.

Exercises 6.5

1. (a) Degree 1.

   (b) Degree 3.

   (c) Degree 2.

Exercises 6.6

1. (a) $P_3^1(x) = \frac{3}{2}(1 - x^2)^{1/2}(5x^2 - 1)$.

   (b) $P_4^2(x) = \frac{15}{2}(1 - x^2)(7x^2 - 1)$.

   (c) $Q_0^2(x) = \dfrac{2x}{1 - x^2}$.

   (d) $P_3^5(x) = 0$.

2. $x(1 - x^2)^{1/2} = \frac{1}{3}P_2^1(x)$.

CHAPTER 7

Exercises 7.1

1. (a) $t^2 = O(e^{kt})$, $k > 0$.
   (b) e^{t^2} is not of exponential order.

2–6. Answers found in Appendix 2.

7. (b) $\dfrac{5!}{s^6}$.

8. (a) $\sqrt{\pi}/2s^{3/2}$.
   (b) $3\sqrt{\pi}/4s^{5/2}$.
   (c) $15\sqrt{\pi}/8s^{7/2}$.

9. (c) $k/(s^2k^2 + 1)$.

10. Fails to exist.

Exercises 7.2

1. (a) $\dfrac{s^2 + 9}{(s^2 - 9)^2}$.

   (b) $\dfrac{18(s^2 - 3)}{(s^2 + 9)^3}$.

3. (a) $\dfrac{2}{(s - 3)^3}$.

   (b) $\dfrac{6s^2 + 2}{(s^2 - 1)^3}$.

4. (a) $4e^t - 3e^{2t}$.
   (b) $-\frac{2}{5}e^t + \frac{7}{5}\cos 2t - \frac{14}{5}\sin 2t$.

5. $y(t) = -e^{-2t} + e^{-t}$.

6. $y(t) = \dfrac{2A + B}{5}\, e^{3t} + \dfrac{3A - B}{5}\, e^{-2t}$.

7. $y(t) = \dfrac{3}{2}\sin t - \dfrac{t}{2}\cos t$.

8. $y(t) = \dfrac{3B - 3A + 1}{3}\, e^{2t} + \dfrac{4A - 2B - 1}{2}\, e^t + \dfrac{1}{6}\, e^{-t}$.

9. $y(t) = \frac{1}{2}(e^t + \cos t - \sin t)$.

10. $y(t) = e^{t/2}\left[\cos \sqrt{15}\,\dfrac{t}{2} + \dfrac{1}{\sqrt{15}}\sin \sqrt{15}\,\dfrac{t}{2}\right]$.

11. $y(t) = 1 - t^2 e^{-t}$.

12. $y(t) = 2\sin t$.

Exercises 7.3

1. (a) $\dfrac{1}{(s+3)^2}$.

(b) $\dfrac{s-2}{(s-2)^2+9}$.

2. (a) $\dfrac{e^{-3s}}{s^2-1}$.

(b) $\dfrac{se^{-2s}}{s^2-1}$.

3. (a) $\dfrac{e^{-s}+e^{-2s}-2e^{-3s}}{s}$.

(b) $\dfrac{e^{-2s}}{s^2}$.

(c) $\dfrac{e^{-\pi s}}{s^2+1}$.

(d) $\dfrac{1-e^{-s}}{s}$.

4. $\dfrac{e^{-2s}}{s-1}$.

5. (a) $3(t-1)U(t-1)$.

(b) $U(t-3)\sin 2(t-3)$.

(c) $\frac{1}{2}U(t-3)\sin 2(t-3)$.

7. (a) $\dfrac{s+e^{-s}-e^{-2s}(2s+1)}{s^2(1-e^{-2s})}$.

(b) $\dfrac{-s(e^{-2\pi s}+e^{-\pi s})}{(s^2+1)(1-e^{-2\pi s})}$.

(c) $\dfrac{1-e^{-(s-1)}}{(s-1)(1-e^{-s})}$.

8. (a) $y(t) = \sin t + U(t-\pi)[1-\cos(t-\pi)]$.

(b) $y(t) = \frac{1}{2}U(t-2)[1+e^{-2(t-2)}-2e^{-(t-2)}]$.

(c) $y(t) = -e^{2t}+2e^{t}+\frac{1}{2}U(t-2)[1+e^{2(t-2)}-2e^{t-2}]$.

(d) $y(t) = \frac{1}{2}U(t-\pi)[\sin(t-\pi)-(t-\pi)\cos(t-\pi)]$.

9. (a) $y(t) = 3U(t-1)e^{-2(t-1)}+e^{-2t}$.

(b) $y(t) = \cos 2t + \frac{1}{2}U(t-\pi)\sin 2(t-\pi)$.

(c) $y(t) = e^{-t}[\cos t+\sin t]+4U(t-1)e^{-(t-1)}\sin(t-1)$.

(d) $y(t) = [1+U(t-\pi)]\cos t$.

Exercises 7.4

1. (a) $\dfrac{1}{s(s^2+1)}$.

(b) $\dfrac{s}{(s+1)(s^2+1)}$.

2. (b) $\dfrac{1-\cos 2t}{2}$.

(c) $\dfrac{2t-\sin 2t}{4}$.

(d) $\dfrac{e^{2t}-2\sin t-\cos t}{5}$.

(e) $\dfrac{(e^{2(t-2)}-1)U(t-2)}{2}$.

(f) $\dfrac{\sin t - t\cos t}{2}$.

3. (b) $\dfrac{1}{(s^2+1)}$.

(c) $\dfrac{1}{s(s-1)}$.

(d) $\dfrac{\cosh 3t - 1}{9}$.

(e) $\frac{1}{2}(e^{2(t-1)}-1)U(t-1)$.

4. (b) $\dfrac{e^{kt}-kt-1}{k^2}$.

5. (c) $\log\dfrac{s+m}{s+k}$.

8. (a) $y(t)=\dfrac{2e^t-\sin 2t-2\cos 2t}{10}$.

(b) $y(t)=e^t\displaystyle\int_0^t (t-\tau)e^{-\tau}f(\tau)\,d\tau$.

(c) $y(t)=\displaystyle\int_0^t [1-\cos(t-\tau)]f(\tau)\,d\tau$.

9. (a) $y(t)=t$.

(b) $y(t)=\dfrac{1+e^{2t}}{2}$.

(c) $y(t)=e^{-t}$.

(d) $y(t) = \dfrac{2t - 1 + e^{2t}}{4}$.

(e) $Y(s) = \dfrac{F(s)}{1 - B(s)}$.

14. (a) $x(t) = \dfrac{2 + e^{-3t}}{3}$, $\quad y(t) = \dfrac{1 - e^{-3t}}{3}$.

(b) $x(t) = \dfrac{\sinh \sqrt{2}t}{\sqrt{2}}$, $\quad y(t) = \cosh \sqrt{2}t + \dfrac{\sinh \sqrt{2}t}{\sqrt{2}}$, $\quad z(t) = -\dfrac{\sinh \sqrt{2}t}{\sqrt{2}}$.

(c) $x(t) = 1 + e^t$, $\quad y(t) = 2 - e^{-t}$.

Exercises 7.5

1. $y(x, t) = 2(t - 2x)^2 U(t - 2x) - t^2$.

2. $y(x, t) = \dfrac{1}{2} U(t - 2x)[4(t - 2x)^2 + (t - 2x)] - t^2 - \dfrac{t}{2} + x$.

3. $y(x, t) = U(t - 2x)[(t - 2x)^2 + 2(t - 2x)] - 2t$.

4. $y(x, t) = U(t - 4x)[(t - 4x)^2 + (t - 4x)] - 2t$.

Exercises 7.6

3. (a) $\dfrac{L}{n\pi} [1 - (-1)^n]$. $n \neq 0$.

(b) $\dfrac{(-1)^{n+1} L^2}{n\pi}$, $n \neq 0$.

(c) 0, $n \neq 0$.

(d) $\dfrac{L^2}{n^2 \pi^2} [(-1)^n - 1]$, $n \neq 0$.

(e) $\dfrac{kL^2[(-1)^n e^{kL} - 1]}{k^2 L^2 + n^2 \pi^2}$.

(f) $\dfrac{Ln\pi[(-1)^{n+1} e^{kL} + 1]}{k^2 L^2 + n^2 \pi^2}$.

(g) $\dfrac{2\pi i (-1)^n}{n}$, $n \neq 0$.

(h) 0, $n \neq \pm 1$.

Exercises 7.7

2. $F_s(\alpha) = \dfrac{2\alpha}{(1 + \alpha^2)^2}$.

4. $\displaystyle\int_0^\infty \sin t \sin \alpha t \, dt$ fails to converge.

8. $\displaystyle f(t) = \frac{2}{\pi t}\left[1 - \frac{\sin t}{t}\right].$

9. $\displaystyle f(t) = \frac{2t}{\pi(1 + t^2)}.$

13. $F_e(\alpha - c).$

14. $\displaystyle \frac{2}{1 + \alpha^2}.$

16. $\displaystyle \frac{1}{1 - i\alpha}.$

18. $y(t) = Ke^{-t^2/2}.$
19. $y(t) = Ke^{t - t^2/2}.$

Exercises 7.11

1. (b) $F_M(s + \alpha).$

(c) $\displaystyle \frac{-F_M(-s/h)}{h}.$

(f) $a^{-s}\Gamma(s).$

(h) $\displaystyle \Gamma(s) \cos\left(\frac{\pi s}{2}\right).$

CHAPTER 8

Exercises 8.1

1. $\displaystyle y(x, t) = \sum_{n=1}^\infty A_n \sin\frac{n\pi x}{2} \sin\frac{n\pi at}{2}$ where $A_n = \frac{2}{n\pi a}\int_0^2 g(x) \sin\frac{n\pi x}{2} \, dx.$

2. $\displaystyle y(x, t) = \frac{k\pi}{2} - \frac{4k}{\pi}\sum_{n=1}^\infty \frac{\cos(2n - 1)x \cos(2n - 1)at}{(2n - 1)^2}.$

3. $y_{tt} = a^2 y_{xx}, \quad 0 < x < L, \; t > 0$

$$y(x, 0) = f(x) = \begin{cases} 0.02x & \text{if } 0 \leqslant x \leqslant L/2 \\ 0.02(L - x) & \text{if } L/2 \leqslant x \leqslant L \end{cases}$$

$y_t(x, 0) = 0, \quad y(0, t) = y(L, t) = 0$

Solution: $\displaystyle y(x, t) = \frac{0.08L}{\pi^2}\sum_{n=1}^\infty \frac{(-1)^{n+1}}{(2n - 1)^2} \sin\frac{(2n - 1)\pi x}{L} \cos\frac{(2n - 1)\pi at}{L}.$

4. $y_{tt} = a^2 y_{xx}$, $0 < x < 2$, $t > 0$.

$y(0, t) = y(2, t) = 0$, $t \geq 0$

$y_t(x, 0) = g(x) = \begin{cases} 0.05x & \text{if } 0 \leq x \leq 1 \\ 0.05(2 - x) & \text{if } 1 \leq x \leq 2 \end{cases}$

$y(x, 0) = 0$, $0 \leq x \leq 2$

Solution: $y(x, t) = \dfrac{0.8}{\pi^3 a} \sum_{n=1}^{\infty} \dfrac{(-1)^{n+1}}{(2n-1)^3} \sin \dfrac{(2n-1)\pi x}{2} \sin \dfrac{(2n-1)\pi a t}{2}$.

5. $y(x, t) = 0.05 \sin \dfrac{4\pi x}{L} \cos \dfrac{4\pi a t}{L}$.

6. $y(x, t) = \dfrac{1}{4\pi} \sin 2\pi x \sin 4\pi t$.

8. $y(x, t) = \sum_{n=1}^{\infty} B_n \sin \dfrac{(2n-1)\pi x}{2L} \cos \dfrac{(2n-1)\pi a t}{2L}$.

where $B_n = \dfrac{2}{L} \displaystyle\int_0^L f(x) \sin \dfrac{(2n-1)\pi x}{2L} \, dx$.

9. $\theta(x, t) = \dfrac{2kL}{\pi} \sum_{n=1}^{\infty} \dfrac{(-1)^{n+1}}{n} \sin \dfrac{n\pi x}{L} \cos \dfrac{n\pi a t}{L}$.

10. $\theta(x, t) = \dfrac{8kL}{\pi^2} \sum_{n=1}^{\infty} \dfrac{(-1)^{n+1}}{(2n-1)^2} \sin \dfrac{(2n-1)\pi x}{2L} \cos \dfrac{(2n-1)\pi a t}{2L}$.

Exercises 8.2

1. $u(x, t) = \exp\left[-\dfrac{9\pi^2 a^2 t}{L^2}\right] \cos \dfrac{3\pi x}{L}$.

2. $u(x, t) = \sum_{n=1}^{\infty} B_n \exp\left[-\dfrac{(2n-1)^2 \pi^2 a^2 t}{4L^2}\right] \sin \dfrac{(2n-1)\pi x}{2L}$ where

$B_n = \dfrac{2}{L} \displaystyle\int_0^L f(x) \sin \dfrac{(2n-1)\pi x}{2L} \, dx$.

3. (a) $v(x, t) = 5t(1 - x) + \frac{85}{3}x + \frac{5}{2}x^2 - \frac{5}{6}x^3$.

(b) $w(x, t) = \dfrac{160}{\pi^3} \sum_{n=1}^{\infty} \left[\dfrac{3(-1)^n + 1}{n^3}\right] \exp\left(-\dfrac{n^2 \pi^2 t}{16}\right) \sin \dfrac{n\pi}{4} x$.

(d) 70.84.

(e) 70.98.

4. $u_t = a^2 u_{xx}$, $0 < x < L$, $t > 0$; $u_x(0, t) = u(L, t) = 0$, $t \geq 0$;

$u(x, 0) = \cos \dfrac{5\pi x}{2L}$. Solution: $u(x, t) = \exp\left[-\dfrac{25\pi^2 a^2 t}{4L^2}\right] \cos \dfrac{5\pi x}{2L}$.

5. $u(x, t) = \dfrac{16}{\pi^2} \displaystyle\sum_{n=1}^{\infty} \dfrac{(-1)^{n+1}}{(2n-1)^2} \exp\left[-\dfrac{(2n-1)^2\pi^2 t}{8}\right] \sin \dfrac{(2n-1)\pi x}{4}$.

6. $u(x, t) = \displaystyle\sum_{n=1}^{\infty} B_n \exp\left[-\dfrac{(2n-1)^2\pi^2 t}{16}\right] \cos \dfrac{(2n-1)\pi x}{4}$ where

$B_n = \dfrac{16}{(2n-1)^2\pi^2}\left[\cos\dfrac{(2n-1)\pi}{4} - \cos\dfrac{(2n-1)\pi}{2}\right]$.

Exercises 8.3

1. $u_{xx} + u_{yy} = 0$, $0 < x < 1$, $0 < y < 2$, $u(0, y) = u(1, y) = u(x, 2) = 0$;
$u(x, 0) = f(x)$.

Solution: $u(x, y) = \displaystyle\sum_{n=1}^{\infty} B_n \dfrac{\sinh n\pi(2-y) \sin n\pi x}{\sinh 2n\pi}$

where $B_n = 2\displaystyle\int_0^1 f(x) \sin n\pi x \, dx$.

2. $u(x, y) = \displaystyle\sum_{n=1}^{\infty} B_n \sin \dfrac{n\pi x}{2} \sinh \dfrac{n\pi y}{2}$

where $B_n = \dfrac{1}{\sinh n\pi} \displaystyle\int_0^2 f(x) \sin \dfrac{n\pi x}{2} \, dx$.

3. $u(x, y) = \displaystyle\sum_{n=1}^{\infty} \left[A_n \exp\left(\dfrac{n\pi y}{a}\right) + B_n \exp\left(\dfrac{-n\pi y}{a}\right)\right] \sin \dfrac{n\pi x}{a}$ where

$A_n = \dfrac{1}{a \sinh (n\pi b/a)} \displaystyle\int_0^a \left[g(x) - \exp\left(\dfrac{-n\pi b}{a}\right)f(x)\right] \sin \dfrac{n\pi x}{a} \, dx$,

$B_n = \dfrac{1}{a \sinh (n\pi b/a)} \displaystyle\int_0^a \left[\exp\left(\dfrac{n\pi b}{a}\right)f(x) - g(x)\right] \sin \dfrac{n\pi x}{a} \, dx$.

4. $u_{xx} + u_{yy} = 0$, $0 < x < 2$, $0 < y < 2$; $u(0, y) = u_x(2, y) = 0$; $u(x, 0) = 0$,

$u(x, 2) = \sin \dfrac{3\pi x}{4}$.   Solution: $u(x, y) = \dfrac{1}{\sinh (3\pi/2)} \sinh \dfrac{3\pi y}{4} \sin \dfrac{3\pi x}{4}$.

5. Selected values, $u(x, y)$, $x = 2, 4, 6, 8$

   $y = 2$: 2.12   4.24   6.06   6.67
   $y = 4$: 4.24   8.79   13.33   16.61
   $y = 6$: 6.06   13.33   21.88   30.42
   $y = 8$: 6.67   16.61   30.42   47.21

6. $u(x, y) = \dfrac{4}{\pi} \displaystyle\sum_{n=1}^{\infty} \dfrac{\sinh ny \sin nx}{(2n-1) \sinh [2(2n-1)\pi]}$.

7. $u(x, y) = -\dfrac{1}{\sinh \pi} \cosh (\pi - y) \cos x.$

8. $u(x, y) = \dfrac{2}{\sinh \pi} \cos y \cosh x.$

Exercises 8.4

1. $u(r, z) = 2 \displaystyle\sum_{k=1}^{\infty} \left[\dfrac{1}{J_1^2(\alpha_k)} \int_0^1 \xi f(\xi) J_0(\alpha_k \xi) \, d\xi \right] J_0(\alpha_k r) \dfrac{\sinh (2 - z)\alpha_k}{\sinh 2\alpha_k}.$

2. $u(r, z) = 2T_0 \displaystyle\sum_{k=1}^{\infty} \dfrac{\sinh \alpha_k (h - z) J_0(\alpha_k r)}{\alpha_k J_1(\alpha_k) \sinh \alpha_k h}.$

3. $u(r, z) = \dfrac{2}{a^2} \displaystyle\sum_{k=1}^{\infty} \dfrac{J_0(\alpha_k r) \sinh \alpha_k z}{J_1^2(\alpha_k a) \sinh \alpha_k h} \int_0^a \xi f(\xi) J_0(\alpha_k \xi) \, d\xi.$

4. $u(r, z) = \displaystyle\sum_{n=1}^{\infty} B_n \dfrac{\sinh \alpha_n (2 - z)}{\sinh 2\alpha_n} J_0(\alpha_n r)$

where $B_n = \dfrac{2\alpha_n^2}{[k^2 + \alpha_n^2] J_0^2(\alpha_n)} \displaystyle\int_0^1 r f(r) J_0(\alpha_n r) \, dr.$

5. $u(r, t) = \displaystyle\sum_{k=1}^{\infty} B_k [Y_0(\alpha_k p) J_0(\alpha_k r) - J_0(\alpha_k p) Y_0(\alpha_k r)] \exp (-\alpha_k^2 a^2 t)$

where $B_k = \dfrac{\displaystyle\int_p^q r f(r) [Y_0(\alpha_k p) J_0(\alpha_k r) - J_0(\alpha_k p) Y_0(\alpha_k r)] \, dr}{\displaystyle\int_p^q r [Y_0(\alpha_k p) J_0(\alpha_k r) - J_0(\alpha_k p) Y_0(\alpha_k r)]^2 \, dr}.$

Exercises 8.5

1. $u(\rho, \phi) = 15 + 15 \displaystyle\sum_{n=0}^{\infty} [P_{2n}(0) - P_{2n+2}(0)] \left(\dfrac{\rho}{a} \right)^{2n+1} P_{2n+1} (\cos \phi).$

2. $\dfrac{3T_0 \rho P_1(\cos \phi)}{2a} - \dfrac{7T_0 \rho^3 P_3(\cos \phi)}{8a^3} + \cdots.$

3. For a sphere radius 1, upper half of surface has temperature $u(1, \phi) = T_0$; temperature on lower half kept at 0.

4. $v(\rho, \phi) = -V_0 \rho P_1(\cos \phi) + V_0 \rho^{-2} P_1(\cos \phi).$

6. $u(\rho, \phi) = \frac{1}{3} P_0(\cos \phi) - \frac{4}{3} \rho^2 P_2(\cos \phi).$ Steady state temperature problem in a spherical shell.

CHAPTER 9

Exercises 9.1

1. $-5 \cos 5t$.
2. $-5U(t-2) \sin 5(t-2)$.
3. $\frac{1}{8}e^{-3t}(4 \cos 4t + 3 \sin 4t)$.
4. $\frac{1}{4}(-3 \sin 8t + t \sin 8t)$.

5. $Q(t) = \dfrac{1 - e^{-100t}}{10}$, $I(t) = 10e^{-100t}$.

6. $Q(t) = \dfrac{1 - e^{-100t} - (1 - e^{-100(t-10)})U(t-10)}{10}$,

   $I(t) = 10(e^{-100t} - e^{-100(t-10)}U(t-10))$.

7. $Q(t) = 2(\cos 40t - \cos 50t)$, $I(t) = 2(50 \sin 50t - 40 \sin 40t)$.

8. $Q(t) = \dfrac{4}{5}\left[1 - e^{-5t/2}\left(\cos \dfrac{5\sqrt{19}t}{2} + \dfrac{\sin(5\sqrt{19}t/2)}{\sqrt{19}}\right)\right]$,

   $I(t) = \dfrac{40}{\sqrt{19}} e^{-5t/2} \sin \dfrac{5\sqrt{19}t}{2}$.

Exercises 9.2

1. $u(x, y, t) = 0.04 \displaystyle\sum_{m=1}^{\infty} \sum_{n=1}^{\infty} \dfrac{(-1)^{n+m}}{nm} \sin nx \sin my \cos a\sqrt{n^2 + m^2}\,t$.

2. $u(r, t) = \dfrac{1}{2} \displaystyle\sum_{k=1}^{\infty} \left[\dfrac{1}{J_1^2(2\alpha_k)} \int_0^2 \xi f(\xi) J_0(\alpha_k \xi)\, d\xi \right] J_0(\alpha_k r) \cos \alpha_k at$.

3. $u(r, \theta, t) = \displaystyle\sum_{k=0}^{\infty} \left\{ \sum_{j=1}^{\infty} M_{kj} J_k(\alpha_j r) \cos \alpha_j at \right\} \cos k\theta$

   $\qquad\qquad + \displaystyle\sum_{k=0}^{\infty} \left\{ \sum_{j=1}^{\infty} K_{kj} J_k(\alpha_j r) \cos \alpha_j at \right\} \sin k\theta$ where

$$M_{kj} = \frac{2}{\pi c^2 J_{k+1}^2(\alpha_j c)} \int_0^c \int_0^{2\pi} rf(r, \theta) J_k(\alpha_j, r) \cos k\theta\, dr\, d\theta, \quad \text{if } k \in \mathbf{N}$$

$$M_{0j} = \frac{1}{\pi c^2 j_1^2(\alpha_j c)} \int_0^c \int_0^{2\pi} rf(r, \theta) J_0(\alpha_j r)\, dr\, d\theta, \quad \text{if } k = 0$$

$$K_{kj} = \frac{2}{\pi c^2 J_{k+1}^2(\alpha_j c)} \int_0^c \int_0^{2\pi} rf(r, \theta) J_k(\alpha_j r) \sin k\theta\, dr\, d\theta, \quad k \in \mathbf{N}$$

4. $u_{tt} = a^2\left[u_{rr} + \dfrac{1}{r}u_r + \dfrac{1}{r^2}u_{\theta\theta}\right]$, $0 < r < 2$, $0 < \theta < \dfrac{\pi}{2}$, $t > 0$;

$u(2, \theta, t) = u(r, 0, t) = u(r, \pi/2, t) = 0$; $u_t(r, \theta, 0) = 0$, $u(r, \theta, 0) = f(r, \theta)$.

Solution: $u(r, \theta, t) = \displaystyle\sum_{j=0}^{\infty} \sum_{n=1}^{\infty} A_{nj} J_{2n}(\alpha_j r) \sin 2n\theta \cos \alpha_j at$ where

$$A_{nj} = \frac{2}{\pi J_{2n+1}^2(2\alpha_j)} \int_0^2 \int_0^{\pi/2} rf(r, \theta) J_{2n}(\alpha_j r) \sin 2n\theta \, d\theta \, dr.$$

5. $y(x, t) = \dfrac{0.4}{\pi} \displaystyle\sum_{n=1}^{\infty} \frac{1}{2n-1} e^{-kt/2}\left[\cos \frac{\sqrt{n^2\pi^2 - k^2}}{2} t\right.$

$\left. + \dfrac{k}{2} \sin \dfrac{\sqrt{n^2\pi^2 - k^2}}{2} t\right] \sin \dfrac{n\pi x}{2}$.

6. $y(x, t) = \dfrac{gx}{2a^2}(x - \pi) + \dfrac{2g}{a^2\pi} \displaystyle\sum_{n=1}^{\infty} \left[\frac{1 - (-1)^n}{n^3}\right] \sin nx \cos nat$.

7. $y(x, t) = 2k\pi \displaystyle\sum_{n=1}^{\infty} \frac{[1 - (-1)^n e^{-c}]n}{c^2 + n^2\pi^2} \sin \frac{n\pi x}{c} \cos \frac{\sqrt{hc^2 + n^2\pi^2 a^2 t}}{c}$.

8. $y(x, t) = \dfrac{0.4}{\pi^2} \displaystyle\sum_{n=1}^{\infty} \frac{(-1)^{n+1}}{(2n-1)^2} \sin \frac{(2n-1)\pi x}{2} \cos \sqrt{1 + \frac{(2n-1)^2\pi^2}{4}} t$.

9. $y(x, t) = \dfrac{k}{\sqrt{2}} e^{-t/2} \sin \sqrt{2} t \cos \dfrac{3x}{2}$.

10. $y(x, t) = \displaystyle\sum_{n=1}^{\infty} A_n \sin \frac{n\pi x}{L} \cos \frac{n^2\pi^2 at}{L^2}$ where $A_n = \dfrac{2}{L} \displaystyle\int_0^L f(x) \sin \frac{n\pi x}{L} \, dx$.

11. $u(x, t) = \dfrac{2}{\pi} \displaystyle\sum_{n=1}^{\infty} \left[\int_0^\pi f(\tau) \sin n\tau \, d\tau\right] \cos nt \sin nx$.

Exercises 9.3

1. $u(x, t) = 10(x + 1) + \dfrac{40}{\pi} \displaystyle\sum_{n=1}^{\infty} \frac{1 + (-1)^n}{n} \exp\left[-\frac{n^2\pi^2 t}{4}\right] \sin \frac{n\pi x}{4}$.

2. $u(x, t) = \dfrac{T_0 x}{\pi} + \dfrac{2T_0}{\pi} \displaystyle\sum_{n=1}^{\infty} \frac{(-1)^n}{n} e^{-n^2 t} \sin nx$.

3. $u(x, t) = x - 3 - \dfrac{8}{\pi} \displaystyle\sum_{n=1}^{\infty} \left[\sin \frac{(2n-1)\pi}{2} + 1\right] \exp\left[-\frac{(2n-1)^2\pi^2 t}{4}\right]$

$\times \cos \dfrac{(2n-1)\pi x}{2}$.

4. (a) $u(x, t) = \dfrac{A_0}{2} e^{-ht} + \displaystyle\sum_{n=1}^{\infty} A_n \exp\left[-\left(h + \dfrac{n^2\pi^2 a^2}{L^2}\right)t\right] \cos \dfrac{n\pi x}{L}$

where $A_n = \dfrac{2}{L} \displaystyle\int_0^L f(x) \cos \dfrac{n\pi x}{L} \, dx.$

(b) $v(x, t) = \dfrac{A_0}{2} + \displaystyle\sum_{n=1}^{\infty} A_n \exp\left[-\dfrac{n^2\pi^2 a^2 t}{L^2}\right] \cos \dfrac{n\pi x}{L}$ where

$A_n = \dfrac{2}{L} \displaystyle\int_0^L f(x) \cos \dfrac{n\pi x}{L} \, dx.$ Note: $u(x, t) = e^{-ht} v(x, t)$ is the

solution (a).

5. $u(x, t) = \dfrac{k}{a^2} [1 - e^{-x} + (e^{-1} - 1)x] + \dfrac{2k}{a^2\pi} \displaystyle\sum_{n=1}^{\infty} \dfrac{(-1)^n e^{-1} - 1}{n}$

$\times \exp[-n^2\pi^2 a^2 t] \sin n\pi x.$

6. $u(x, t) = k \sin x + (1 - k)e^{-t} \sin x.$

7. $u(x, y) = 2 \displaystyle\sum_{n=1}^{\infty} \dfrac{\sin \alpha_n}{\alpha_n(1 + \sin^2\alpha_n)} e^{-\alpha_n y} \cos \alpha_n x.$

8. $u(\rho, t) = 2 \displaystyle\sum_{n=1}^{\infty} \dfrac{(-1)^{n+1}}{n} e^{-n^2\pi^2 a^2 t} \dfrac{\sin n\pi\rho}{\rho}.$

9. $u(x, t) = \dfrac{2}{\pi} \displaystyle\sum_{n=1}^{\infty} A_n(\tau)e^{-n^2(t-\tau)} \sin nx + \displaystyle\sum_{n=1}^{\infty} B_n(0)e^{-n^2 t} \sin nx$ where

$A_n = \displaystyle\int_0^\pi h(x, t) \sin nx \, dx, \, B_n(0) = \int_0^\pi f(x) \sin nx \, dx.$

Exercises 9.4

2. Odd.

3. (a) $\alpha = b, \, \beta = a;$

(b) $\xi = b.$

5. (a) $u(x, t) = \dfrac{1}{\pi} \displaystyle\int_0^\infty f(\xi) \int_0^\infty e^{-\alpha^2 a^2 t}[\cos \alpha(\xi - x) - \cos \alpha(\xi + x)] \, d\alpha \, d\xi.$

Exercises 9.5

1. (a) Selected values, $u(x, t)$, $x = 2, 4, 6, 8, 10$

$t = 0$: 0.00000 0.00000 0.00013 0.05399 0.39894
$t = 1$: 0.00001 0.00057 0.01600 0.11825 0.23033
$t = 4$: 0.00384 0.01800 0.05467 0.10648 0.13298
$t = 9$: 0.01906 0.03602 0.06018 0.08240 0.09153

(b) Selected values, $u(x, t)$, $x = 2, 4, 6, 8, 10$

$t = 1$: 0.13929 0.64104 0.96452 0.99942 0.99999
$t = 4$: 0.32228 0.57436 0.81315 0.94389 0.97912
$t = 9$: 0.45812 0.58365 0.72848 0.83724 0.87680

Exercises 9.6

1. (a) $y(x, t) = \dfrac{1}{\pi} \displaystyle\int_0^\infty \cos \alpha at \int_{-\infty}^\infty f(\tau) \cos \alpha(\tau - x)\, d\tau\, d\alpha.$

2. $y(x, t) = \dfrac{0.04}{\pi} \displaystyle\int_0^\infty \dfrac{\alpha}{(1 + \alpha^2)^2} \cos \alpha at \sin \alpha x\, d\alpha.$

3. (a) $v(x, y) = \dfrac{1}{\pi} \displaystyle\int_0^\infty e^{-\alpha y} \int_{-\infty}^\infty f(\xi)[\cos \alpha\xi \cos \alpha x + \sin \alpha\xi \sin \alpha x]\, d\xi\, d\alpha.$

7. (a) $u(x, y) = \dfrac{2T_0}{\pi} \displaystyle\int_0^\infty \left[\dfrac{\cos \alpha - 1}{\alpha} \cos \alpha x \right] e^{-\alpha y}\, d\alpha.$ Steady   state   heat problem.

(b) $u(x, y) = \dfrac{2T_0}{\pi} \displaystyle\int_0^\infty \dfrac{e^{-\alpha y} \sin \alpha \sin \alpha x}{\alpha}\, d\alpha.$

8. (a) $v(x, y) = \dfrac{2}{\pi} \displaystyle\int_0^\infty \dfrac{\sin \alpha}{\alpha \cosh \alpha} \cosh \alpha x \cos \alpha y\, d\alpha.$

(b) $v(x, y) = \dfrac{2}{\pi} \displaystyle\int_0^\infty \dfrac{\cosh \alpha x \cos \alpha y}{(1 + \alpha^2) \cosh \alpha}\, d\alpha.$

9. $v(x, y) = \dfrac{1}{\pi} \displaystyle\int_0^\infty \int_{-\infty}^\infty \dfrac{f(\tau) \cos \alpha(\tau - x) \sinh \alpha y}{\sinh \alpha}\, d\tau\, d\alpha.$

APPENDIX 1

SELECTED INTEGRALS

Constants of integration are omitted

1. $\displaystyle\int u\,dv = uv - \int v\,du$

2. $\displaystyle\int \sin^2\alpha x\,dx = \frac{x}{2} - \frac{\sin 2\alpha x}{4\alpha}$

3. $\displaystyle\int \cos^2\alpha x\,dx = \frac{x}{2} + \frac{\sin 2\alpha x}{4\alpha}$

4. $\displaystyle\int \sin \alpha x \cos \alpha x\,dx = \frac{\sin^2\alpha x}{2\alpha}$

5. $\displaystyle\int x \sin \alpha x\,dx = \frac{\sin \alpha x}{\alpha^2} - \frac{x \cos \alpha x}{\alpha}$

6. $\displaystyle\int x \cos \alpha x\,dx = \frac{\cos \alpha x}{\alpha^2} + \frac{x \sin \alpha x}{\alpha}$

7. $\displaystyle\int x^2 \sin \alpha x\,dx = \frac{2x \sin \alpha x}{\alpha^2} + \frac{2 \cos \alpha x}{\alpha^3} - \frac{x^2 \cos \alpha x}{\alpha}$

8. $\displaystyle\int x^2 \cos \alpha x\,dx = \frac{2x \cos \alpha x}{\alpha^2} - \frac{2 \sin \alpha x}{\alpha^3} + \frac{x^2 \sin \alpha x}{\alpha}$

9. $\displaystyle\int \sin \alpha x \cos \beta x\,dx = -\frac{1}{2}\left[\frac{\cos(\alpha-\beta)x}{\alpha-\beta} + \frac{\cos(\alpha+\beta)x}{\alpha+\beta}\right], \quad \alpha^2 \neq \beta^2$

10. $\displaystyle\int \cos \alpha x \cos \beta x\,dx = \frac{1}{2}\left[\frac{\sin(\alpha-\beta)x}{\alpha-\beta} + \frac{\sin(\alpha+\beta)x}{\alpha+\beta}\right], \quad \alpha^2 \neq \beta^2$

11. $\displaystyle\int \sin \alpha x \sin \beta x\,dx = \frac{1}{2}\left[\frac{\sin(\alpha-\beta)x}{\alpha-\beta} - \frac{\sin(\alpha+\beta)x}{\alpha+\beta}\right], \quad \alpha^2 \neq \beta^2$

12. $\displaystyle\int xe^{\alpha x}\,dx = \frac{e^{\alpha x}[\alpha x - 1]}{\alpha^2}$

13. $\displaystyle\int x^2 e^{\alpha x}\, dx = \frac{e^{\alpha x}[\alpha^2 x^2 - 2\alpha x + 2]}{\alpha^3}$

14. $\displaystyle\int e^{\alpha x} \sin \beta x\, dx = \frac{e^{\alpha x}[\alpha \sin \beta x - \beta \cos \beta x]}{\alpha^2 + \beta^2}$

15. $\displaystyle\int e^{\alpha x} \cos \beta x\, dx = \frac{e^{\alpha x}[\alpha \cos \beta x + \beta \sin \beta x]}{\alpha^2 + \beta^2}$

16. $\displaystyle\int x e^{\alpha x} \sin \beta x\, dx = \frac{x e^{\alpha x}[\alpha \sin \beta x - \beta \cos \beta x]}{\alpha^2 + \beta^2}$

$$- \frac{e^{\alpha x}[(\alpha^2 - \beta^2) \sin \beta x - 2\alpha\beta \cos \beta x]}{(\alpha^2 + \beta^2)^2}$$

17. $\displaystyle\int x e^{\alpha x} \cos \beta x\, dx = \frac{x e^{\alpha x}[\alpha \cos \beta x + \beta \sin \beta x]}{\alpha^2 + \beta^2}$

$$- \frac{e^{\alpha x}[(\alpha^2 - \beta^2) \cos \beta x + 2\alpha\beta \sin \beta x]}{(\alpha^2 + \beta^2)^2}$$

18. $\displaystyle\int e^{\alpha x} \sin \beta x \cos \gamma x\, dx = \frac{e^{\alpha x}[\alpha \sin (\beta - \gamma)x - (\beta - \gamma) \cos (\beta - \gamma)x]}{2[\alpha^2 + (\beta - \gamma)^2]}$

$$+ \frac{e^{\alpha x}[\alpha \sin (\beta + \gamma)x - (\beta + \gamma) \cos (\beta + \gamma)x]}{2[\alpha^2 + (\beta + \gamma)^2]}$$

19. $\displaystyle\int e^{\alpha x} \cos \beta x \cos \gamma x\, dx = \frac{e^{\alpha x}[\alpha \cos (\beta - \gamma)x + (\beta - \gamma) \sin (\beta - \gamma)x]}{2[\alpha^2 + (\beta - \gamma)^2]}$

$$+ \frac{e^{\alpha x}[\alpha \cos (\beta + \gamma)x + (\beta + \gamma) \sin (\beta + \gamma)x]}{2[\alpha^2 + (\beta + \gamma)^2]}$$

20. $\displaystyle\int e^{\alpha x} \sin \beta x \sin \gamma x\, dx = \frac{e^{\alpha x}[\alpha \cos (\beta - \gamma)x + (\beta - \gamma) \sin (\beta - \gamma)x]}{2[\alpha^2 + (\beta - \gamma)^2]}$

$$- \frac{e^{\alpha x}[\alpha \cos (\beta + \gamma)x + (\beta + \gamma) \sin (\beta + \gamma)x]}{2[\alpha^2 + (\beta + \gamma)^2]}$$

APPENDIX 2

TABLE OF LAPLACE TRANSFORMS

$$F(s) = \mathcal{L}\{f(t)\} = \int_0^\infty e^{-st}f(t)\,dt, \quad f(t) = \mathcal{L}^{-1}\{F(s)\} = \int_{c-i\infty}^{c+i\infty} e^{st}F(s)\,ds$$

	$f(t)$	$F(s)$
1.	1	$1/s$
2.	t^n	$n!/s^{n+1}$
3.	$t^\alpha,\ \alpha > -1$	$\Gamma(\alpha+1)/s^{\alpha+1}$
4.	$t^{-1/2}$	$(\pi/s)^{1/2}$
5.	e^{kt}	$1/(s-k)$
6.	$\sin kt$	$k/(s^2+k^2)$
7.	$\cos kt$	$s/(s^2+k^2)$
8.	$\sinh kt$	$k/(s^2-k^2)$
9.	$\cosh kt$	$s/(s^2-k^2)$
10.	$e^{kt}t^n$	$n!/(s-k)^{n+1}$
11.	$e^{kt}\sin mt$	$m/[(s-k)^2+m^2]$
12.	$e^{kt}\cos mt$	$(s-k)/[(s-k)^2+m^2]$
13.	$t\sin kt$	$2ks/(s^2+k^2)^2$
14.	$t\cos kt$	$(s^2-k^2)/(s^2+k^2)^2$
15.	$\sin kt - kt\cos kt$	$2k^3/(s^2+k^2)^2$
16.	$U(t-k)$	e^{-ks}/s
17.	$U(t-k)f(t-k)$	$e^{-ks}F(s)$

	$f(t)$	$F(s)$
18.	$\delta(t - k)$	e^{-ks}
19.	$f(kt)$	$(1/k)F(s/k)$
20.	$(1/t)f(t)$	$\int_s^\infty F(\beta)\,d\beta$
21.	$\int_0^t F(\alpha)\,d\alpha$	$(1/s)F(s)$
22.	$e^{kt}f(t)$	$F(s - k)$
23.	$(1/k)f(t/k)$	$F(ks)$
24.	$f^{(n)}(t)$	$s^n F(s) - s^{(n-1)}f(0) - \cdots - f^{(n-1)}(0)$
25.	$(-t)^n f(t)$	$F^{(n)}(s)$
26.	$\int_0^t f(t - \tau)g(\tau)\,d\tau$	$F(s)G(s)$
27.	$f(t)$ periodic (p)	$[1/(1 - e^{-ps})]\int_0^p e^{-st}f(t)\,dt$
28.	$f(t)$ if $f(t + p) = -f(t)$	$[1/(1 + e^{-ps})]\int_0^p e^{-st}f(t)\,dt$
29.	$(1 - e^{-t})/t$	$\ln(1 + 1/s)$
30.	$(1/t)\sinh kt$	$(1/2)\ln[(s + k)/(s - k)]$
31.	$(1 - \cosh kt)/t$	$(1/2)\ln(1 - k^2/s^2)$
32.	$(1 - \cos kt)/t$	$(1/2)\ln(1 + k^2/s^2)$
33.	$(e^{kt} - e^{mt})/t$	$\ln[(s - m)/(s - k)]$
34.	$(1/t)\sin kt$	$\arctan(k/s)$
35.	$\operatorname{erf}(\sqrt{t})$	$1/(s\sqrt{s + 1})$
36.	$\operatorname{erf}(t/(2k))$	$e^{k^2 s^2}[1 - \operatorname{erf}(ks)]/s$
37.	$1 - \operatorname{erf}(kt^{-1/2}/2)$	$e^{-k\sqrt{s}}/s$
38.	$(1/\sqrt{t})\cos 2\sqrt{kt}$	$\sqrt{\pi/s}\,e^{-k/s}$
39.	$\sin 2\sqrt{kt}$	$\sqrt{k\pi/s^3}\,e^{-k/s}$
40.	$\sinh 2\sqrt{kt}$	$\sqrt{k\pi/s^3}\,e^{k/s}$
41.	$(1/\sqrt{t})\cosh 2\sqrt{kt}$	$\sqrt{\pi/s}\,e^{k/s}$
42.	$t^{-3/2}\exp(-k^2/(4t))$	$(2\sqrt{\pi}/k)e^{-k\sqrt{s}}$
43.	$J_0(kt)$	$1/\sqrt{s^2 + k^2}$
44.	$J_0(2\sqrt{kt})$	$(1/s)e^{-k/s}$
45.	$J_\alpha(kt),\ \alpha > -1$	$k^{-\alpha}(\sqrt{s^2 + k^2} - s)^\alpha(s^2 + k^2)^{-1/2}$
46.	$(1/t)J_\alpha(kt),\ \alpha > 0$	$(\sqrt{s^2 + k^2} - s)^\alpha/(\alpha k^\alpha)$

APPENDIX 3

TABLES OF FINITE FOURIER TRANSFORMS

FINITE SINE TRANSFORMS

$$F_S(n) = S_n\{f(x)\} = \int_0^L f(x) \sin\left(\frac{n\pi x}{L}\right) dx , \quad f(x) = \frac{2}{L} \sum_{n=1}^{\infty} F_S(n) \sin\left(\frac{n\pi x}{L}\right)$$

	$f(x)$	$F_S(n)$
1.	1	$L[1 - (-1)^n]/(n\pi)$
2.	x	$(-1)^{n+1}L^2/(n\pi)$
3.	e^{kx}	$Ln\pi[1 - (-1)^n e^{kL}]/[(n\pi)^2 + (kL)^2]$
4.	$\cos kx$	$Ln\pi[1 - (-1)^n \cos kL]/[(n\pi)^2 - (kL)^2], \, n\pi \neq kL$
5.	$\sin kx$	$(-1)^n[Ln\pi \sin kL]/[(kL)^2 - (n\pi)^2], \, n\pi \neq kL$
6.	$f'(x)$	$-(1/L)n\pi F_C(n)$ (Cosine Transform)
7.	$f''(x)$	$n\pi[f(0) - (-1)^n f(L)]/L - (n\pi)^2 F_S(n)/L^2$

FINITE COSINE TRANSFORMS

$$F_C(n) = C_n\{f(x)\} = \int_0^L f(x) \cos(n\pi x/L)\, dx\,,$$

$$f(x) = \frac{F_C(0)}{L} + \frac{2}{L} \sum_{n=1}^{\infty} F_C(n) \cos(n\pi x/L)$$

	$f(x)$	$F_C(n)$
1.	1	L if $n = 0$, 0 if $n \neq 0$
2.	x	$L^2/2$ if $n = 0$, $L^2[(-1)^n - 1]/(n\pi)^2$ if $n \neq 0$
3.	e^{kx}	$kL^2[(-1)^n e^{kL} - 1]/[(n\pi)^2 + (kL)^2]$
4.	$\cos kx$	$(-1)^n kL^2 \sin kL/[(kL)^2 - (n\pi)^2]$, $n\pi \neq kL$
5.	$\sin kx$	$kL^2[(-1)^n \cos kL - 1]/[(n\pi)^2 - (kL)^2]$, $n\pi \neq kL$
6.	$f'(x)$	$(1/L)n\pi F_S(n) - f(0) + (-1)^n f(L)$ (Sine Transform)
7.	$f''(x)$	$-(n\pi/L)^2 F_C(n) - f'(0) + (-1)^n f'(L)$

The finite Fourier transform pairs adopted for the above tables originate directly from the sine and cosine *half range* series (3.7), (3.8), (3.9), and (3.10). For more tabular entries see Sneddon [43, pp. 529–530]. As one may observe, the use of $L = \pi$ simplifies the recording of these tables. If we adopt $L = \pi$, then

$$F_S(n) = \int_0^L f(x) \sin nx\, dx\,, \quad f(x) = \frac{2}{\pi} \sum_{n=1}^{\infty} F_S(n) \sin nx$$

and

$$F_C(n) = \int_0^L f(x) \cos nx\, dx\,, \quad f(x) = \frac{F_C(0)}{\pi} + \frac{2}{\pi} \sum_{n=1}^{\infty} F_C(n) \cos nx$$

Tables corresponding to these formulas are available in Churchill [14, pp. 467–470].

APPENDIX 4

TABLES OF FOURIER TRANSFORMS

EXPONENTIAL FOURIER TRANSFORMS

$$F_e(\alpha) = E_\alpha\{f(x)\} = \int_{-\infty}^{\infty} f(x)e^{-i\alpha x}\, dx\,, \quad f(x) = \frac{1}{2\pi}\int_{-\infty}^{\infty} F_e(\alpha)e^{i\alpha x}\, d\alpha$$

	$f(x)$	$F_e(\alpha)$
1.	$U(x)$	$1/(i\alpha)$
2.	$xU(x)$	$(-1)/\alpha^2$
3.	$f(-x)$	$F_e(-\alpha)$
4.	$f(x-k)$	$e^{-i\alpha k}F_e(\alpha)$
5.	$f(kx)$	$(1/|k|)F_e(\alpha/k)$
6.	$e^{ikx}f(x)$	$F_e(\alpha - k)$
7.	$e^{-|x|}$	$2/(1+\alpha^2)$
8.	$xf(x)$	$iF_e'(\alpha)$
9.	$x^2 f(x)$	$-F_e''(\alpha)$
10.	$x^n f(x)$	$i^n F_e^{(n)}(\alpha)$
11.	$f'(x)$	$i\alpha F_e(\alpha)$
12.	$f''(x)$	$-\alpha^2 F_e(\alpha)$
13.	$f^{(n)}(x)$	$(i\alpha)^n F_e(\alpha)$
14.	$xf'(x)$	$[\alpha F_e(\alpha)]'$
15.	$xf''(x)$	$-i[\alpha^2 F_e(\alpha)]'$
16.	$x^2 f''(x)$	$[\alpha^2 F_e(\alpha)]''$

For additional entries see Erdelyi [19, Vol. I].

FOURIER SINE TRANSFORMS

$$F_S(\alpha) = S_\alpha\{f(x)\} = \int_0^\infty f(x) \sin \alpha x \, dx , \quad f(x) = \frac{2}{\pi} \int_0^\infty F_S(\alpha) \sin \alpha x \, dx , \quad x > 0$$

	$f(x)$	$F_S(\alpha)$
1.	$1 - U(x - k)$	$[1 - \cos k\alpha]/\alpha$
2.	$x[1 - U(x - k)]$	$[\sin k\alpha - k\alpha \cos k\alpha]/\alpha^2$
3.	$1/x$	$\pi/2$
4.	$e^{-kx}, k > 0$	$\alpha/(\alpha^2 + k^2)$
5.	$(1/k)f(x/k)$	$F_S(k\alpha)$
6.	$xf(x)$	$-F_C'(\alpha)$ (Cosine Transform)
7.	$x^2 f(x)$	$-F_S''(\alpha)$
8.	$f'(x)$	$-\alpha F_C(\alpha)$
9.	$f''(x)$	$\alpha f(0) - \alpha^2 F_S(\alpha)$

FOURIER COSINE TRANSFORMS

$$F_C(\alpha) = C_\alpha\{f(x)\} = \int_0^\infty f(x) \cos \alpha x \, dx , \quad f(x) = \frac{2}{\pi} \int_0^\infty F_C(\alpha) \cos \alpha x \, d\alpha , \quad x > 0$$

	$f(x)$	$F_C(\alpha)$
1.	$1 - U(x - k)$	$(1/\alpha) \sin k\alpha$
2.	$x[1 - U(x - k)]$	$[k\alpha \sin k\alpha + \cos k\alpha - 1]/\alpha^2$
3.	$2k/[\pi(x^2 + k^2)]$	$e^{-k\alpha}, k > 0$
4.	$e^{-kx}, k > 0$	$k/(\alpha^2 + k^2)$
5.	$(1/k)f(x/k)$	$F_C(k\alpha)$
6.	$xf(x)$	$F_S'(\alpha)$ (Sine Transform)
7.	$x^2 f(x)$	$-F_C''(\alpha)$
8.	$f'(x)$	$-f(0) + \alpha F_S(\alpha)$
9.	$f''(x)$	$-f'(0) - \alpha^2 F_C(\alpha)$

Additional entries may be found in Erdelyi [19, Vol. I].

INDEX

*Plural formed by adding s.